SYSTE
DESIG
BEHAVIORAL PERSP
ON DESIGNER
TOOLS, AND ORGAN

SYSTEM DESIGN

BEHAVIORAL PERSPECTIVES ON DESIGNERS, TOOLS, AND ORGANIZATIONS

Edited by

William B. Rouse
Search Technology, Inc.
Norcross, Georgia
Georgia Institute of Technology
Atlanta, Georgia

Kenneth R. Boff
Armstrong Aerospace Medical Research Laboratory
Wright-Patterson Air Force Base, Ohio

NORTH-HOLLAND SERIES IN
SYSTEM SCIENCE AND ENGINEERING
Andrew P. Sage, *Editor*

North-Holland
New York • Amsterdam • London

The U.S. Government shall have the right to reproduce material from the book for internal purposes, but this shall not include the right to grant permission to others to reproduce any portion of the book or to distribute any reproduction of this material outside U.S. Government agencies. In all cases in which a U.S. Government employee reproduces a portion of the book, a copyright notice as it appears in the book shall be included on the first page of the reproduction, accompanied by a suitable credit to the book.

Elsevier Science Publishing Co., Inc.
52 Vanderbilt Avenue, New York, New York 10017

Distributors outside the United States and Canada:
Elsevier Science Publishers, B.V.
P.O. Box 211, 1000 AE Amsterdam, the Netherlands

© 1987 by Elsevier Science Publishing Co., Inc.

This book has been registered with the Copyright Clearance Center, Inc.
For further information, please contact the Copyright Clearance Center, Inc., Salem, Massachusetts.

Library of Congress Cataloging in Publication Data

System design.

(North-Holland series in system science and engineering)
Includes index.
1. System design—Psychological aspects. I. Rouse, William B. II. Boff, Kenneth R. III. Series.
QA76.9.S88S9523   1987   003'.019   87-18472

ISBN 0-444-01230-3

Current printing (last digit)
10 9 8 7 6 5 4 3 2 1

Manufactured in the United States of America

CONTENTS

	Preface *William B. Rouse and Kenneth R. Boff*	ix
1	**Introduction and Overview** *Fredric F. Doppelt*	1
2	**Workshop Themes and Issues:** **The Psychology of System Design** *William B. Rouse and Kenneth R. Boff*	7
3	**Effects of Technological and** **Organizational Trends on** **System Design** *Robert S. Cooper*	19
4	**Designers, Tools, and Environments:** **State of Knowledge, Unresolved** **Issues, and Potential Directions** *William B. Rouse and Kenneth R. Boff*	43
5	**An Experimental View of the** **Design Process** *Joseph M. Ballay*	65

6	**The Tower of Babel Revisited: On Cross-Disciplinary Chokepoints in System Design** *Kenneth R. Boff*	83
7	**Psychology or Reality** *Paul R. Chatelier*	97
8	**Some Intellectual Requirements for System Design** *Anthony Debons*	103
9	**The Changing Nature of the Human-Machine Design Problem: Implications for System Design and Development** *Robert G. Eggleston*	113
10	**Designing in Virtual Space** *Thomas A. Furness III*	127
11	**The Difficulties of Design Problem Formulation** *Ruston M. Hunt*	145
12	**The Role of Man in the System Design Process: The Unresolved Dilemma** *Edgar M. Johnson*	159
13	**Analytical Versus Recognitional Approaches to Design Decision Making** *Gary A. Klein*	175
14	**Unified Life Cycle Engineering** *Bernard Kulp and Anthony Coppola*	187
15	**Information Technology and Other Factors in System Design** *L. W. Lassiter*	199
16	**On Nature of Design and an Environment for Design** *Larry Leifer*	211

17	**Toward a More Systematic, Efficient Design Process: The Potential Impact of Intelligent Design Aids** *Edward A. Martin*	221
18	**A Cognitive Theory of Design and Requirements for a Behavioral Design Aid** *David Meister*	229
19	**Designing for User Acceptance of Design Aids** *Nancy M. Morris*	245
20	**Engineering Design Support Systems** *Robin Popplestone, Tim Smithers, Jonathan Corney, Anastasia Koutsou, Karl Millington, and Gideon Sahar*	257
21	**Designers, Decision Making, and Decision Support** *William B. Rouse*	275
22	**Knowledge, Skills, and Information Requirements for Systems Design** *Andrew P. Sage*	285
23	**Intuition by Design** *J. MacGregor Smith*	305
24	**The Nature of Design (and the Designer)** *Edward J. Zagorski*	319
	Index	331

PREFACE

This volume emerged from our wrestling, independently initially and later mutually, with trying to understand individual and organizational aspects of information utilization in system design. In particular, we both were concerned with why much of the information that results from basic and applied research does not appear to be utilized by designers. This concern led us to review the literature on the psychology of system design, organize a workshop on the topic, prepare this volume, and most recently perform a series of field studies of aerospace design organizations. We now feel that we have a reasonable grasp of the problems of supporting information utilization in design—our current goal is to develop and evaluate design support systems based on this understanding.

We got to this point by somewhat different paths. One of us has a systems engineering background (WBR) while the other's background is experimental psychology (KRB). Both of us are fascinated by the behavioral aspects of human-machine interaction in the operation, maintenance, management, and design of complex systems.

One of the paths to this point (WBR's) involved many years of studying human problem solving, first in static tasks (e.g., troubleshooting) and later in dynamic situations (i.e., aircraft, ships, process plants, and communications networks). The interest in human problem solving in the context of system design grew out of the realization that information utilization is particularly intriguing when this utilization is discretionary. It can be very difficult to predict (or even explain) humans' information utilization behaviors when the amount and variety of information available far exceeds the minimum necessary to perform the task of interest. However, as designers of information systems, we have to make this prediction (at least implicitly), or resort to simply providing everything imaginable and potentially overwhelming the user. This dilemma suggests that we need to develop a practical theory of information access and utilization. This theory must explain why designers (and people in general) value some types of information and not others, as well as how these preferences change in time and with situations. It is clear from the discussions in this volume that this theory will have to be based on both cognitive and organizational psychology.

As a Senior Scientist at the Human Engineering Division of the Air Force's Armstrong Aerospace Medical Research Laboratory, KRB has been concerned, over the past ten years, with resolving issues and problems in the design of control and display interfaces in military crew systems. The principal thrust of this activity is to develop a theoretical and empirical basis for matching the designed capability of crew system controls and displays with the sensory, perceptual, and cognitive characteristics of the human operator. While a good deal of potentially useful human performance data exist, these have had little direct impact on the design of crew system interfaces. Though the nature and availability of these data are a key part of this problem, the problem can also be attributed to the basic skills and inclinations of designers, limitations in the available support environment, and to constraints imposed by the system design and acquisition processes. Therefore, enhancing utilization of technical information in system design depends on the development of a practical access technology based on the factors which influence the "perceived" value and cost of information by system designers.

While we have orchestrated the production of this volume, we were certainly not the sole or most important performers. Listed below are all of the participants in the Workshop on the Psychology of System Design, most of whom are also authors of chapters in this volume. We are indebted to these individuals for their stimulation, enthusiasm, and meeting our many editorial and scheduling demands.

Joseph M. Ballay	Robert G. Eggleston
Leslie E. Beavers	Thomas A. Furness III
William L. Blackwood	David P. Glenn
Kenneth R. Boff	William W. Hartsfield
Gian M. Cacioppo	Ruston M. Hunt
Paul R. Chatelier	Edgar M. Johnson
William J. Cody	Gary A. Klein
Robert S. Cooper	Bernard A. Kulp
Anthony Debons	L.W. Lassiter
Fredric F. Doppelt	Larry J. Leifer

Edward A. Martin	William B. Rouse
Kathy L. Martin	Andrew P. Sage
David Meister	J. MacGregor Smith
Nancy M. Morris	Erik K. Vermulen
Lee R. Penick	Edward J. Zagorski
Robin Popplestone	

We are grateful for the support and sponsorship of the workshop by the Human Engineering Division of the Armstrong Aerospace Medical Research Laboratory at Wright Patterson Air Force Base under Contract Number F33615-82-C-0513 with MacAulay-Brown, Inc. Gian Cacioppo, MacAulay-Brown Program Manager, and Kathy Martin, Task Coordinator, made many suggestions beneficial to this effort.

We are also very pleased to acknowledge several important administrative, editorial, and clerical contributions to this effort. Earl Alluisi of the Office of the Undersecretary of Defense for Research and Advanced Technology generously aided our editorial effort. Gwyn Sheridan coordinated the workshop, Gabrielle Cacioppo served as technical editor for this volume, Joel Dickerson did most of the artwork, and Anita Smith typed the entire volume, both as preprints and as final galley proofs. The strong support of these individuals was the key to producing a quality volume in a timely manner.

William B. Rouse
Kenneth R. Boff

Norcross, Georgia
Wright-Patterson Air Force Base, Ohio

March 1987

Chapter 1

INTRODUCTION AND OVERVIEW

Fredric F. Doppelt
Major General, USAF, MC

Aerospace Medical Division
Brooks Air Force Base, Texas

I. INTRODUCTION

Our capability to produce "effective" defense systems is critical to national security. Similarly, in the commercial sector, the ability to develop "effective" industrial equipment and processes is both a contribution to and a by-product of our nation's economic health. In practical terms, effective systems are those which operate *efficiently* (i.e., satisfactorily perform the functions for which they were designed), *reliably* (in terms of low failure rate and ease of supportability), and *competitively* (in terms of the cost to acquire, operate, and maintain over the life of the system). For systems in which the human is a vital operational component, *operability* also becomes a major factor in system effectiveness.

The design of effective man-machine systems requires a practical understanding of the relationships among these factors. Also required is the ability to resolve tradeoffs among these factors to meet the different needs of customers, end-users, and designers. On the one hand, customers and end-users define requirements and set expectations. They also control the training, operations, and maintenance environments which influence the perception of the effectiveness of a given system long after it has been designed. On the other hand, designers are expected to formulate and resolve individual design tradeoffs which eventually culminate in the synthesis of a given system design.

The "goodness" of these decisions is, in turn, dependent on the designer's understanding of and ability to factor appropriate information into this decision process. Hence, the acquisition and use of information becomes central to the effectiveness of a given design.

The interdependence among the myriad variables which contribute to the use of information in the design of complex man-machine systems (discussed in Chapter 2) makes it difficult to quantify the influence of any single factor on a given design. The design of a complex system, such as a military aircraft, is dependent on the participation and contributions of many individuals. This typically includes researchers, engineers, draftsmen, managers, and others whose quantitative, analytic, analogic, oracular, and creative talents are guided, melded, and constrained by the sequence of steps, stages, and procedures collectively described as the "design process." This process is, in turn, regulated by standards, legal statutes, political constraints, and whims, while bounded by limitations of time and resources. Given the complexity of this multitude of requirements, dependencies, and behavioral intangibles (such as creativity, foresight, etc.), it is not surprising that the synthesis of a successful system design—a remarkably understated achievement—is generally not well understood.

The likelihood that a given system design will be effective is, in large part, dependent on accessible experiential knowledge, available methods for predicting the value of identifiable contributing factors to a given design, and the ability to resolve design tradeoffs appropriately. The support and amplification of these capabilities of designers should be the driving force in the development of design support systems. An improved understanding of the designer and the design process, coupled with the emergence of powerful, low-cost computing systems and machine intelligence (discussed in Chapter 3), presages major advances in design support technology including "intelligent" information access, interpretation, and decision aiding. Quite appropriately, this "vision" and its various implications are the focus of the discussions, speculations, and positions which comprise this volume.

II. TRANSFER OF TECHNOLOGY: RESEARCH TO DESIGN APPLICATIONS

A multitude of laboratories across the military services collect and disseminate data regarding the human operator's physical and mental performance as an element in complex systems. Within the Air Force Aerospace Medical Division, which it is my good fortune to currently command, it is the mission of the Armstrong Aerospace Medical Research Laboratory (AAMRL) to conduct behavioral and biomedical research to assess the effect on human performance of environmental, physiological, and cognitive stimuli and stressors associated with aerospace operations. Results of scientific efforts supported by these laboratories are published as technical reports, articles in scientific journals, handbooks, military specifications, and occasionally as academic textbooks. The intent of these research and development activities is to provide a technology basis for advanced engineering developments.

A review of AAMRL's accomplishments over the past fifty years reveals many scientific achievements, breakthroughs, and milestones which have had significant potential to impact subsequent human-system designs. Over the same period of time, the academic and industrial research literature has burgeoned with data equally suggestive of applications to the human-system design problem. Indeed, there have been many significant applications of biotechnology in the design of aerospace crew systems. The problem is that a staggering volume of potentially applicable research data remains to be exploited.

In other words, it appears to me that much research data exist that could contribute to system effectiveness, but these data are not presently factored into design decision making. This problem appears to be due to a variety of factors, ranging from the practical usefulness of the data (Rouse, 1985), the choice of format and media by which it is typically communicated (Boff, Calhoun, and Lincoln, 1984), the personal habits and styles in the use of information by design engineers (Allen, 1977; Rouse, 1986a; Boff, Chapter 6 of this volume), and the perceived costs/benefits of acquiring and usefully implementing these data (Rouse, 1986b). A large measure of this problem may also be that designers are already deluged with managing available information and have little time or resources to invest in searching, discriminating, and acquiring more, albeit potentially useful, information.

Enhancing the transition of basic research findings to application is dependent on dramatically increasing the ability of the individual designer to access, interpret, and track greater amounts of useful information and to factor it effectively into the appropriate design decisions. The design of support systems which amplify and assure the effectiveness of the designer's ability to design is, in turn, critically dependent on an accurate perspective and characterization of designers, the tools and methods useful to them, and the environments in which they perform. This volume provides this perspective and characterization.

III. SYSTEM DESIGN: A BEHAVIORAL PERSPECTIVE

This volume grew out of the three-day Workshop on the Psychology of System Design which was held in March of 1986. As noted in the Preface, the workshop included 30 invited participants from a wide variety of disciplinary and organizational backgrounds. As background material, each participant received a draft of Chapter 2 of this volume, "Workshop Themes and Issues," which Bill Rouse and Ken Boff prepared in the course of planning the workshop. Each participant then prepared a draft position paper on a subset of

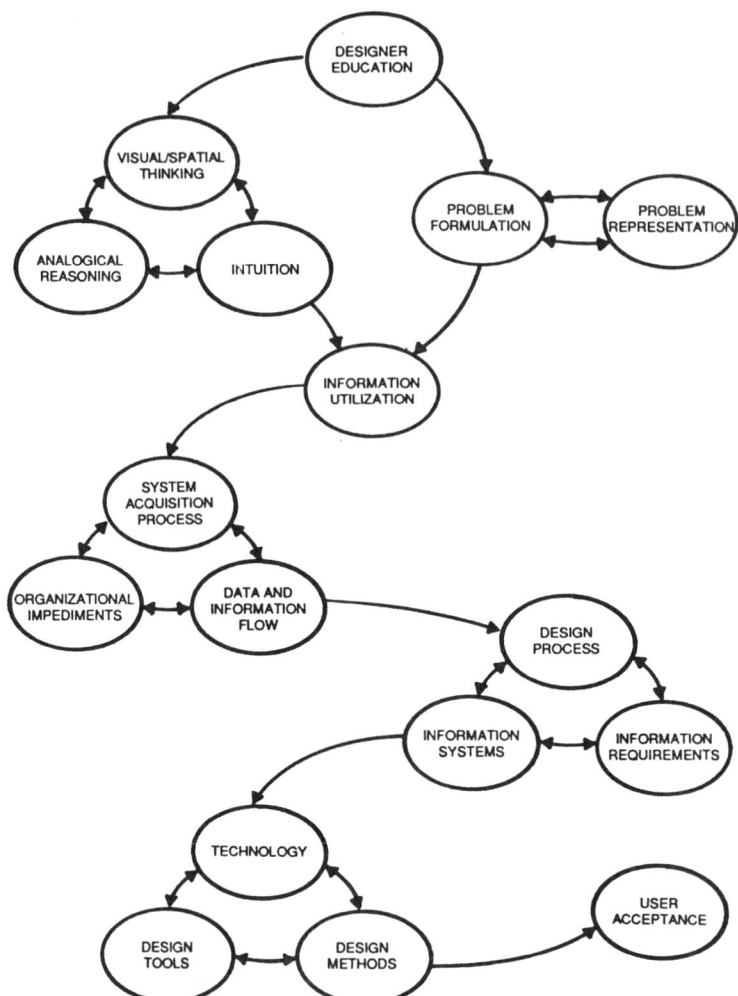

Figure 1. Issues in the Position Papers.

the issues in this chapter. The resulting set of 20 position papers served as the preprints for the workshop. Subsequent to the workshop, all position papers were reviewed and authors were asked to revise their papers on the basis of the many comments and suggestions emerging from this review process—as a result, many papers were substantially revised. The final versions of the position papers appear in this volume as Chapters 5 through 24.

Figure 1 is a representation of the relationships among the issues discussed in the position papers. In general, this figure flows from individual considerations at the top, through organizational issues and the design process, and finally to technology for designer support. While there are many important themes underlying the elements in this figure, the dominant theme of Chapters 5 through 24 is information utilization and flow, as well as information systems and the technology upon which these systems are based. This theme is also an important element of the other chapters in this volume.

As explained in Chapter 4, "Designers, Tools, and Environments," the participants received additional background material during the workshop in terms of three keynote presentations. This introductory chapter was drawn from my presentation as the initial keynote speaker. Chapter 3 is based on Bob Cooper's comprehensive and insightful perspectives on technological and organizational trends and their likely impact on system design.

In Chapter 4, Bill Rouse and Ken Boff have integrated the results of the three days of intensive discussions and debate by the designers, researchers, and managers who participated in the workshop. This integrative effort resulted in reasonably crisp conclusions regarding the state of knowledge, unresolved issues, and potential directions for increasing our understanding of the design of complex systems and developing design information systems.

To summarize the overview of this volume, Chapters 1 through 3 served as background material (provided either before or during the workshop); Chapter 4 represents the direct results of the workshop; Chapters 5 through 24 both impacted the workshop (in draft form as preprints) and were impacted by the workshop (in revised form as they appear here). Considered as a whole, this volume represents a comprehensive and impressive integration of a wide variety of points of view on the many important issues associated with designers, tools, and organizations.

IV. EPILOGUE

Meetings such as the Workshop on the Psychology of System Design can be a heady experience—a small group of interesting and intelligent people wrestling with important and challenging issues in a relaxed and supportive environment. Furthermore, volumes such as this can be quite gratifying—seeing the deliberations of the workshop integrated and nicely packaged. Such experiences and products are definitely rewarding.

In addition to this feeling of satisfaction, I find myself also feeling cautious. This volume may prove to be seminal in outlining the roles of designers, tools, and organizations in the design of complex systems; however, this volume does not provide comprehensive and detailed solutions for the many problems identified. While it is true that understanding a problem is the first step toward solving it, we must be careful to avoid resting on the laurels of understanding. Thus, my message to Bill Rouse, Ken Boff, and the other participants is: *great job, but don't quit yet—use your understanding to help improve system design.*

REFERENCES

Allen, T.J. (1977). *Managing the flow of technology: Technology transfer and the dissemination of technological information within the R&D organization.* Cambridge, MA: The MIT Press.

Boff, K.R., Calhoun, G.L., and Lincoln, J. (1984). Making perceptual and human performance data an effective resource for designers. *Proceedings of the NATO DRG Workshop* (Panel 4). Royal College of Science. Shrivenham, England.

Rouse, W.B. (1985). On better mousetraps and basic research: Getting the applied world to the laboratory door. *IEEE Transactions on Systems, Man, and Cybernetics, SMC-15*(1), 2-8.

Rouse, W.B. (1986a). A note on the nature of creativity in engineering: Implications for supporting system design. *Information Processing and Management,* 22(4), 279-285.

Rouse, W.B. (1986b). On the value of information in system design: A framework for understanding and aiding designers. *Information Processing and Management,* 22(2), 217-228.

Chapter 2

WORKSHOP THEMES AND ISSUES: THE PSYCHOLOGY OF SYSTEM DESIGN

William B. Rouse

Search Technology, Inc.
Norcross, Georgia

Kenneth R. Boff

Human Engineering Division
Armstrong Aerospace Medical Research Laboratory
Wright-Patterson Air Force Base, Ohio

I. INTRODUCTION

The degree of attention presently directed at automating the process of design suggests the need for a thoughtful reevaluation of the factors which contribute to design decision making and tradeoffs. Factors such as the expertise and experience of the designer; the nature of the design problem; the constraints of time, resources, work environment, and organization; and the availability and support effectiveness of design aids and tools should be considered in the development of advanced design support systems.

These factors are depicted as themes in Figure 1 and are decomposed in terms of underlying issues in the discussion which follows. The discussion is formulated around a prospective outline of salient issues and questions selected with the intent of provoking further discussion, debate, and perhaps even dissent among the workshop participants.

II. NATURE OF DESIGN

Design has frequently been naively portrayed as an unstructured, creative process of dreaming up new light bulbs or better mousetraps. In fact, while creativity is an extremely important element of design, it is certainly not the only important element. Typically, there is a process that explicitly or implicitly

Figure 1. Workshop Themes.

underlies design. While a variety of characterizations of this process have been proposed (Rouse, 1986), all represent versions of the following:

o   Formulation of problems,
o   Generation (or synthesis) of alternative solutions,
o   Evaluation (or analysis) of alternatives, and
o   Selection (or optimization) among alternatives.

Different types of support are needed during each step or phase. For example, during formulation and generation, designers tend to be "artists," while during evaluation and selection, designers are definitely "analysts." The artist versus analyst distinction has clear and important implications for how the structure and flexibility of the support should vary across design phases (Rouse, 1986).

The design of complex systems entails consideration and integration of technical information into tradeoffs and decisions at each stage of the process (Boff, Calhoun, & Lincoln, 1984). Given the constraints of time, available resources, and the resident expertise and experience available to designers, it is likely that the information considered is not the best information objectively available. Hence, it is probable that for a given design problem, technical information exists that may contribute to the effectiveness of design decisions but is not factored into the design process.

While design is often viewed as concentrating solely on new products, there is actually a much wider range of design problems. In fact, new product design is probably much less frequent than design associated with improving or remedying problems with existing products, or modifying existing products for new uses. For example, it is difficult to imagine starting "from scratch" when designing the next generation of aircraft, diagnostic test equipment, or infantry rifle. The implication is that design is often very constrained by past practices and products; a support system that is premised on designing "from scratch" is unlikely to be successful.

The above discussion suggests several important questions:

1. To what extent is any top-down, systematic view of design unrealistic relative to the heterarchical chaos that typically underlies creative activities such as design?

2. To what extent might design be enhanced if appropriate tools could foster a more systematic, orderly process?

3. To the extent that "new" designs are heavily based on past experiences, is it likely that useful support systems will require very substantial investments in data base development and maintenance?

4. To what extent is design a multiperson, multidisciplinary activity, and how does this mix vary across design domains?

5. To what extent does the success of a new design depend on prior successful designs? To what extent do designers purposively seek analogical matches between old and new design requirements and specifications?

III. NATURE OF DESIGNERS

When designing an aircraft crew station or a piece of test equipment, one should consider the characteristics of the operators and maintainers who will utilize these systems. Similarly, the design of a support system for designers should be based on an understanding of the nature of designers. Three attributes are of particular interest:

- o Knowledge and skills,
- o Inclinations and biases, and
- o Information processing characteristics.

Issues within the category of knowledge and skills include the *content of knowledge*, ranging from basic principles, to specific practices and procedures, to general heuristics. A major issue of concern under knowledge content is the domain of expertise of the designer, particularly with regard to how well it matches the technical knowledge domains germane to the design problem. This has implications for design problem reduction and the formulation of objectives for information search and retrieval (Boff, 1982; Boff, Calhoun, & Lincoln, 1984). A related issue is the *form of mental representation* of knowledge, with the most useful distinction being between spatial and verbal forms. Also of interest are *procedural skills* which include abilities to manipulate representations and use computer-based systems. Useful questions related to knowledge and skills are:

6. Which types of knowledge are most central to successful design?

    a. Basic principles (e.g., principles of structural mechanics or theories of human information processing).
    b. Applied practices and procedures (e.g., system simulation languages or human engineering guidelines).
    c. Available and foreseeable materials, components, hardware, and software technology.

7. To what extent must the requisite knowledge be drawn from multiple disciplines within engineering, behavioral science, and other areas?

8. To what extent must the requisite knowledge be represented in particular ways (e.g., spatially) in order for designers to find this knowledge accessible and useful?

9. To what extent do the types and forms of the requisite knowledge depend on the creative, analytical, and computer skills of designers?

Within the category of inclinations and biases, an important issue is design *problem solving*. By this is meant the strategy used to decompose or reduce a design requirement to performance and equipment considerations embodied as either functional or detailed specifications (Boff & Martin, 1980). Another somewhat amorphous factor can be *cognitive style*, which loosely describes how individuals prefer to approach problems. While differences in cognitive style may reflect a variety of inclinations, one fairly common characteristic across all styles is a *tendency to "satisfice"* in the sense of not being willing to invest the effort required to determine the "best" solution. Beyond inclinations, there are also *judgmental biases* which relate to natural tendencies to be affected inordinately by recent experiences, somewhat arbitrary reference points, and the imaginability of outcomes. Important questions concerned with inclinations and biases include:

10. Is it important to provide support that matches designers' individual styles?

11. Have "style" and related constructs been sufficiently researched to provide a basis for such customization?

12. Is the tendency to satisfice a serious intrinsic problem, or merely an artifact of awkward and inefficient information retrieval systems?

13. Can designers be "debiased" by appropriate transformation and formatting of technical information?

Human information processing characteristics can be important determinants of how designers should or should not be supported. One of the more salient concerns here is the distinction between *recognition* and *recall*; humans are excellent at recognizing displayed patterns, but relatively poor at recalling previously observed patterns. Somewhat related is the issue of *working memory*, which limits humans from keeping track of more than a few things simultaneously without being aided. Also of concern is the possibility of *stress* and *fatigue* due to almost continual interaction with any computer-based system. Within the area of human information processing characteristics, questions of interest include:

14. How might information retrieval systems be designed to take advantage of humans' pattern recognition abilities while also minimizing demands on recall and working memory?

15. To what extent is the design of such systems likely to depend on the expertise of the system user (e.g., menus for novices and command languages for experienced users)?

16. To what extent is the design of such systems likely to depend on individual differences in abilities, limitations, and inclinations?

17. Are stress and fatigue, due to prolonged interaction with CRTs (visual fatigue) and keyboards (muscle fatigue), likely to lead to problems?

This section has outlined a variety of important issues relative to characterizing designers' abilities, limitations, and inclinations. These characteristics are of interest to the extent that they interact with designers' work environments. The following section considers research issues in context.

IV. EFFECTS OF ORGANIZATION

Design problem solving always occurs in some organizational context. This context usually has very strong effects upon the problem solving process.

Formal and informal objectives constraints, practices, and procedures affect the alternative solutions considered and how selections are made. It has also been suggested that goals, criteria, and approaches intrinsic to design organizations can strongly influence the utilization of technical information (Rouse, 1985b).

While the nature of most organizations is influenced by the markets they serve, these effects are perhaps strongest for organizations whose primary market is the Federal Government in general, and the Department of Defense in particular. Many organizational practices and procedures are explicitly or implicitly dictated by procurement and cost accounting regulations. Furthermore, the character of relationships between technical project monitors and contractors influences design decisions. Thus, both internal and external organizational contexts can affect the design process.

Organizational attributes can strongly affect the practicality and acceptability of alternative approaches to enhancing the use of technical information in design. It is important to understand the points in the organization, as well as in the design process, where changes are most likely to be allowed and effective. Further, it is useful to know how broader organizational changes (e.g., in procurement regulations) might facilitate local changes that would otherwise be impossible.

The above discussion suggests the following important questions:

18. To what extent are designers constrained by organizational goals, criteria, and approaches?

    a. Do the effects depend on the type of organization?
    b. Do the effects depend on the domain of design?
    c. Do designers perceive these constraints as inhibiting?

19. To what extent does the procurement process of the Government dictate both the specifications for systems *and* the possible solutions?

20. To what extent are these effects due to Government personnel synthesizing requirements, and to what extent are they due to contractor personnel marketing particular solutions?

21. Do these effects constrain innovation, limit system effectiveness, and/or substantially increase costs?

22. Should the locus of any changes in the overall process be in Government practices and procedures, or in contractors' ways of doing business?

23. Where in the design process, either within the Government or

contractor organizations, are changes in information access and utilization most likely to be effective?

V. DESIGN SUPPORT SYSTEMS

A variety of systems to support designers have been envisioned, many of which are available in the guise of CAD/CAM/CAE systems. Most of these systems can be characterized in terms of three attributes:

o   Type of information,
o   Type of manipulation, and
o   Function of support.

Types of information include facts, fantasies, and feedback (Rouse, 1986). *Facts* include the data, models, and algorithms that are traditionally central to CAD/CAM/CAE systems. *Fantasies* are provided by mechanisms which enable generation, integration, and elaboration of solution concepts. Excellent examples of such mechanisms are the many rapid prototyping tools now available. *Feedback* includes mechanisms for monitoring, evaluation, and advice. An example of such a mechanism is an expert system for evaluating display formats. Salient questions related to type of information include:

24. To what extent do traditional, fact-oriented information retrieval systems provide substantive support to designers?

25. What are the advantages and disadvantages of rapid prototyping in terms of quickly seeing alternative solutions, prematurely focusing on initial ideas, etc?

26. To what extent are on-line expert systems likely to be of value for different types of design problem and phases of design?

27. In what ways might current CAD/CAM/CAE systems be enhanced, within current technology constraints, to provide facts, fantasies, and feedback?

Types of manipulation range from retrieval to management to transformation (Rouse, 1985a; Reitman, Weischedel, Boff, Jones, & Martino, 1985). Information *retrieval* involves providing the designer with information that he or she is unaware of or, more likely, information that will only be used if readily available. Information *management* can, in effect, provide designers with a spread-sheet program, a support package analogous to Visicalc, Multi-Plan, or

Lotus 1-2-3. Finally, information *transformation* can enable the designer to perform calculations and simulations that would be impossible without computational support. This type of support includes those analytical and computational tools that are often viewed as the essence of CAD/CAM/CAE systems. Important questions related to manipulation of information include:

28. To what extent are designers' information seeking inclinations related more to ease of access than knowledge of availability of information?

29. To what extent do designers take advantage of existing commercial or in-house computer-based information retrieval systems?

30. Are information management problems an impediment to the efficiency and/or effectiveness of the design process, particularly for large-scale, complex systems?

31. To what extent do designers make use of models, simulations, etc? Does usage depend on the type of design problem and phase of design?

"Function of support" refers to the role of the support system relative to the designer. At the lowest level, the system might function as a *clerk* in the sense of performing bookkeeping, message handling, etc. A higher level would involve functioning as a *consultant* by answering questions (perhaps via transformation) and providing advice. Another high-level function would be that of *coach*, both in terms of tutoring and providing strategic advice. These characterizations of levels of support suggest the following important questions:

32. Is it reasonable to conceptualize a support system in terms of human-like characteristics (e.g., coach), or might such terminology tend to be misleading?

33. Is it likely that designers will be confused and/or dissatisfied by a support system whose role is adapted to the nature and state of their design problem solving?

34. To what extent might a support system serve as both a training and aiding device, and thereby allow designers to "bootstrap" themselves into becoming system users?

35. What other roles can be imagined for a design support system, and when might these roles be invoked?

By combining the different types of information, types of manipulation, and functions of support, it is easy to envision a variety of support system concepts.

For example, CAD/CAM/CAE systems usually function as clerks or consultants in the retrieval and transformation of facts. Considering the range of concepts possible within this framework, it is easy to imagine that the most appropriate concepts depend on the phases of design and/or types of designer.

VI. ACCEPTANCE OF SUPPORT SYSTEMS

An understanding of the factors associated with acceptance of support systems is crucial to system design. A system is accepted to the extent that it is favorably received and willingly used as intended (Rouse & Morris, 1985). It can be argued that automation will be accepted if the benefits of the system are perceived as exceeding the costs of utilization.

Within the context of the workshop, benefits can be viewed in terms of the value of information provided. While the value of information is, in itself, a perhaps surprisingly complicated concept (Rouse & Rouse, 1983), it is possible to be reasonably concrete within the domain of design. It has been argued that the value of information in this context can be defined as having three attributes: 1) *reduction of uncertainty*, 2) *task relevance*, and 3) *appropriateness of form*.

More specifically, to be valuable, information must reduce uncertainty by either informing the recipient of something previously unknown, or reminding the recipient of something previously known but forgotten. Further, information must be relevant to the tasks of interest; this precludes information that is not yet relevant or about to be relevant, no longer relevant, or categorically irrevelant. Finally, information must be in an appropriate form to be valuable; either it can be in a form that is natural to use or in a form that is transformable. It can be shown that the benefits of alternative support systems for different phases of design can be characterized in terms of these three attributes of the value of information (Rouse, 1986).

The perceived costs of accepting a support system include straightforward considerations such as the difficulty of interacting with a system that is not "user friendly" in the sense of having an understandable dialogue structure, etc. For support systems that function as consultants or coaches, the personality or demeanor of the support can be crucial; an assertive support system may be readily accepted initially, but viewed as a "nag" after experience is gained. More subtle and pervasive are factors that can lead to alienation. Lack of understanding, devaluation of skills, and perceptions of loss of control are factors that can undermine acceptance (Rouse & Morris, 1985). From this perspective, it is important to understand the nature of designers and how they view their roles in the design process.

The above discussion suggests the following important questions:

36. To what extent is the proposed definition of the value of information appropriate across different technology domains and phases of design?

37. Can a similar or complementary construct be developed for the cost of access and utilization?

38. Is there a significant risk that designers will view "intelligent" support systems as threats to their "turf"?

39. Will the emergence of networked design workstations lead to alienation due to decreased face-to-face interaction among members of design teams?

VII. CONCLUSION

The purpose of the preceding discussion was to outline briefly a host of issues which underlie the key factors influencing design. While it is not reasonable to expect workshop participants to formulate a monolithic approach to the resulting complex network of related issues and questions, it is important that partial, incremental answers and solutions be consistent with an integrated perspective of these issues. As a result, partial solutions will be complementary rather than competing, and the impact of limited resources can be optimized.

REFERENCES

Boff, K.R. (1982). Integrated perceptual information for designers. *Proceedings of the National Aerospace and Electronics Conference.* Dayton, OH.

Boff, K.R., Calhoun, G.L., & Lincoln, J. (1984). Making perceptual and human performance data an effective resource for designers. *Proceedings of the NATO DRG Workshop (Panel IV).* Shrivenham, England.

Boff, K.R., & Martin, E.A. (1980). Aircrew information requirements in simulator display design. *Proceedings of the Second Annual Interservice/Industry Training Equipment Conference,* pp. 355-362.

Reitman, W., Weischedel, R., Boff, K., Jones, M., & Martino, J. (1985). *Automated information management technology: A technology investment strategy* (AFAMRL-TR-85-042).

Rouse, W.B. (1982). On models and modelers: N cultures. *IEEE Transactions on Systems, Man, and Cybernetics, SMC-12*(5), 605-610.

Rouse, W.B. (1985a). *Computer-aided crew station design: An approach to supporting engineering judgment in function allocation.* Norcross, GA: Search Technology, Inc.

Rouse, W.B. (1985b). On better mousetraps and basic research: Getting the applied world to the laboratory door. *IEEE Transactions on Systems, Man, and Cybernetics, SMC-15*(1), 2-8.

Rouse, W.B. (1986). On the value of information in system design: A framework for understanding and aiding designers. *Information Processing and Management*, 22(2), 217-228.

Rouse, W.B., & Morris, N.M. (1985). Understanding and avoiding potential problems in implementing automation. *Proceedings of the 1985 International Conference on Systems, Man, and Cybernetics*, pp. 787-791.

Rouse, W.B., & Rouse, S.H. (1983). Human information seeking and design of information systems. *Information Processing and Management*, 20, 129-138.

Chapter 3

EFFECTS OF TECHNOLOGICAL AND ORGANIZATIONAL TRENDS ON SYSTEM DESIGN

Robert S. Cooper

Atlantic Aerospace Electronics Corporation
Arlington, Virginia

ABSTRACT

This chapter is based on an edited transcript of Dr. Cooper's keynote presentation at the Workshop on the Psychology of System Design. Dr. Cooper's presentation focused on high technology trends that are affecting system design in general and design support systems in particular. He elaborated on developments in microelectronics, rapid prototyping, computer graphics, computer networking, machine intelligence, and his views of the prospects for a "Designer's Associate." He also discussed a variety of organizational issues affecting system design, including industrial and government attitudes toward risk and increasing regulatory pressures. Dr. Cooper is a former Assistant Secretary of Defense for Research and Technology, former Director of the Advanced Research Projects Agency, former Director of NASA's Goddard Spaceflight Center, and has held a variety of other high-level R&D management positions.

I. INTRODUCTION

I am delighted to have this opportunity to talk with you about system design and the possible role of technology in future system design. Technology has been my life. I got into the technology business at MIT in my student days, and I have spent my entire career doing advanced research and development of one sort or another in just about every setting imaginable. In academia, I was a faculty member for a number of years and also worked in a research laboratory at MIT. In defense management at the Office of the Secretary of Defense, I managed very high technology space and laser weapon research programs. I also worked extensively in the space program as Director of NASA Goddard Spaceflight Center. As Vice President of Engineering for a satellite communications company, I was the architect of, and designed and built, a very advanced digital satellite communication system which formed the backbone of Satellite Business Systems' network. For the last four years, I had the best R&D job in the world as Director of ARPA, the premiere playground for technologists.

Based on this background and experience, I would like to relate some of my impressions of the way systems get designed, as well as what the influence of technology is likely to be on system design in the future. I also would like to discuss some of the problems the Department of Defense (DoD) has getting technology into its systems, and how I view this defense problem more broadly as a national problem. For the last ten years, the industry in this country has had an incredibly difficult time taking advantage of the technology which was generated in the 1960s and 1970s, putting this technology into products, and making those products competitive around the world. I would like to talk about the problems that designers face in being able to use the best technology, to get their ideas implemented and tested, and to see real products emerge.

II. DESIGN TRENDS

The process of creation in design takes place over many orders of magnitude. Design is pursued at scales ranging from molecular to cosmic. As a result, if you look at what is possible in the way of engineering systems these days and how the design process might be supported, whatever ideas you come up with that are truly kernel ideas have to apply over an incredibly dynamic range.

For instance, over the past several years I have been exposed to molecular engineers who are able to use recombinant DNA processes to create molecules of 100,000 atoms or more—that is, to make living material that is literally a new man-made life form. This is now a controlled process. Engineers can design organic molecules of extraordinary complexity and, in many cases, predict the characteristics of those molecules.

One scale up from that, circuit design engineers are now placing a million electronic gates on a single 16-square-millimeter chip of silicon in patterns which have vast computational power over which designers can have almost complete control. Not only can this be done, but it can also be done with reasonable facility. All of the essential design and testing aids, at least first-

Effects of Technological and Organizational Trends

generation versions, are there to accomplish this. Unfortunately, these aids are still too expensive. Later on, I will talk some more about technology trends in that area and how these trends are likely to affect the design business.

One scale up from that, electronic and aeronautical engineers using design workstations of extraordinary power (attached to class six computers, which are among the fastest computers) are now able to design half-million-pound aircraft that have excellent flying qualities when they are first flown. These aircraft have the additional qualities of little or no radar backscatter when illuminated by a radar source. This is an incredible juxtaposition of electromagnetic and aeronautical designs in the same large scale design integration. Ten years ago, this was completely impossible.

One scale up from that, in the next year, the DoD will orbit a system of 24 new design navigation satellites and will begin to deploy almost 30,000 receiver-processor systems. The receiver-processors are electronic boxes on the earth that will make it possible to navigate in the near earth space, up to a few thousand nautical miles in altitude, in three dimensions with precisions on the order of a few meters. To know exactly where you are in space to within a few meters has been possible for over ten years. However, it is an incredibly complex task to design a system that can work to a part in 10^9 or a part in 10^{10} of the radius of the earth, and position the receiver accurately ten times a second in a very dynamic environment. To engineer a system of that complexity and precision twenty years ago would have been totally out of the question.

Now one scale up from that, defense engineers are developing the technology today that could provide us with system concepts for creating a globe-girdling, space-based defense against ballistic missiles and against a ballistic missile attack of enormous proportions. Such a system would consist of thousands of elements, containing literally millions of communication channels, with capabilities to engage tens of thousands of objects in time periods on the order of hours. A system of that complexity clearly would not have even been conceivable in the 1960s when we started the Apollo program, which was about the most complex system conceivable in that period.

Look at the orders of magnitude of scale size of the design conception that human beings are now able to cope with using the tools at hand today. It is truly daunting to think of the capability to operate in that kind of design environment, to pursue problems at all scales ranging from the molecular level all the way up to the cosmic level.

III. DESIGN TOOLS

The tools being made available to design engineers are increasing in complexity themselves. Essentially, it is a bootstrapping process. A system has to be designed in order to develop tools which will aid designers of such systems. Now, there is some kind of magic in the ability to bootstrap in that way. It is not just a first order effect. It has an exponential influence on the productivity of engineers.

I had an interesting experience at ARPA. When I came to ARPA, the organization had not changed its internal way of operating since it was created twenty years ago. The way ARPA operates is very simple. There is no bureaucracy, which is great, and that is one of the reasons why it is such a powerful research organization.

ARPA has program managers who are really good and who do the creative work. Then there is the Director who is basically the fellow with all the money. The Director doles the money out to the program managers who have the good ideas. Essentially, two people have to come to an agreement to do something at ARPA, and it can be done on a large scale. Thus, $100 million commitments can be made by agreements between two people, a program manager and the Director of ARPA. I know of nowhere else in the bureaucracy where that can happen.

However, the paperwork process to make these agreements legal was based on a scheme that was set up in the 1960s by the first Management Office Director at ARPA. It really had not changed in twenty years. The papers generated are called ARPA orders, and they had grown in complexity and size through the years.

One of the things I decided to do while I was at ARPA was to get rid of all the paper and make ARPA completely electronic. Now, you would think that an advanced research organization could handle something like that, particularly people who are working on the cutting edge of technology doing space-based laser weapon programs and getting machine intelligence technology cooking at a tremendous pace. Well, it took four years for me to essentially depaperize ARPA. I had to drag some of those high-technology program managers kicking and screaming to the well. The way I finally did it was to require that no money leave the system except by completion of the electronic ARPA order. This approach was risky because some of the program managers were computer scientists who were very clever at jimmying the electronic data processing systems. It worked, and eventually it multiplied the productivity of the program managers by manyfold. Later, they recognized this, and even those who were most resistant to the change said, "I've now got more time to do the things that I need to do to make my programs better, and I can do more because I don't have the burden of the paperwork." Thus, there is resistance to using the tools which can make us more effective and efficient.

The tools that can be made available today are not taking hold in the engineering community as rapidly as they could. One of the things that has struck me over the past several years in my peripatetic journeys throughout the U.S., talking to people in ARPA programs that are doing high-technology work, is that those who are most productive and innovative in their abilities are the ones who have found ways to grab on to the latest technical capability to enhance their power to create new designs and capabilities. Thus, the new technology that is coming has the potential to exponentially improve the creative powers of engineering design people, and it will do that for the best designers. The people who hang on to the past way of doing things are the ones who are plodding behind and are not at the cutting edge of capability. Nevertheless, we have to recognize that there is resistance to change and encourage the acceptance of new technology as design tools.

Of course, new design tools are just one part of the story. New technology will also play a role in the actual design of systems. Later on, I will come back to this possibility and the barriers to that kind of transition.

IV. DESIGN EXPERIENCES

Some of the views I have about what can be done and what will be done in the future are obviously biased by my experiences. Let me note some of the design problems with which I have been involved. When I was at MIT, I ran a faculty group, including several students, that was working on plasma stability problems in experiments designed to heat hydrogen gas to atomic fusion temperatures. The group's goal, in part, was to create magnetic structures which could confine very hot plasmas and heat them to high temperatures in search of the holy grail of fusion. More specifically, the objective was to achieve a product of plasma temperature and density equal to about 10^{14}. We "hacked" our way toward our goal by building interesting magnetic plasma confinement structures, by creating plasmas within them, and then firing electron guns and ion beams into them. The difficult part was designing complex magnetic field configurations in three dimensions. We used an IBM-709, pencil and paper, and a lot of chalk on the blackboard. One design took us the better part of a year to complete. Today, at Princeton University where the TOCAMAC magnetic fusion reactor is being designed, the design aids being used include advanced graphics workstations attached to incredibly powerful computers which are able to model magnetic structures in a few minutes. Using these tools, they have designs at hand which will meet the temperature-density criterion many times over. This means that magnetic confinement fusion is probably on the horizon, largely due to the design aids that are available today which were not previously available. I think we knew almost all the physics that is known today and used as the basis for the design. However, we did not have the power to rapidly move through the alternatives and seek out the optimal designs.

My next interesting design problem grew out of work I was doing at Lincoln Laboratory which involved looking at radar scattering from the ionized wakes of reentry vehicles. We had a small group there that essentially solved the ballistic missile reentry discrimination problem. This problem involved determining which one of the pieces of junk coming down from an ICBM attack was the "twin with the Toni." In other words, the concern was how to pick the ICBM out of the decoys and chaff. An Anti-Ballistic Missile (ABM) system uses this information to aim an interceptor at the ICBM. By looking at radar phenomenology from ICBMs that were fired at the Kwajalein Missile Range and observed with radars, we discovered ten separate criteria by which reentry vehicles could be discriminated from junk. We implemented these criteria in hardware in a very clever processor design, probably one of the most complex processors that was ever built up to that time (1968). This design was incorporated into the Missile Site Radar, which was the terminal defense radar system of the ABM system deployed in the early 1970s. That research and design effort was an interesting part of my life and was my first experience in the electronic design world where electronic components and their inherent limitations had to be dealt with.

I moved to another kind of radar project and became one of the principals in building the first laser radar for making images of low earth orbiting spacecraft. This is done through a process of Doppler imaging. Radars can make two-dimensional images of things that move by you. The process involves computing each pixel in a two-dimensional image from a one-dimensional signal. The approach is employed on a wide variety of wideband radars today. The interesting thing about creating images this way with a laser is that the resolution of the images does not depend on the aperture of the telescope or, in the case of a microwave radar, on the size of the antenna as it does for a real aperture system. Instead, the resolution is related to the wavelength of the illumination. Since laser radars could be made in the 10-micron range very easily in those days, we made a laser radar with resolution that far exceeded the seeing conditions for the atmosphere. This was due to the fact that resolution mainly depends on the bandwidth of the radar which can be made very broad for a laser. The bandwidth depends on the precision with which the Doppler velocity, due to the motion of the body as it moves by you, can be resolved which, in turn, depends on the wavelength—the smaller the wavelength, the better it is. We were able to make incredibly good images of low earth orbiting spacecraft and found out some interesting things about the nature of what the Soviets had in orbit. It was an experimental system, but it was a complex system that was built out of state-of-the-art components. Every designer knows that "state-of-the-art" is synonymous with components that do not work very well!

My interest in laser radar took me into the high-energy laser weapon business, and I began experimenting with some of the early high-energy laser weapons that were developed by companies in the United States. They were initially gas-dynamic lasers, with emitted optical powers on the order of two or three magnitudes beyond anything that had been worked with before. There were many questions about nonlinear atmospheric optical effects in these lasers that were of interest. I worked on very complex instrumentation systems to diagnose the characteristics of high-energy laser weapons. Next, I moved to the DoD to manage the national high-energy laser program for three years. During that time, I became familiar with the design problems associated with space weapon systems because I had the added duties of overseeing all of the space programs in the DoD and coordinating activities with NASA. One of the programs I started during that time was the Global Positioning System, a precision navigation satellite system. It was a system of extraordinary proportion for a satellite system in those days. There were many, many conceptual design problems, particularly at the architectural level.

I went on to become the Director of NASA's Goddard Space Flight Center for a number of years. During that time, in a four year period, I had an opportunity to be involved in the initial design, development, and placing in orbit of approximately 15 satellites involved in space science and space applications missions. I found the process that NASA uses to create new systems to be an incredibly interesting process. Both the management and detailed engineering design processes were very well integrated, and, I think, provide a model for the best way to manage a one-of-a-kind, high-technology project. There are probably variations on NASA's management techniques that would work in almost any environment.

One of the major programs I started as I came through the door at Goddard was the tracking and data relay satellite system, which was an advanced communication satellite system meant to communicate with all of the satellites NASA had in low earth orbit. At any one time, there were about 30 satellites in low earth orbit sending back scientific data, engineering data, earth observing data, weather data, and so on. Goddard managed a series of about 14 tracking sites around the earth to collect data from those satellites. Those tracking sites were located in various countries. Each tracking site had a limited horizon and could only see the satellites about 15% of the time. Only when a satellite came over the horizon could data be collected from it. The concept for avoiding this limitation involved putting satellites up in synchronous orbit that could look down on the earth and see the low earth orbiting satellites all the time. This would enable shutting down the ground sites, which were very expensive to operate, and collecting data nearly 100% of the time from these satellites in synchronous orbit. This concept was developed, the system design was completed, and a contract was let to have the satellites built and service provided to NASA. Although I went out the door before the system got into orbit, it was a very interesting system and got me into the billion-dollar system business. While the complexities of the design of a billion-dollar system were extraordinary, the complexities of the contract procedures and of managing the system were its nemesis. I will discuss some of these kinds of problems when I talk about management issues in more detail.

That experience, along with an interest in sending my two boys to college, prompted me to go work at Satellite Business Systems for a few years and to build another advanced space communication system. This system was a spread-spectrum, burst, time division multiple access format communication satellite operating in the Ku band. All of the previous communication satellites were done in C band. We were moving up in frequency and down in wavelength with a number of attendant design problems related to weather, placement in orbit, new components in the satellite, and new problems associated with controlling the satellite. I was completely absorbed in designing this system for about three years, eighteen hours a day, seven days a week. I probably learned as much as I will ever want to know about system design from that. I may never go back to the system design business, but that was a really interesting business, and it was done without benefit of advanced state-of-the-art design aids now available in the community.

V. TECHNOLOGY TRENDS

What are some of the kinds of advanced technology that are going to revolutionize or re-revolutionize the advanced design business?

A. Microelectronics

Electronics technology is still moving forward at an incredible pace. The single most important effect this is going to have is to reduce the cost of electronic aids for engineering designers over the next ten years. The indication is that cost is going to be reduced by an order of magnitude. First of all, during the last seven years, silicon circuitry has gone from essentially 5-micron features to 1.25-micron features. In that reduction in feature size, the number of elements has gone from small-scale integrated circuitry to very large-scale integrated circuitry. Not only that, but the process by which integrated circuits are developed has achieved a level of control at the 2-micron level which creates yields for the electronic circuitry that are far higher than ever anticipated seven or eight years ago—up in the 80% to 85% yield factors. I would suspect that at 1.25-microns we will approach similar kinds of yields at some time within the next two or three years, although some new changes in process control, probably automation of process control, will be required to accomplish this.

Another trend evolving at a tremendous rate is design automation—the ability to create circuit designs. The Daisys, Mentors, and other companies are creating design workstations and design packages that provide the capability to design new circuits that, in turn, make it possible to design new workstations which will make engineers even more creative in that area. The avalanche effect is working in the electronic design industry.

If you look at what is happening in the United States' microelectronics industry today, the big players are moving away from competition with Japanese commodity electronics. They are moving toward applications-specific integrated circuits (ASICs). Instead of trying to take one design, like a microprocessor, and make a million or ten million ICs out of that one design, they are trying to get a thousand designs and make ten thousand ICs each out of those designs and have those ten thousand designs be application specific.

Effects of Technological and Organizational Trends

Thus, instead of a general-purpose processor for all jobs, these processors will be more powerful for specific jobs. To do this, the cost of designing those thousand circuits must be reasonable. Again, the issue of design automation comes in; that is, you need to make the individual designer more efficient.

Three years ago, I went around the country to talk to the microelectronics folks. At Intel, I talked with Bob Norris, who is one of the senior vice presidents and founders of Intel. He told me that Intel, at that time, was contemplating building a new microprocessor. Furthermore, he said that the Japanese had just "reverse engineered" Intel's latest microprocessor in about 95 days. In other words, 95 days from the time Intel put their chip on the market, there was a Japanese equivalent floating around—perhaps not selling yet—but floating around. The Japanese distributed a few of them to people to show they could make them. Intel had a 95-day lead on the Japanese from the reverse engineering standpoint, and he told me that it was going to cost him 18 months and $50 million to design the next microprocessor. Eighteen months and $50 million for a microprocessor design that had probably somewhere around 600,000 gates on the chip. The question he was contemplating was whether or not Intel should spend the $50 million. If the Japanese reduced their reverse engineering time to 30 days, the investment would be wasted.

The problem for Intel was that the 18 months and $50 million would be invested in a very large team of designers who were trying to work together with old design aids; they did not have state-of-the-art design aids. The whole idea of design automation in the electronics business is crucial to this country's ability to maintain its electronics industry. It is also crucial to being able to create the next generation of design aids. I see that process working right now in this country. I think people are moving rapidly toward advanced design aids. They are moving rapidly toward design automation, toward creating more IC design engineers by making the training requirements less grueling than they were before. They are doing this by putting more of the smarts into design aids and requiring less of the designer.

If you look at what is happening elsewhere in the electronics industry, much of the industry is involved in making moderate-scale integrated circuit boards that can be found everywhere. They can be bought for PCs, home electronics equipment, and so on. They cost anywhere from $100 to $350. The ICs, which now all come from Japan, that plug into the boards are medium-scale and large-scale integrated circuits. The designers who design these boards are basically moderately skilled designers. They are logic designers who can send their designs to a circuit board house where they are stuffed and returned. They are cheap and work reasonably well, and the process is limited to a certain extent. Those boards can now be easily reduced to a chip and the question is: Can a design package be built that allows the circuit board designer, with no additional skills, to take the logic diagrams he created before and create a chip out of it instead?

There is a technology now available to do that. This technology includes design aids as well as management schemes and arrangements that can provide access to this technology for a person who has no real connection with the integrated circuit industry. ARPA was one of the first organizations to lead the charge in the direction of trying to make this "impedance match" with the integrated circuit industry. It did so by creating a project called the Metal Oxide Semiconductor Implementation System (MOSIS) Project at the University of Southern California in the Information Sciences Institute (ISI).

Today it is very expensive to try out an integrated circuit design. Say that you are a circuit designer, and you have designed an integrated circuit. You want to get it fabricated to see if it works. Well, you take it to an IC house, a merchant semiconductor shop, and they do not really want to talk to you because what they like to do is make a million chips of one kind. What you really want is only one that works so you can test it to see if you want to buy some more. Even if it tests out right, you probably only want to buy a hundred because you want to put them into a new experimental computer that you are working on. The IC house really does not want to talk, but if you bring $50,000, they will give you back one of those chips. You may have to wait around a long time for it because they will try to get you into the queue sometime when it does not interrupt their process of making a million chips of one kind.

B. Rapid Prototyping

The idea of the MOSIS Project was to take say 1,000, 5,000, or 10,000 designers who were designing chips that were relatively mundane. Each designer would put his design into a mill and every Thursday he would go to the IC house and say, "I want another one. Here's my $50,000." Whoever had his design there by Thursday would have his design put through a semiconductor foundry. It would be reproduced at enough die sites on a silicon wafer so that the yield times the number of die sights would yield at least a few parts that were workable which would get sent back through the mail. The question was how fast can that be turned around? Could it be turned around in four weeks...three weeks...two weeks?

The whole concept of rapid prototyping of integrated circuits and the idea of multiple designs on a single wafer was born at ARPA and was implemented in this project at ISI. Today the MOSIS Project runs 5-micron, 3-micron, and 2-micron complementary metal oxide semiconductor (CMOS) circuits. It runs, on an experimental basis, a 1.25-micron CMOS which is state of the art. It has a turnaround time of approximately four weeks.

Remember that Bob Norris at Intel was facing an 18-month period to get his new microprocessor. The idea of reducing that time and making it possible for a single designer to do the job, rather than a team of 50 designers working for a year and a half, was the contribution of the MOSIS Project. This process now works, and it works very well. The industry is moving toward application-specific individual designs. It is kind of like what compiling computer programs was back in the 1960s and 1970s. You wrote your computer program on punch cards, gave it to the computation section, and they gave you back your compiled program. The next day you checked and debugged it, and gave them back the debugged program deck. They ran the program and returned the answers to you. Now you do it in silicon. You send your design in; they compile it in silicon and send it back. You debug the IC you get back, send back the new design, and they send you back a few hundred or a few thousand parts that work.

This concept of rapid prototyping was introduced by Mead and Conway at Cal Tech in the 1970s. Lynn Conway came to work at ARPA for a couple of years. She worked in the machine intelligence program and has now moved to the University of Michigan where she is Dean of Engineering. She has the idea of trying to move the rapid prototyping that was accomplished in the integrated circuit business to mechanical rapid prototyping. What if you could turn around the mechanical design of something in four weeks and prototype it so that you did not have to get it all right the first time? You would not have to put your design into a model shop and have little tool makers diddle away at it for two, three, or four months in a queue. Instead, you could get it back from a numerically controlled machine or some automated process in a few weeks. What would that do to mechanical design? Would it change mechanical design in the same way that rapid prototyping is radically changing the integrated circuit industry today?

All of the things I have talked about tend to drive down the cost. In the case of the MOSIS Project, the $50,000 per lot does not change. However, if there are 100 designers who have 100 designs on that wafer, then for each designer, the cost is $50,000 divided by 100 or $500. If a design could be done at 1.25 microns for $500 bucks, that is a real bargain! What it means is that the cost of the design has been reduced from 18 months of several highly trained people who, fully burdened, cost $150,000 per year, to maybe three or four months of one $150,000 person working for a short period of time. Furthermore, the cost of actually prototyping and testing the chip is trivial compared to this individual's salary. That is an amazing reduction in the potential costs for producing specific kinds of integrated circuits.

I believe there are going to be amazing reductions in the cost of hardware that can be used to create inexpensive new design automation systems. What you find in the electronic design industry is that it is difficult to justify investing more than about two or three times the cost of a personal computer in a design workstation for an electronics designer. Capitalization beyond that gets to be problematic. The goal in the IC industry is to get a complete workstation and design package that will put a moderately capable design engineer at the state of the art for $50,000 to $100,000. If you can do that, then you have a product that will sell and something that will be incredibly powerful in creating a new capability for the electronics designer, with which he can bootstrap his technology.

What can chips be built for and what kind of systems can be built with this capability? Let me give you some examples. There were several chips that were built at companies under ARPA sponsorship using the MOSIS Project rapid prototyping. These chips were either controllers for computers or complete computers. One was a restricted limited instruction set architecture computer called MIPS that was built at Berkeley. This chip had about 50,000 gates on it and a reduced instruction set which was carefully tailored to handle certain kinds of problems. It was a general-purpose computer and worked remarkably well. That whole chip, which was essentially the functional equivalent in its computer power to the $50 million chip that Bob Norris told me about, probably costs something like $100,000 for ARPA to produce. The cost has been reduced from $50 million to $100,000. The time period was not much different; it did take 18 months, but that was because it was done in a university. It was not done in an industrial environment where it would have been prosecuted on a much more pronounced schedule.

That is one indicator of investment cost. The recurring cost is probably not all that much. However, if you only buy thousands or tens of thousands, instead of millions, the recurring cost has to be weighted by the cost of design and development. You have got to keep the design and development costs down if you want to buy in limited numbers. That is a generic problem in the DoD. The DoD never buys many of anything, and the preponderance of the cost of every unit, whether it is an aircraft or a ship or whatever, is in the R&D costs. It is not in the recurring cost of stamping out new copies.

Let me tell you about two other chips. One chip is another computer developed at Carnegie-Mellon called the systolic array. The computer is essentially a multiprocessor architecture computer with one chip as the basic design element. The design and development was done at a very low cost and will be produced by a company named ESL, I believe, in the coming year or so for defense purposes. Another interesting design, by Bolt Beranek and Newman, is the switching element in their new multiprocessor computer called the Butterfly. An upgraded version of that element will go into the new super Butterfly which, of course, is called the Monarch. That chip probably cost 10% of what it would have cost if done by the merchant semiconductor industry in the past. So, we have, through the prototyping design process, reduced the cost of producing very large scale integrated (VLSI) hardware by an order of magnitude.

Some of the multiprocessor computers that I see coming on the market these days are beginning to drive down the cost per mega instruction per second (i.e., cost per MIP) into ranges where we may all be able to afford supercomputers. To give you some examples, a Cray computer currently will run at a sustained rate of perhaps 30 to 80 MIPs and at peak rates of 240 MIPs for a quad Cray machine. Because the price of a Cray is over $10 million, that boils down to around $50,000 per MIP. In contrast, there are multiprocessor machines now built out of Reduced Instruction Set Computer (RISC) architecture chips, that have been developed at low cost, and which are in the range of $5,000 per MIP.

I can see the day when we may be able to buy computers, particularly multiprocessor computers, at less than $1,000 per MIP. That means a 500-MIP machine would cost $500,000, which is about what a super-mini costs today. This is a lot of money, particularly for a design station, but think of what you could do with a design workstation that has 500 MIPs of computer power.

C. Computer Graphics

Graphics can be done that are comparable to what Tom Furness (1986) describes in this volume. Furness has the designers surrounded by incredible representations of reality. The movie industry is picking up that technology at a terrific pace and the DoD could do more with that. ARPA had a project in this area which was used to make one of the science fiction movies. We funded some work at Boeing to use a certain kind of mathematics called fractal geometry to create images symbolically. In other words, a symbolic representation is created of an image which is very sparse. The information required to represent the image symbolically may be a thousandth, a ten-thousandth, or a hundred-thousandth of what is needed if the image is stored completely. Then, fractal geometry is used to create that image, and the result is not only the image but also the textures of the image normally missing from symbolic representations of images in the past. It does it in a way that makes the images look very realistic, like a photograph. The problem is that it takes enormous computer power to do that because the images have to be put together in real time at rates of 30 times a second or so to keep the flicker rate down to allow motion in the images. We have taken the algorithms to do that and have converted those algorithms to VLSI circuits.

There are application-specific chips to do this that are now in the works, and they will make it possible to create the ability to fly a cruise missile down a valley and to see everything in that valley exactly the way it would be seen by the cruise missile. It would be seen with very high fidelity by storing the symbology of the image at full bandwidth. It would be stored at an incredibly reduced bandwidth. It may be possible someday to put the images that one can see with reconnaissance photography into this symbolic representation, store these images in a cruise missile, and make it possible for the cruise missile to see differences between what the images were before and what they are now as a basis for en route planning.

Think of what you could do with that technology if you used it in a design workstation. You could take theoretical representations of almost anything, create a symbolic representation of it, use the fractal image generator to generate an instantaneous image of it, and then rotate or change the image in specified ways. To the extent that Furness is right—that design is a visual process—such images could be very powerful, and there are systems that produce these images at low bandwidth today. For example, they are used in the aerospace industry to help design structures within aircraft design.

D. Computer Networking

Thus far, I have talked about computational power and rapid prototyping, both for electronic and other kinds of equipment, and how these developments are going to reduce the cost of electronics and make more powerful design aids possible. Let me talk now about networking and the way design is done. When I was doing designs at MIT in the magnetic confinement fusion lab, we had a small group of people who worked together as designers. Some of us were good at the theoretical things that happened to plasmas when they get hot and confined. Some of us were good at designing magnetic fields. I remember Ward Getty could, in a few days, design a new electron gun that would fit well with the particular thing that we had in mind. Another couple of persons in that crowd were good in creating the electronics that would control the gun in a certain way to get the temperature to rise exactly right. We would work as a team to design a particular configuration. Then we would send Louis Smullin off to the National Science Foundation to see if enough money could be drummed up to actually create that design. People still do designs in teams, but it is a different kind of a process when using design aids.

For instance, ARPA had a program called the X29. It was an advanced, forward swept wing, integrated technology aircraft that was built by Grumman. Grumman had several engineers working on the design and they used advanced design aids. In fact, one of the goals of the program was to develop design aids that could be used later for other advanced fighter aircraft. Design workstations were networked, and a design integration software package was built that functioned like the supervisor of the design group. This idea was a bit counterculture for the engineers because typically they had worked together on a personal basis in the past. They had done their detailed designs in their individual design areas, and they would get together for weekly meetings to integrate their individual designs. Instead of this process, the design automation system allowed them to interact more quickly on a day-by-day, hour-by-hour, even minute-by-minute basis. The supervisory software kept track of how the design was changing, kept everyone apprised of what was happening and why, and off-loaded the burden of information management from any one particular person.

I think this idea has incredible power. It is an idea that has been tried at ARPA before, the most salient example of which is the ARPANET, which is a netted communication system that builds technical networks. A technical network is a group of people who have a common interest and who have both a communication channel which connects them and a set of software aids that make it possible for them to communicate easily through this network. In this way, they are able to pass information to one another in a way that is friendly to their normal mode of communication. In the past, this information transfer has been done by journal publications, by visiting professorships, by personal letter writing and communication, and by spending time at each others' homes. In contrast, through the ARPANET, we have seen groups of people building relationships that are essentially electronic relationships. These relationships can be as strong and powerful in terms of providing cross-pollination of ideas as the more familiar techniques have been in the past.

That which has been accomplished on a national basis through the ARPANET is now being tried in local area networks to build a close local community of interest with a supervisory routine which makes it possible for the design to move more rapidly. Nowhere is this type of design aid more powerful than in the area of software systems design. Software system design is a black art. It is variously described as a management problem, as no problem, or as an impossible problem. Whatever it is, there is certainly no science to software system development. It is an art and needs to be handled by a close coupling mechanism, particularly as the scale size of software systems grows. And, it is growing—the software system for a program like the space shuttle, for instance, is an enormous problem. The software system that will control the space station will be an even larger system. The software system that might be associated with a future strategic defense system boggles the mind. No one knows how to create such a software system today because the design tools are not available to make it possible for the literally thousands of people who would have to participate in that design.

Clearly, executive routines, or these close collaboration networks with a supervisory routine, is a design concept which is going to be developed and will be very powerful. Characteristic of that kind of system is one I saw a couple of years ago being developed at Xerox Corporation Research Laboratory in Palo Alto, California. The system is called CEDAR and is an attempt to make it possible for a group of software system designers to work together electronically. A noncommercial product was designed for Xerox to build this experimental network. Designers can, in principle, work together on a software project of sizable proportion. Xerox had succeeded in avoiding the need for face-to-face communication through experiments with a supervisory routine which could supervise and coordinate the activities of 10 to 15 programmers. Although these programmers were at Xerox and they had offices scattered around the building, they could have been thousands of miles apart. If this concept could be generalized and made so that a thousand people or even ten thousand people could contribute in a meaningful way to a sizable software project, then some day it may be possible to build a command and control system for a strategic defense system of the future.

E. Machine Intelligence

Another thing that is coming down the pike at a pretty good clip, if Graham-Rudman-Hollings does not stop it cold, is the ARPA machine intelligence program. Machine intelligence is going to have a substantial impact on the design automation business in the future. Things are happening at a rate which is both gratifying and frightening for those of us who are somewhat resistant to change.

Let me outline a few of the advanced technologies designers are going to have available to them in the not too distant future. First of all, the ARPA program is a substantial program; it was billed as a billion-dollar program. Actually, it builds up in three steps from the first step of $50 million to the third step of $150 million per year for three years. It is scheduled to continue at the $150-million rate for seven years, which would put it in the billion-dollar category. The principal objective of the program is to create intelligent machines, that is, computing machines that can emulate many of the processes that go on in human interactions with their surroundings. Some of the key technologies that are being worked on are speech understanding, image understanding, and natural language processing. The natural language effort will concentrate on English first, but foreign languages will come later. A natural fallout of this research should be a translator. There is a major push to create multiprocessor architectures that will be able to support image understanding, speech understanding, signal processing, and symbolic processing of various kinds including running expert systems. Of course, the whole issue of expert systems and how to create software packages that emulate human thought processes is also a major part of the machine intelligence program.

Multiprocessor architectures are the road to low-cost computation. If the combination of the electronic cost reductions I mentioned before are proliferated in multiprocessor architectures, it is conceivable that computer powers can be driven up to 1,000 to 10,000 times the processing power of the most powerful computers that are available today. It is conceivable, but there are theoretical barriers.

There is an old saw that says that any discipline that has the word science in it probably isn't, and that probably goes ten times over for computer science. Computer science is a phenomenologically based discipline. It has almost no theoretical underpinnings—a terrible circumstance for the world to find itself in. As a matter of fact, during the approximately 25 years that large-scale computers have been available in the western world, one would expect that somebody would step forward, a sort of Einstein character, who could take all of the computer programming experiments of the past 25 years and create a statistical mechanics of computation or a relativity theory of computation. That smart person has not stuck his head out of the woodwork so far, and computer science continues to evolve as an experimental discipline, with few overarching principles bringing it all together.

I still hope that it will be the other way, but I spent four years at ARPA trying to interest some of the bright people in the computer science community to think of computer science as a real science, to find out its foundations and where its limits of computation are, but the community is too wrapped up in its own fun things. Speech, vision, and natural language are primary examples.

They get ARPA to buy a big computer. Then, they go off and do heuristic experiments in these fields and it is great fun. It is not great science.

Nevertheless, I think we will make incremental progress, particularly because there are so many people working in this area. ARPA provides $150 million every year for work on machine intelligence technology and, as a result, it is going to move forward and make a big difference. For instance, I believe that by the early 1990s, in the area of speech, you will be able to talk to your computer and it will be able to talk back to you very well. It will be able to do as well as the average MIT undergraduate, which is fine as long as it is on a topic where it has the necessary contextual knowledge. It will not be as erudite as Truman Capote, nor will it be as knowledgeable in some narrow discipline as a fellow like Herb Simon. However, in a particular discipline area, such as fighter aircraft design, you will be able to communicate with a machine in that context and hardly know you are not talking to a person. Of course, I am a technology freak and an optimist.

For machine vision, I think it is going to be a lot longer before we get systems that will actually be able to understand images well. There are a lot of interesting things happening these days, particularly in the DoD, concerning automatic target acquisition. There are infrared images of a target area and an automatic device for looking in that area to pick out things that look like they might be targets. Much progress is being made there, but I think in terms of being able to understand images in the way that a human being understands, and in the kind of detail possible for humans, machine vision is further off and may require a lot more technology in the sensor. It would be nice to have a sensor as simple as the eye and the ability to get as much understanding out of the image as the mind gets out of the image the eye presents.

Natural language processing is part of the speech activity. The issue of having a machine talk back in ways that are understandable is really an issue of reasoning and planning. When we talk, we plan what we are going to say. What are the points to be made and how are they best expressed so that the audience can understand? This is a very complicated process and may be the toughest part of speech understanding. Just turning connected words into script is a tough problem. It is, as yet, an unsolved problem in general, but people are closing in on it by taking advantage of the context of the discourse.

F. Designer's Associate

We have multiprocessors and associated architectures. The next task is to take problems and divide them so that several processors can work on them efficiently and simultaneously. Let us say that by the early 1990s, the machine intelligence program has solved many of these problems or substantially solved them. What does that mean for the 1990s design aid? Well, it means that the concept, at least today, of a smart Designer's Associate, which is the functional equivalent of the person I had in my design group at MIT, is reasonable.

Remember, this person could design those magnetic circuits so well because he had this incredible capability to see how to put those strips of copper around those pancakes we built to confine the plasma. He could see the magnetic field and exactly where to put the copper bars in order to make the magnetic field he wanted. He was a Design Associate and could whip up a design in a few days. It is that kind of capability that can be seen in a machine in the early 1990s, and it can be seen in such a way that the designer can talk to the machine and get the design out of the machine in the same way I got it out of my friend at MIT. That is, we would discuss the nice features we would like to get and what it would take to get those features. He would then do the design. Then he would come back and say, "Here it is." At that time, we would have to build it and test it because the IBM-709 we had could not simulate it. It would take a year to design it, build it, and test it in the lab if NSF would give us the money. The Designer's Associate may well be able to do the testing also, as it is likely to have the necessary simulation capability built into it—particularly if 500 MIPs can be bought for $500,000.

The Designer's Associate may also be able to maintain a data base, a sort of corporate memory of what has happened in the past. One nice thing that I found about the folks around MIT was that they were a repository of all the things that had gone wrong before. The same mistakes were never made twice. In contrast, at ARPA the technical people come and go about every three and a half years; thus, the same mistakes may occur but nobody remembers them. It would be nice to have a data base which is a corporate memory. It is possible to conceive of a Designer's Associate that remembers things archivally and learns from experience. In other words, it has the requisite experience and the ability to express this experience to you in the same way as a human associate expresses his experience in a particular design discipline.

Machines using expert system technology available today can both respond to "what if" questions and can explain, in considerable detail, the reasoning behind the process that led to the conclusion of the "what if" question. It would be extremely useful to have an associate that could explain how it arrived at a solution, and give you the intermediate steps in the process and some of the intermediate results. A real-time simulation of a preliminary design arrived at in conjunction with a Designer's Associate would also be a powerful capability. Of course, modeling reality comprehensively costs computer cycles, and that costs money. Thus, the real issue is how much can the cost of computer cycles be reduced in the future. As I emphasized, this cost has been decreasing at an incredible rate over the last ten years. If it would just continue for another 20 years, you really will be able to model reality and completely avoid doing experiments. The trend is certainly in that direction. For instance, at NASA's Ames Research Center, they are building a national aerodynamic simulation facility with a concentration of class six computers that rivals the capability of the national weapons laboratories. This capability will allow them to model complete aircraft and to simulate the aerodynamic flow around these aircraft.

Effects of Technological and Organizational Trends

This allows several design iterations where all of the testing is done numerically in the machine and only, perhaps, the last version would be put into a wind tunnel. Thus, real-time simulation will give designers an even more powerful way to seek optimal designs. Instead of "satisficing," you go for optimal. With enough computer power, the designer can afford to expend the effort to find an algorithm which allows him to work his way down to that extremum of performance that every designer really desires.

That is my concept of the Designer's Associate. It is where things can go if electronic technology, and some of the other technologies I have talked about, really work.

VI. ORGANIZATIONAL ISSUES

A. Attitudes Toward Risk

Now I would like to turn to what I believe is the most discouraging aspect of engineering design today. The fact is that in many cases, designers are not allowed to let their creative capabilities really go. This is because both commercially and militarily, we are wedded to incrementalism. We take concepts we had in the past and make them a little bit better. Now that is not always bad. If you take a jet engine or turbojet engine and make it 5% better every three or four years, you end up with an F100 engine after about twenty years. It is not a bad engine; it "goes like 60" and it powers a lot of high performance aircraft. But that is not the "be all and end all" of creative design. There is a high road that takes the power of advanced technology and asks: What if I try to make that giant leap forward? The problem is that the system is not rigged up these days, either in the commercial industry or the defense industry, to actually take these leaps of design and faith. There are not enough chances to do the far out design, to test it, to try it, or to prototype it. The reasons for this are expense and risk.

The DoD ought to spend some substantial fraction of its resources on R&D. We are not doing that now. The basic research budget has been relatively level, but that is not where the problem is in the DoD or industry. The problem is at that point where you have something that you have perfected somewhat in an exploratory development program, and you really need to try it out on a substantial scale. That is where the bottleneck occurs. The bottleneck is due to the cost which ranges upward from tens of millions of dollars to hundreds of millions of dollars for testing integrated systems. You have to be ready, willing, and able to make that level of investment on speculation. In other words, you do not really have a requirement for it in the hard sense of the word, but you can see that if you had it and it worked, it really could make a difference, although you would have to figure out how to use it after you proved it.

This approach is perceived as a risky business by defense managers, and they do not like to accept risks like that. It is even worse for commercial industry and that is why the Japanese are overtaking IBM so quickly, just as they did with General Motors. General Motors has not spent a risk dollar in creating new technology for its industry for 50 years. Even the electronics industry is getting to the point where it is not willing to take flyers. The flyers are being taken by venture capital and even venture capital has had some second thoughts recently. We have what I believe is a national crisis in our unwillingness to prototype things, to take technology and move it to product. In contrast, the Japanese are taking our technology, moving it to product, and bearing the expense of that prototyping step. This problem is acute in the defense business, and it is where the engineering design industry is being hurt most.

B. Regulatory Pressures

This situation is worsened by the fact that the relationship between the DoD and industry is very poor these days. As I see it, the reason is that ever since the latter part of the 1960s, when there was a general disaffection with the DoD and defense R&D, the lawyers on Capitol Hill piled on laws to control defense procurement, and those laws have resulted in many regulations. These regulations have, in turn, resulted in layers of bureaucracy. These layers of bureaucracy have resulted in many people watching the people who are doing R&D. There are also watchers to watch the watchers. We have created an enormous inspector general bureaucracy, an enormous contract audit organization, an enormous bureaucracy of approvals where every rule and regulation imaginable to man, and some of them unimaginable, are being administered by layer after layer of bureaucracy. It has stymied our ability to make rapid decisions about what is right and what is wrong, to concentrate power and authority to manage projects on a substantial scale, and to drive them through to completion. There are also the fits and starts of interference on a line item-by-line item basis by Congress, with the defense R&D budget making very irregular progress.

You may know that the President's Blue Ribbon Commission on Defense Management, also called the Packard Commission, has just delivered a preliminary report to the President. They have proposed a sweeping reconstruction of the DoD management in almost every one of its aspects, starting from the Secretary's Office and the Joint Chiefs of Staff, all the way down to the basement of the Pentagon. The report is great and sweeping and all done in 26 pages—it is the most fantastic report I have ever seen. Of course, the Pentagon cannot take a recommendation like this and do anything with it. Therefore, the Packard Commission is writing a detailed report, due in three months, to provide a cookbook that the Pentagon will have to adhere to if they want all of the things that are recommended in this report to happen. One of the recommendations in this report is that a new emphasis be placed on the business of technology transition. The DoD and industry are creating technologies that need to be considered in our defense systems. The DoD needs to pay attention to prototyping and feasibility demonstrations.

The 6.3A account, or the advanced technology demonstration account, in the DoD is meant to accomplish this. That account has about $2 billion in it. What happens to all those dollars? Those dollars are used for almost everything imaginable except for transitioning technology from the laboratory into our weapon systems. It pays for overruns in engineering development contracts, which interestingly enough are also included in the combined 6.3 account. It is easy to get the dollars back and forth across this permeable barrier between the 6.3A and 6.3B accounts. The 6.3A account pays for "hobby shops" in laboratories and R&D shops all over the defense establishment in $50,000 and $100,000 chunks. You can imagine how many hobby shops you can buy for $2 billion.

The 6.3A account also pays for some perhaps questionable kinds of things in large projects such as in the strategic defense program. One of the things I have been urging as I go around the country talking to the scientific and engineering community is that the strategic defense program is the President's program, and there is no way that the community is going to get rid of it. The community should find out what is right with it and what is wrong with it, and then give some constructive criticism. One of the things that is wrong is in that 6.3A account. They are wasting money and not using it for technology transition. We ought to demand that money be spent wisely.

C. Prospects for Change

I believe that the focus of attention in the Packard Commission report is in the right place. Its recommendations can, if properly managed in the defense establishment, provide an outlet for engineering design creativity. Its recommendations on prototyping will go a long way toward relieving the technology transition stagnation that currently exists. We need to create more experimental aircraft. We need to fill up the technology pipeline to the feasibility demonstration level. We need to move our technology more rapidly into the area of confidence where a designer knows that an idea can be used because it has been tested. He knows what it can do, knows where its limitations are, and has all the necessary design data. This allows him to decide to use it and make something better out of it. That is where our process is falling down. It is falling down in defense and in the commercial industry, and we need to do more about it.

VII. QUESTIONS AND ANSWERS

Q: What are the prospects for the Packard Commission report?

A: Well, let me give you a little historical background on that. I went to the Pentagon for the first time in 1972. In 1972, you may remember that Nixon was the President, Melvin Laird was the Secretary of Defense, David Packard was the Deputy Secretary of Defense, and Johnny Foster was the Director for Research and Engineering. I worked for Johnny Foster. When I got there, Dave Packard had just received a report from a blue ribbon commission chartered by President Nixon to look at the abominable R&D management practices in the DoD. The blue ribbon panel was chaired by a prominent industrialist known as Fitzhugh, and the panel's report later became known as the Fitzhugh report. It exists in some sizable number of volumes in the DoD archives. The Fitzhugh report contains recommendations similar to those in the Packard Commission report and, of course, Dave Packard was right there. He was the defense official they handed it to. He said thank you very much and he marched with Sancho Panza by his side to try and do something about it. Now I have to say that shortly after that he was called away. He had to resign his job in the Pentagon to give some attention to his business interests. I think just about every Deputy Secretary after that has tried somehow to take some of the Fitzhugh recommendations and put them into practice. The most recent was Frank Calluci, who created 32 initiatives out of what was the grist from the Fitzhugh report to help the Pentagon correct some of its faults.

The Packard Commission report is comprehensive. It goes further than the Fitzhugh report because it reorganizes the Joint Chiefs of Staff; it reorganizes the Office of the Secretary of Defense; it changes the procurement cycle to two years with a five-year plan that is submitted to the Congress; and it does not allow the Congress to touch line items at all—Congress can touch only mission areas. Congress is interested in agreeing to this, partially because Dave Packard has gotten Congressional support. He gave the report to the President on national television. In his letter to the President, he recommended the changes be implemented as soon as possible. The President said he would implement them as soon as he could. Cap Weinberger, later on national television, said he liked that report. I think it is great. Now I would say that all these endorsements bring the probability up to about 5% that something will happen. Of course, you have to realize that I am jaded after having been in Washington so many years.

Q: Earlier in your talk you made the remark that at ARPA, only the program manager and the Director have to agree in order to get a project going, which seems to make ARPA an ideal customer for industry. My company is fortunate in just having been awarded a contract which originated at ARPA. However, this contract is administered by the Air Force Aeronautical Systems Division (ASD), and there was a battle as to whether the Avionics Lab or the Flight Dynamics Lab would control the project. At our kickoff meeting, we had 16 people from ASD and 2 people from ARPA. Would you comment on how it is decided who is going to administer ARPA contracts?

A: It used to be the ARPA program manager. Now, ARPA is getting caught up in the same procurement bureaucratic mess as the rest of the DoD. More and more, the administrative decisions are being removed from ARPA's control. It is not quite as bad as the rest of the DoD for a lot of reasons, but one of the last things I did as a parting shot at the system was to talk to Will Taft, Don Hicks, and a number of people on the Hill about the fact that bureaucratic regulations were closing in on ARPA. There are too many inspector generals running around. There are competition advocates all over the military departments that do not recognize the special arrangements that ARPA has made with the services. Before I left ARPA, I was spending all of my time trying to solve those types of problems rather than trying to run good programs, find good technology, and convince the services they ought to use it. I do not know whether it is going to get better or get worse. If we implement the Packard Commission's recommendations, it is going to get better.

Chapter 4

DESIGNERS, TOOLS, AND ENVIRONMENTS: STATE OF KNOWLEDGE, UNRESOLVED ISSUES, AND POTENTIAL DIRECTIONS

William B. Rouse

Search Technology, Inc.
Norcross, Georgia

Kenneth R. Boff

Human Engineering Division
Armstrong Aerospace Medical Research Laboratory
Wright-Patterson Air Force Base, Ohio

I. INTRODUCTION

As explained earlier, the "Themes and Issues" chapter (see Chapter 2) was provided as background reading for the workshop participants. After receiving and reviewing this background material and responding to a questionnaire keyed to this document, each participant was asked to prepare a position paper on a particular subset of the themes and issues. These position papers were distributed to all participants prior to the workshop. Revised versions of these position papers appear in this volume.

Reviewing briefly the format of the workshop, there were only three formal presentations. These were the one-hour keynote talks by Major General Fredric Doppelt, Commander of the Air Force Aerospace Medical Division; Brigadier General Leslie Beavers, from the Office of the Army Deputy Chief of Staff; and Dr. Robert Cooper, former Director of the Advanced Research Projects Agency. The rest of the three-day workshop was devoted to plenary and small working group discussions, as well as presentations of the results of each working group's deliberations. This chapter integrates and summarizes these results.

A. Objectives

Each working group had the same objectives. First, they were to assess and outline the current state of understanding of the psychology of design. Second, they were to determine where important knowledge is lacking. Finally, they were asked to recommend R&D efforts that would both provide missing knowledge and contribute to improving system design.

B. Method

Four working groups were formed by dividing the 26 participants into groups of six or seven people. Group assignments were counterbalanced in terms of disciplinary background and organizational affiliation, with engineers, psychologists, and industrial designers/architects equally divided among the groups. Similarly, individuals from government, industry, and academia were divided as equally as possible. Each group was chaired, in a nondirective manner, by a participant from one of the organizations of the workshop chairmen.

Each group's assignment was to fill out the matrix shown in Figure 1. As indicated in this figure, the groups worked on one column per day. Approximately one half of each day was devoted to completing this assignment, and the remainder of the day was allocated to background discussions, brainstorming, etc.

Each group was provided on each day with four transparencies for overhead projection. One transparency was similar to Figure 1 except that the column of

Figure 1. The Format of Working Group Assignments.

TOPIC / AGENDA	DESIGNER	TOOLS	ENVIRONMENT
	Designer behavior from both individual and organizational perspectives.	Design support systems from both psychological and technological perspectives.	Implementation and evaluation from both psychological and methodological perspectives.
STATE OF KNOWLEDGE Level of understanding, current practices, and regulatory constraints.			
UNRESOLVED ISSUES Research, technology development, and organizational considerations requiring attention.	1st DAY	2nd DAY	3rd DAY
POTENTIAL DIRECTIONS Benefits, costs, and priorities of alternative R&D investments.			

Designers, Tools, and Environments

interest for the day's assignment was much wider, and the other two columns were very narrow. The remaining three transparencies were blank except for labels: State of Knowledge, Unresolved Issues, and Potential Directions.

At the end of each day, a plenary discussion was held. One member of each working group was selected to present their group's results using the four transparencies. A different individual presented each day, and 12 of the 26 participants made presentations. The resulting 48 transparencies (4 per group per day x 4 groups x 3 days) constituted the products of the groups. This chapter is an integrated summary of the findings documented in the 48 transparencies.

C. Integration of Results

The sections of this chapter are organized according to the columns of Figure 1; subsections are organized according to the rows of this figure. However, there is no attempt to associate particular conclusions with specific days. Similarly, conclusions are not attributed to specific groups.

As might be imagined, a substantial amount of interpretation is required to synthesize 48 transparencies into a coherent chapter. This process was aided by notes taken during the aforementioned plenary presentations, as well as the high degree of consistency among groups. In addition, interpretations were verified by sending a draft of this chapter to all participants for review and comment prior to preparing the final draft for publication. Nevertheless, the editors accept full responsibility for the interpretation, integration, and summary presented in this chapter.

II. THE DESIGNER

Designers of aircraft cockpits, power plant control rooms, automobile interiors, etc. have long realized that understanding users' needs is essential to the design of a successful system. At the very least, the marketing and sales of such systems require an understanding of what users value. With this user-oriented perspective in mind, the working groups devoted the first day of the workshop to discussing the nature of design and designers as a basis for later consideration of design support systems.

A. State of Knowledge

In contrast with aircraft pilots, power plant operators, and automobile drivers, it is surprisingly difficult to determine *who* to study if the target population is identified as designers. Similarly, it can be difficult to delimit the activities of this elusive population. Despite the fact that many of the individuals in the working groups were trained as engineering or industrial designers, these issues were surprisingly difficult for the groups to resolve.

1. Who is the Designer?

To answer this question, one might look for those individuals whose job titles include the word "designer." However, this answer is much too narrow. For example, in the aerospace industry, those with the job title of designer are usually limited in responsibilities to detailed design and drafting. The conceptual design is usually complete before these individuals become involved.

Once the scope of the answer to the "who" question is broadened, it becomes very difficult to identify particular job titles, department names, etc. that qualify to be included. The working groups reached this conclusion quite early. The consensus was that designers are different people at different stages of design. Thus, rather than focus on job titles and so on, it is more useful to consider what designers do.

It was also noted that task requirements, as well as the individuals involved, change as the stages of design progress. The level of detail in the design obviously increases as the process moves from concept to product. The degree of flexibility also decreases, partially due to past decisions and partially due to regulatory factors such as those involved in government procurements.

These considerations imply changing information and knowledge requirements as the design process proceeds. For example, particular information may not yet be valuable because it is more detailed than the current state of the design. Similarly, specific information may no longer be valuable because flexibility has decreased to the point that new information can no longer be incorporated.

2. What Do Designers Do?

By formally characterizing designers' activities, it is likely that approaches to supporting designers can be better identified and integrated. Figure 2 is a list of 11 activities that the working groups felt were characteristic of designers. These activities can be grouped into four tasks as shown on the right of Figure 2.

It is of particular interest to note that synthesizing a solution, which from a traditional perspective might be viewed as the essence of design, is only one of several important tasks and activities. Formulation and understanding the problem or need and fostering acceptance of the solution are also essential. This assertion was easily accepted by the industrial designers and architects at the workshop, probably due to the close designer-user relationships typical for those individuals. In contrast, the importance of this conclusion was not initially obvious to many individuals whose experience was limited to large aerospace design organizations.

The activities in Figure 2 can be viewed as information transformation processes. More specifically, each of these activities involve various degrees of seeking, interpreting, manipulating, and disseminating information. Thus, designers and design organizations might be characterized and analyzed in terms

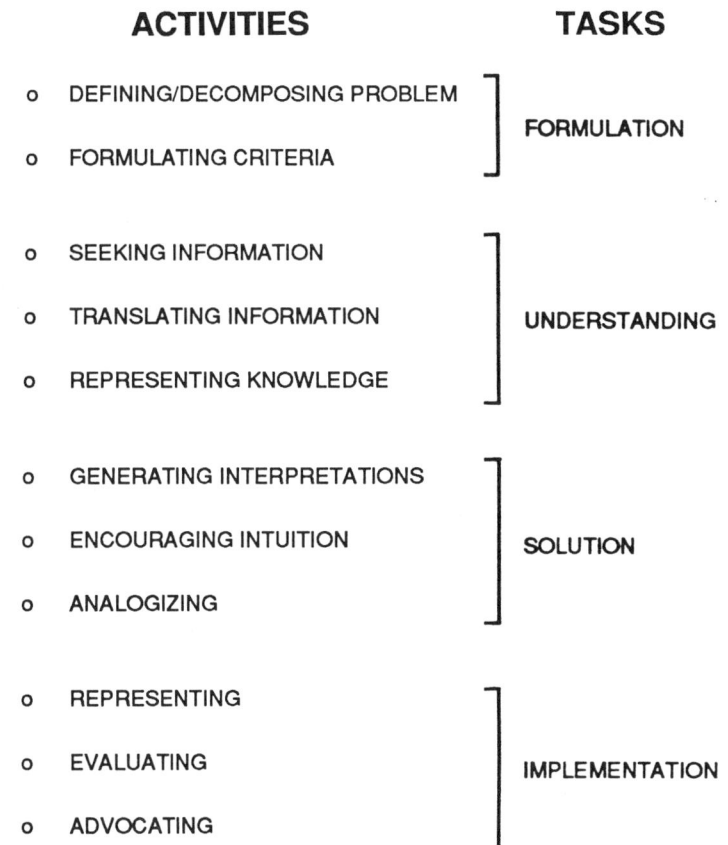

Figure 2. Activities of Designers.

of the information processing involved. To determine the usefulness of this characterization, it is necessary to look at designers' activities in more detail.

3. How Do Designers Pursue these Activities?

This question has been empirically and philosophically addressed by a wide range of commentators, many of whom are referenced elsewhere in this volume. One school of thought describes design as a top-down decomposition of objectives into functional requirements and then physical processes, with a variety of cost versus performance tradeoffs being resolved along the way. This description typifies the *analytical* view. Another school of thought advocates the *artistic* view—designs emerge from bottom-up perceptions of patterns in the context of the designer's understanding and experiencing the whole problem or need.

The analytical school argues that well-done analytical decomposition can

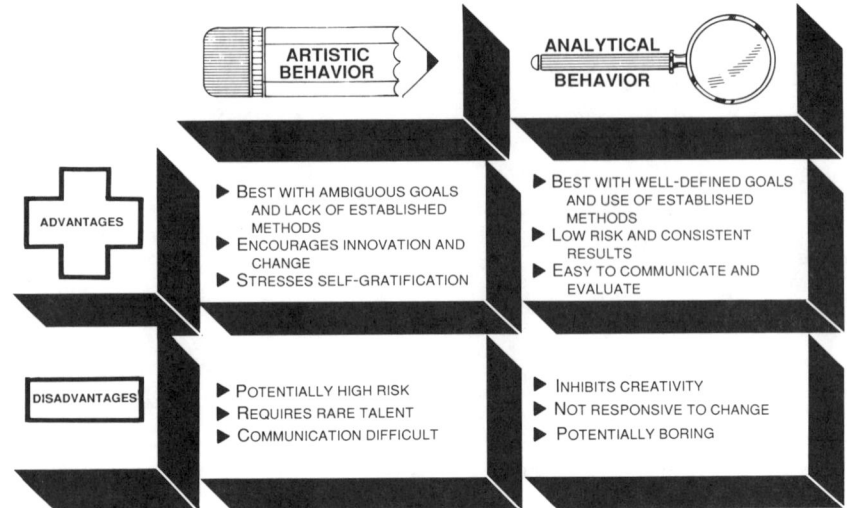

Figure 3. Comparison of Artistic and Analytical Design Behaviors.

make subsequent composition very straightforward, while the artistic school claims that the subtlety and richness of composition is the essence of creative design. To an extent, both of these assertions were accepted as correct. The salient issue concerns the appropriate mix of analytical and artistic behaviors. Figure 3 was produced by one group as they wrestled with this issue.

Appropriate tradeoffs between the advantages and disadvantages shown in Figure 3 depend on the nature of the design problem of interest. For large-scale, well-defined problems, where minimization of risk is essential even if innovation must be compromised, designers' behaviors should be predominantly analytical if appropriate methods are available. A good example of this type of problem is aircraft design.

In contrast, for smaller problems where coordination and integration still exist but are not overriding issues (e.g., all of the design team can sit around a table), and innovation relative to ambiguous goals is paramount, the artistic component of designers' behaviors becomes more important to success. Consumer products such as high-fashion clothes and toys are good illustrations of the type of design problem where purely analytical approaches are unlikely to be sufficient.

Considering the analyst versus artist dichotomy in terms of the activities listed in Figure 2, it seems reasonable to argue that all of the four primary design tasks (i.e., formulation, understanding, solution, and implementation) involve a mixture of decomposition and composition, with more emphasis on decomposition earlier in the process and a shift toward composition later in the process. From this perspective, differences in the analyst-artist tradeoff for complex military systems and consumer products may be due to complex

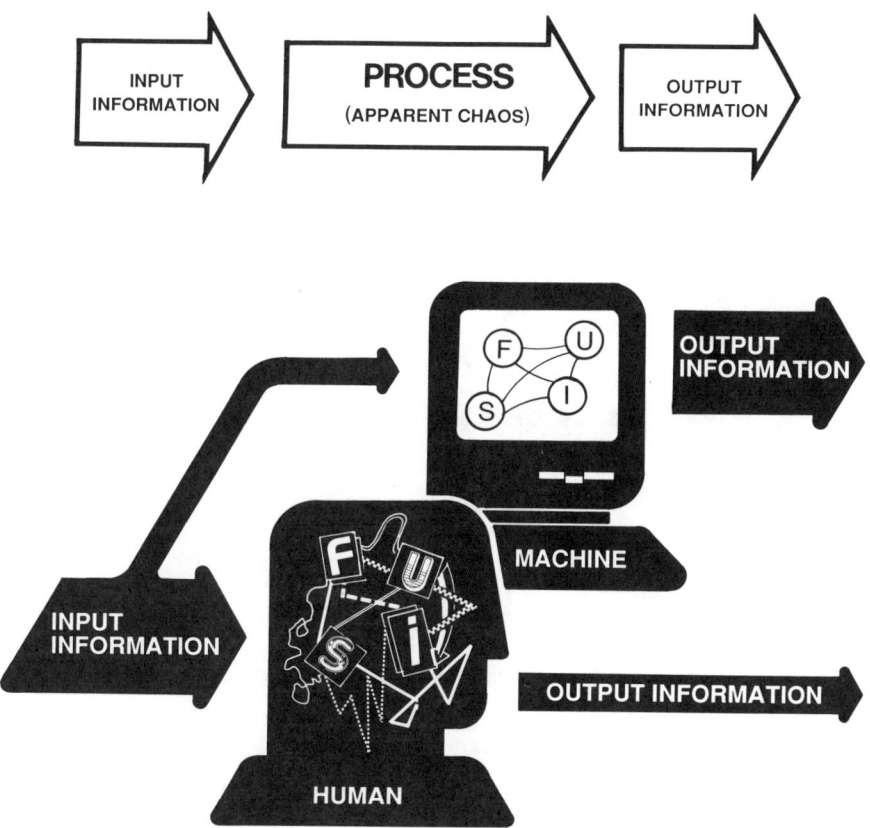

Figure 4. Apparent Chaos of Information Transformation Process.

systems requiring much more formulation and understanding prior to solution, while simpler problems allow the designer to clarify the formulation and develop understanding in the process of composing alternative solutions. Thus, differences in the mix of behaviors observed in different design domains may primarily reflect the natures of the design problems in these domains.

If an outside observer were to characterize designers' behaviors, particularly for complex domains such as aircraft design, it is quite likely that such an observer would conclude that chaos is the most appropriate characterization of design teams at work. Figure 4 was produced by one group in an attempt to capture this confusing image of design. Of course, the apparent chaos is, for the most part, due to the inability of an outside observer to know exactly what is happening.

The simple input-output diagram at the top of Figure 4 represents the apparent chaos as an information transformation process, a characterization that emerged repeatedly as an abstract description of design. From this perspective,

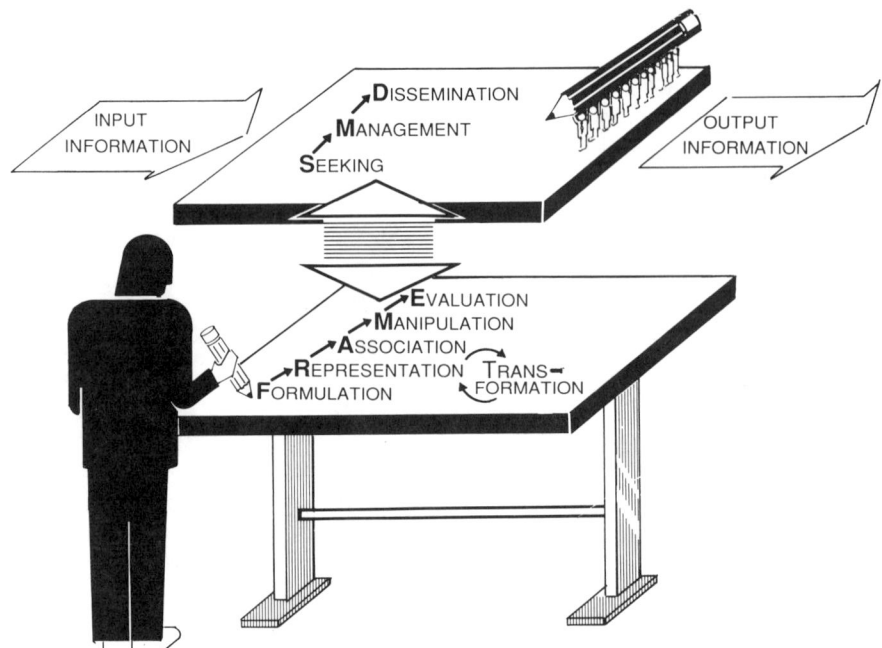

Figure 5. Two-Level Representation of Information Transformation Process.

potential human-machine interaction in design can be viewed as the "mapping" shown in Figure 4 from the apparent random branching among the tasks and activities of formulation (F), understanding (U), solution (S), and implementation (I), to a set of functions for supporting these tasks and activities, perhaps as a more orderly information transformation process.

It is not at all clear how this mapping is to be accomplished unless, of course, the designer is expected to impose order on himself or herself. However, this problem is not as daunting once one explicitly combines the activities listed in Figure 2 with the analyst-artist characterization from Figure 3. The result is a two-level representation of the information transformation process as shown in Figure 5.

The SMD level (Seeking, Managing, and Disseminating information) represents the organizational process whereby an RFP (request for proposal) is decomposed into a SOW (statement of work) via a proposal, and eventually composed into one or more deliverables (e.g., an aircraft). The decomposition and composition is managed by allocating and controlling flows of information (and requisite resources) within the organization's various engineering and production departments, and between the organization and subcontractors and, of course, customers. The SMD level is primarily concerned with the *correctness* of the design product—does it meet performance requirements and is it delivered on schedule and within budget?

The FRAME level (Formulation, Representation, Association, Manipulation,

Designers, Tools, and Environments   53

of detail as the individual activities in Figure 2. Nevertheless, there was a clear consensus that organizational considerations can be very strong determinants of design products.

2. Design Problems

A few issues emerged during the day devoted to "the designer" that did not pertain to the psychology of design in the same manner as the issues discussed in the last section. One of these issues focused on approaches to characterizing the psychology of the user as a basis for aiding designers to produce user-centered designs. Of course, user psychology is mainstream human factors and engineering psychology; as noted earlier, however, this topic was *not* the focus of the workshop.

Another issue that stimulated debate was the correctness versus goodness dichotomy discussed earlier. For the most part, the industrial designers and architects, as well as several other individuals, felt that this distinction was very important. However, another subset of the workshop participants felt that the concept of goodness, despite its general intuitive appeal, had little operational value. Specifically, if goodness was to become a design requirement, how would one measure it and evaluate the extent to which this requirement was satisfied?

Obviously, this is a very difficult question. However, it is not clear that it has to be answered as concisely as some of the workshop participants would like. An alternative perspective is that goodness is not predictable and certainly cannot be proceduralized, and, therefore, cannot be required. However, a design support system should not emphasize correctness to such an extent that goodness is significantly inhibited. Of course, even this more modest goal requires that the nature of design goodness be explicated further.

C. Potential Directions

As discussed earlier, at the end of each day, after pondering the state of knowledge and unresolved issues within the topical area for that day, each working group was requested to produce a set of recommendations for potential R&D directions. The intention of this request was to use the four working groups' results to produce a small set of high-priority directions. To an extent, this goal was accomplished. However, since the groups were instructed to avoid focusing on budgetary, technological, and political constraints, there was a tendency to recommend that all unresolved issues should be pursued. From this perspective, the "potential directions" sections of this chapter tend to be summaries of preceding sections, with a few very important and/or novel ideas highlighted.

Several general recommendations concerned the need to develop a better understanding of designers, with a short-term goal of building better design support systems and a long-term goal of selecting, training, and, as a result, perhaps "optimizing" designers. While this long-term goal may be decidedly unrealistic, the short-term goal obviously represents one of the main themes of the workshop. More specifically, the working groups recommended that more attention be devoted to understanding and supporting the interplay of artistic and analytical behaviors as well as the use of analogies. It was also suggested that the concept of design "goodness" should be better understood, if only to assure that correctness-oriented support systems do not preclude goodness.

Two fairly specific recommendations were concerned with extending, refining, and operationalizing the "design activities" construct summarized in Figure 2. It may be possible to characterize these activities as an exchange between two information structures—the designer and his or her design support system. This leads to the notion of developing a *language* of design activities including a vocabulary, syntax, semantics, and pragmatics. Further, it suggests that observed activities might be "parsed" such that the language of design can be understood by the support system.

A complementary recommendation concerned the observation methods necessary for potential parsing of the information exchange. It was suggested that approaches are needed for "instrumenting" design activities, as a basis for studying design in general, but more specifically for understanding activities, particularly in terms of designers' intentions. The development of a design language, and associated observation and parsing methods, would seem to offer great potential for disambiguating the apparent chaos illustrated in Figure 4.

III. DESIGN TOOLS

Three general topics emerged from the discussion of design tools. One topic was concerned with specific tools that are available, for the most part, as off-the-shelf software packages. The second topic was related to general methods which come from a variety of disciplines and may encompass a range of tools. Finally, the third topic emphasized design data, which includes data from scientific, technical, and regulatory publications as well as data generated during the process of design.

A. State of Knowledge

There are many specific tools available, most of which support context-specific visualization of detailed designs. CAD (computer-aided design) is a burgeoning, but not always thriving, industry which includes well over 1000 CAD vendors. Most of these vendors sell CAD software packages; a few sell hardware and software.

Most of the usage of CAD packages appears to be for detailed design, as opposed to conceptual design. These packages usually support functions such as geometric and volumetric modeling; finite element analysis; and other spatial, algebraic, and tabular manipulations. CAD packages such as SPICE and General Motor's Wire Frame are good examples.

A wide variety of general methods have been developed within several engineering disciplines as well as applied mathematics, economics, and management. Network models, stochastic process models, and mathematical programming techniques have been developed in operations research and management science. Control theoretic techniques for stability analysis, state estimation, and optimal feedback control have emerged from electrical, mechanical, and aeronautical engineering. Cost-benefit analysis and multi-attribute decision making formulations have been developed in management and applied economics.

All of these general methods are used primarily for analysis and interpretation rather than problem formulation. In other words, once the conceptual design has been at least roughed out, these methods can be employed. Typically, these methods are used to estimate the characteristics of system performance (e.g., stability, response time, and throughput) and, if possible, to determine how relevant parameters might be adjusted to optimize performance (e.g., maximize production or minimize cost).

If one compares the capabilities described thus far in this section with the breadth of design activities discussed in the position papers presented in this volume, it is apparent that many design activities are supported by few tools. The nature of this shortfall is pursued in the following section.

B. Unresolved Issues

Three unresolved issues emerged: 1) information needs, 2) current practices, and 3) current methods. The working groups felt that there is a general lack of design data across many disciplines. Even when data exist, it can be difficult to identify and access new data—the same difficulty was noted for new methods. It is also difficult to archive and access data generated during current and past design efforts, which can lead to a rapid depreciation of the corporate experience base.

It was felt that there is usually an inability to extrapolate data to new design problems. The conditions for which data were generated or the methods used are often judged to be inadequate for the new problem and further data collection efforts are pursued. A related difficulty was noted with regard to representing and extrapolating analogies.

Several inadequacies of current design practices were discussed. It was felt that individuals and organizations tend to fixate excessively on their previous designs. Further, it was suggested that they tend to fixate on a particular design concept much too early in the design process. This latter difficulty is understandable and is due, in part, to long turnaround times for developing prototypes. Rapid prototyping techniques are prevalent in software development but, as discussed below, provide a mixed blessing.

Current design methods were thought to be inadequate in terms of both available CAD packages and more general support system concepts. A general complaint was the lack of integration among the plethora of available CAD software and hardware. More specifically, it was felt that CAD design languages are oriented too much around the single discipline from which they emerged. It was also noted that volumetric presentations on most CAD systems are inadequate. Finally, it was suggested that design is often a group activity and, hence, CAD systems should support group interaction in the design process—in fact, one could perhaps argue that communication is a primary function of CAD systems.

From a broader perspective than traditional CAD systems, it was noted that there are few, if any, viable problem formulation tools and model-based management aids (discussed below). There is also a lack of practical methods for generating and managing user-system dialogue. Rapid prototyping emerged repeatedly as both a panacea and a problem. Programming environments, graphics editors, etc. have enabled substantial decreases in the time required to make initial versions of software operational—this lessens the lag in prototyping noted earlier. However, rapid prototyping has also been found, anecdotally at least, to result in people becoming "wedded" to their initial ideas. It is not clear how to balance the benefits of rapid prototyping with the benefits of the slower, more traditional approaches which emphasize more analysis. Of course, this issue is very much related to the earlier discussion of analyst versus artist.

C. Potential Directions

The priorities that emerged from the group discussions emphasized the areas of information needs and support systems. In the area of information needs, the concern was not with needs for new data but rather with accessing and manipulating existing data, as well as increasing the reliability of data bases. Specific recommendations included development of data bases for human factors, maintainability, and related areas; improved methods of data transformation; and approaches to accessing and utilizing analogies.

As noted repeatedly throughout this chapter, information needs include data that are generated during the design process as well as data external to this process. The working groups felt that two related aspects of managing information generation need particular attention. One aspect involves developing knowledge-based information processing methods to help designers keep track of design dependencies (e.g., relationships among interdependent design issues). The second aspect is concerned with improving design team communication. Both of these recommendations relate to the SMD level of Figure 5, particularly for complex systems where design integration tends to be problematic.

Potential directions in the area of design support systems included three recommendations. First, it was felt that CAD systems should be enhanced in the ways discussed in the previous section. Second, it was proposed that "intelligent" workstations be developed to provide balanced support of rapid

Designers, Tools, and Environments

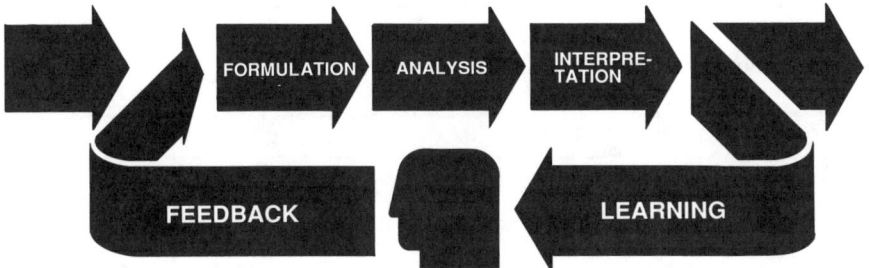

Figure 6. Framework for Model-Based Management Aids.

prototyping and systems integration, while also enabling access and manipulation of the data bases discussed earlier in this section.

The third recommendation concerned model-based management aids, a framework for which is shown in Figure 6. The basic idea is to provide an integrated means for organizing, updating, accessing, and aggregating the many models noted in the earlier discussion of the state of knowledge. In addition, the groups urged the development of aids for problem formulation that would fit within the framework of Figure 6. There was also considerable interest in developing aids for generating and managing user-system dialogues. This interest included aids for design of information presentation methods as well as tools for knowledge acquisition, representation, utilization, and transmission.

IV. DESIGN ENVIRONMENTS

The third and final day of the workshop focused on the organizational environments within which designers work and how these environments may foster or hinder innovation, both in terms of design and the potential acceptance of support systems. Issues of key concern included organizational *complexity*, the role of *creativity* within these environments, and the resistance to *change* that appears to be prevalent.

A. State of Knowledge

The procurement of major equipment systems by the U.S. government is a very complicated process. This appears to be due, in part, to the size of the bureaucracy, which is typical of large organizations in general. It is also due to several characteristics that are perhaps specific to political/governmental organizations whose primary goals include the well-being of its constituents.

Figure 7 summarizes the constraints and points of control in the government's procurement process. Private industry (i.e., the company) has primary control over the design and production of the equipment system of interest. However, the customer (i.e., NASA, the military services, etc.) determines what is needed, when it is needed, and how much the government

CONSTRAINT	POINT OF CONTROL		
	COMPANY	CUSTOMER	POLITICAL
USER REQUIREMENT		●	
RFP		●	
TIME		●	
DOLLARS		●	
TASK COMPLEXITY	●	●	
INFORMATION FLOW	●	●	
SIZE OF DESIGN TEAM	●		
EXPERTISE OF DESIGN TEAM	●		
TEAM ORGANIZATION	●		
TEAM COMMUNICATION	●		
DESIGN AIDS	●		
CORPORATE CULTURE	●		
POLITICAL ENVIRONMENT	●		●

Figure 7. Constraints and Points of Control in Design Environments.

may be willing to pay for it. The customer also affects the SMD level of the design process by influencing task complexity (via functional requirements) and information flow (via briefings and documentation requirements). Finally, of course, the political process including Congress, the media, PACs (political action committees), etc. affect the political environment. For example, the current adversarial nature of industry-government relations is at least partially due to the highly politicized nature of the procurement process.

Thus, there are a variety of "stakeholders" in the process and, as shown in Figure 8, several important factors affect these individuals and organizations. From this figure, it is clear that professional interactions within one's technical peer group is only one of several crucial aspects of design environments. Both industrial and government organizations are affected by their members, the relevant regulatory considerations, and the political environment. The government as a whole is affected by regulations, politics, and cultural norms regarding fair play, social justice, etc. Finally, the eventual user of the equipment system is affected by his or her user peer group (e.g., aircraft pilots) and cultural factors related to roles, aspirations, etc.

As noted earlier, there are a few characteristics of government practices that appear to differ substantially from other large and bureaucratic, but nongovernment, organizations. Particularly problematic is the fact that the

FACTORS \ PEOPLE	PEERS	ORG.	GOV'T	USER
PROFESSIONAL	●	●		●
REGULATORY		●	●	
POLITICAL		●	●	
CULTURAL			●	●

Figure 8. Important Factors Affecting Design Environments.

government's information management system has evolved (as opposed to being designed) in a patchwork manner that tends to be inconsistent and incomplete. While procurement is fairly proceduralized, management of information within the government is very loose and highly unstructured. As a result, while mechanisms exist for identifying government programs that potentially relate to one's interests, it is difficult to determine the state of these programs and access detailed information.

Another characteristic of government practices concerns the emphasis on correctness of processes (as opposed to results) and the use of these processes to accomplish social goals (e.g., equal opportunity) much broader than the reasons for which the processes were established. The penchant for correctness has resulted in greatly inflated costs of doing business. This appears to be due to the inherent great difficulty in judging the intrinsic value of the systems procured and, hence, the emphasis is solely on documenting and auditing the design, development, and implementation process to assure that everything "goes by the book."

The use of procurement to further social goals is an understandable use of one of the government's primary areas of leverage. While there may be pros and cons relative to this process, the effect is to increase costs (perhaps acceptably) to foster equal opportunity, affirmative action, etc. The end result of pursuing these social goals, as well as focusing on correctness, is that equipment systems cost the taxpayer more than they would otherwise. Of course, to the extent that desired social goals are achieved, the taxpayer is also getting more than just an equipment system.

The complexity outlined thus far in this section has important implications for the efficiency and effectiveness of the design process. Industrial organizations often invest substantial levels of creativity and technical expertise in developing proposals for government contracts, in part due to the fact that the proposal for the new project is much more important to success than the company's "portfolio" of results on past projects. Once the contract is awarded, uncreative and even mediocre performance is acceptable since it will not hinder future opportunities, at least not in the short term. In fact, at least one group argued that it is very difficult to foster creativity in large industrial organizations such as those that dominate government procurement unless the corporate culture, including the CEO (Chief Executive Officer), is totally committed to innovation. As a result of the prevalence of this technical and intellectual conservatism (partially instilled by the focus on correctness), new ideas and technologies appear to take roughly 20 years to be implemented in production equipment systems.

This situation can be explained in part by humans' typical resistance to change, which is particularly problematic for large organizations where many people must accept the change. One group suggested that the following factors affect resistance to change:

o extent to which there is agreement that a deficiency exists,
o credibility of the evidence that improvement is possible,
o motivation of the individuals involved to pursue the improvement,
o compatibility of potential improvements with existing subsystems that will remain unchanged,
o types of individual involved in terms of prior training, aspirations, etc.,
o extent of training provided in anticipation of change and during implementation,
o extent of support provided such as operating procedures, consultants, etc., and
o explicitness of the boundary conditions of improvement.

The above factors have indirect implications for system design but, more saliently, these factors are of great importance in the process of introducing a new system. Thus, they should be considered as one proceeds to introduce design support systems such as those discussed throughout this volume.

B. Unresolved Issues

The discussion in this area focused on information needs and organizational considerations. In general, it was felt that there is an information system mismatch in terms of format and content between government and industry, as well as within governmental units and industrial organizations. At the very least, this mismatch imposes enormous overhead costs and severely undercuts *efficiency*. In light of usual cost constraints, this extra overhead also absorbs resources that otherwise could have been invested to improve system *effectiveness*.

As discussed earlier, overhead costs are also inflated by the government's penchant for assuring the "correctness of" the design process. This orientation can be explained, in part, by the lack of auditable design criteria for operability, maintainability, etc. Without such criteria, it requires an enormous amount of judgment and, hence, risk to decide that a product is "good"—a program manager is on much firmer ground if he or she only promises that the product will be "correct."

This observation leads quite naturally to the working groups' conclusions that equipment system users, program managers, Congress, and contractors view design and procurement *very* differently. The perspectives range from trying to do a difficult job better (e.g., piloting a fighter aircraft) to trying to get as large a share as possible of the federal pork barrel (e.g., congressional appropriations). With these different perspectives come varied and often conflicting agendas.

The lowest common denominator among these agendas is *incrementalism*. Industrial, military, and political careers can be built on mediocre increments to the status quo—truly innovative and generic solutions entail risk which can be the bane of most bureaucrats' aspirations. As a result, credible (realistic) demonstration projects involving risky technologies seldom occur. In fact, the reward structures within most governmental and industrial bureaucracies provide few incentives for real innovation. Thus, as one group put it, amidst the avalanche of correctness-oriented requirements and procedures, it is not clear who is responsible for "carrying the spirit of the design."

C. Potential Directions

The above discussions of the state of knowledge and unresolved issues can be viewed as painting a rather bleak picture. However, the workshop participants, perhaps by nature rather than reason, exhibited a cautious measure of optimism for the possibility of improvement. Two changes were advocated. First, there is a great need for increased and improved government-industry interaction, particularly as a catalyst for the credible technology demonstrations necessary for expediting innovation. There is also a need for more interaction in terms of "marketing" design innovation throughout the procurement/political process. Finally, innovation should be explicitly emphasized by requiring it in government RFPs *and* paying for the value added beyond the basic requirements. In other words, the government should encourage and pay for "goodness" rather than simply requiring the least-cost approach to "correctness."

The second type of change suggested concerns the nature of both governmental and industrial organizations. It was felt that the number of levels of decision making has gotten out of hand—an ideal of *two* levels was advocated. It was also strongly felt that there are too many regulations; a periodic and automatic purge might be a good idea. It was suggested that government-industry relations have become much too adversarial in nature with lawyers having disproportionate representation and leverage in the process. Finally, it was argued that fewer levels of decision making, fewer regulations, and fewer lawyers might result in a much more visible design and procurement process that would enable more interaction and earlier resolution of design problems as well as more room for innovation.

V. CONCLUSIONS

Rather than reiterate the many conclusions summarized in this chapter, it is more useful to present the overall perspective that emerged from their consolidation.

A. Design of Complex Systems

System design is much more complicated than a typical undergraduate textbook might lead one to conclude. Analysis is *not* designers' predominant activity—it is just one of the many tasks involved in the job of design. Also, many people are usually involved in the design of a complex system, and they interact within organizational structures that significantly constrain their discretion or latitude in designing approaches to satisfy system requirements. Organizational constraints are useful to the extent that they foster development of a consistent and coherent product—such constraints are counterproductive when they hinder innovation and foster mediocre incrementalism.

In order to improve the efficiency and effectiveness of system design, it is necessary to appreciate and hopefully understand the nature of designers and design organizations. This volume documents much of what is known today. However, we need to move beyond the "naturalist" (or anecdotalist) stage and seriously study design. The goal should be determining the individual and organizational behaviors that are inherent to design endeavors, and those behaviors that are undesirable artifacts of current practices. We should then focus our R&D resources on the changeable—the points where design effectiveness can best be leveraged.

One theme that was common to all of the working groups was the extent to which information access and manipulation is the basis of all design activities. By viewing design as an information transformation process, with inputs both from statements of requirements and each individual FRAMEr, it may be that the best leverage point has already been identified. It is quite likely that design information systems can provide the vehicle, as well as part of the solution, for integrated design support systems.

B. Design Information Systems

The objectives of a design information system should include retrieval and manipulation of external information necessary for design decision making (e.g., scientific, technical, and regulatory information), as well as internal information that is generated in the process of design. To achieve this functionality, it will be necessary to embed support mechanisms within the information system to aid designers in framing their queries for information, as well as manipulating and interpreting the resulting information. Further, it will be desirable if support mechanisms can monitor the flow of information, both within and across organizational boundaries, and prompt designers with potentially relevant and useful information.

The obviously central issue is how one could design an information system with this functionality. Of great importance also is the evaluation of prototype design information systems and "spinning off" of partial, but useful, systems along the long road to a comprehensive system. In general, what is needed is a design and evaluation methodology for developing design support systems.

C. Possible Futures

Considering the perspective presented by Dr. Robert Cooper (see Chapter 3), it is fairly easy to imagine that the technology necessary to design information systems will be increasingly available and decreasingly expensive. However, while the technological future may appear unlimited, the organizational future is more uncertain. Will government-industry relations improve and become less adversarial? Will innovation replace incrementalism as the modus operandi of government and industry? The answers to these questions will determine whether or not the many potential innovations discussed in this volume have an opportunity to enhance the effectiveness of system design.

Chapter 5

AN EXPERIMENTAL VIEW OF THE DESIGN PROCESS

Joseph M. Ballay

Center for Art & Technology
Carnegie-Mellon University
Pittsburgh, Pennsylvania

ABSTRACT

Patterns of design problem solving are revealed by experiments with designers using traditional methods and computer-aided systems to solve a typical industrial design problem. Comparing a model of the design process with the products of designers' problem solving suggests that the early stages of the process are particularly important to the production of successful designs. The early stages of the design process are defined from three viewpoints—as an ill-defined construction task, as a visual task, and as a series of information transactions. These viewpoints provide a context for considering several patterns: patterns of planning, routine processes, information management, and the use of visual representations. In particular, the process of sketching visual representations is examined as it fits within the larger context of design problem solving.

I. INTRODUCTION

In this paper, I intend to convey some insights about the design process that I and my colleagues gained during a recent research project. We were trying to understand how the design process, particularly the design of mass-produced, durable products, was affected by the use of CAD systems. Our aim was, and still is, to improve the usability of computer systems in designing, especially during the most creative phases of the process. As a part of the research project, we defined a design scenario. This design scenario is typical of industrial design tasks and is portable, which allowed us to observe designers working on it in a traditional design environment as well as on a CAD system.

The scenario required designers to design *Easybanker*, a hypothetical but plausible product which allows the payment of credit card bills from home through the telephone system to the bank. It was imagined to be about the size of a typical telephone modem and included a card reader, a keypad, an LCD display, and a miniaturized voice recognition cell. In addition to the product concept, the designers were given a package of information including functional requirements for *Easybanker*, specifications on alternative components, component costs, and production constraints.

The design sessions were videotaped and observed. The designers were asked to verbalize their actions as they went along. Significant events were then organized into a protocol of each designer's session.

The insights reported here are based on the observation of approximately fifty designers, design students, engineers, and draftsmen. Of these, eighteen were the subjects of detailed protocol analysis, and one subject was studied in-depth for approximately twenty hours of design activity.

II. MODELING THE DESIGN PROCESS

For the purpose of organizing our observations, we agreed on a tentative model of the design process. It was based on existing literature (Jones, 1970) about the design process and on a preliminary analysis of the design process as we observed it. The model had five subtask segments which were assumed to run more or less sequentially:

1. *Criteria Formulation* - collecting and analyzing information to establish design criteria.
2. *Space Organization* - volumetric decisions concerning fit, function, ergonomics, and aesthetics.
3. *Details and Structure* - engineering refinement of previously defined concepts.
4. *Appearance Decisions* - decisions that merge concepts with aesthetic criteria.
5. *Release Package* - producing and assembling final documents for release.

In comparison to this model, the design process which was actually observed turned out to be mostly sequential, with some parts of the segments running in parallel and with only a little backtracking. In order to evaluate the designs produced by this process, we established five categories of product

An Experimental View of the Design Process

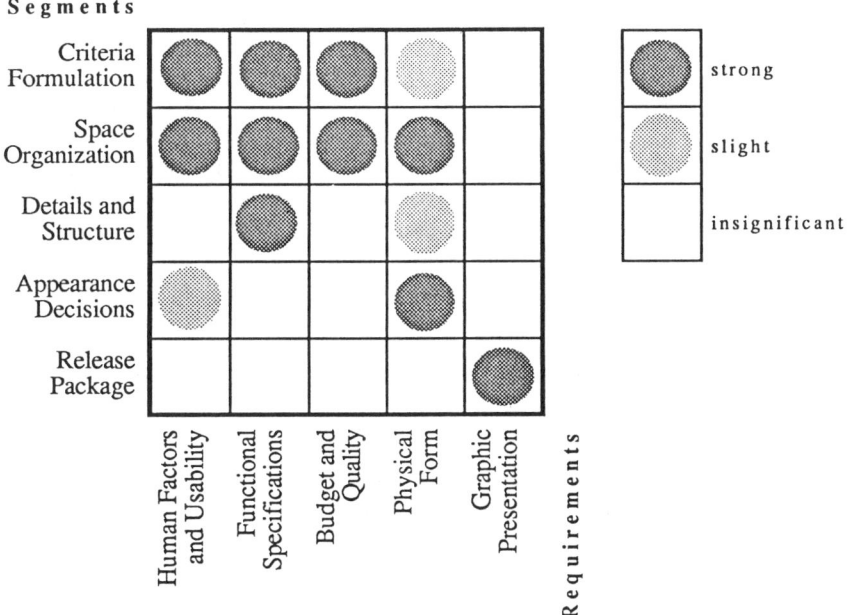

Figure 1. Degree of Segment's Influence on Meeting Product Requirements.

requirements. The designers were told that their designs would have to respond to these requirements, but no priority or weighting was given. These categories were the following:

- *Human Factors and Usability* - meeting the needs of the user.
- *Functional Specifications* - meeting the engineering requirements for production and functional performance.
- *Budget and Quality* - achieving an appropriate balance of cost/performance in the use of alternative components.
- *Physical Form* - assuring size, shape, and appearance are appropriate to the environment in which it will be used.
- *Graphic Presentation* - assuring the release package is appropriate to professional standards.

One analysis of the observed design process involved comparing the five segments of the sequential model against the five categories of product requirements in a simple interaction matrix (see Figure 1). Each cell of the matrix represented the degree (strong, slight, or insignificant) to which design decisions made during that segment influenced the eventual success in meeting that product requirement. In other words, when in the design process were the significant decisions made which influenced the eventual success of the design?

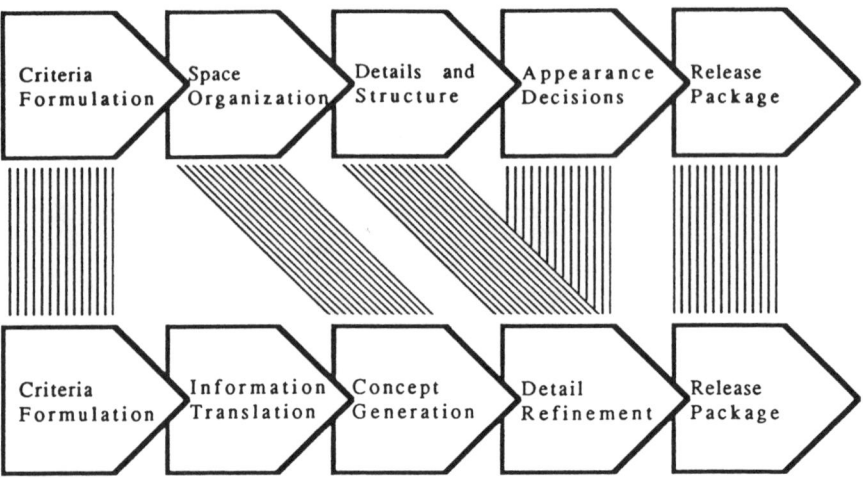

Figure 2. Revised Sequential Model.

The cells of the matrix were assigned degrees of significance based on reviews of the session protocols. It was immediately clear that the early segments were the most significant. Decisions made during the Criteria Formulation and Space Organization segments, in particular, were of prime significance in influencing the design with respect to the Human Factors and Usability, Functional Specifications, and Budget and Quality requirements. The Physical Form requirement was influenced throughout the design process, and the Graphic Presentation requirement was influenced only at the end.

This exercise confirmed our beliefs that the early stages of the design process deserved the most detailed study if we were to affect the ability of designers to design better products on computer systems.

As observational data accumulated, it became clear that our sequential model would have to be revised (see Figure 2). The Details and Structure segment and the Appearance Decisions segment ran so consistently in parallel that they were combined into a single segment named Detail Refinement. Space Organization was renamed Concept Generation to more accurately reflect the scope of activities which occurred during that segment. Most importantly, it became clear that the model needed a new segment named Information Translation. It reflects the designer's significant subproblem of translating the information he receives, such as written briefs, drawings, and specifications, into a format or mode that is useful in the problem solving process. For example, several designers in the traditional environment converted alternative components—part of the information package—into crude models made of

An Experimental View of the Design Process 69

plastic foam, paper, or foamboard. Designers in the CAD environment would convert these drawings into CAD drawing files.

This new Information Translation segment and the recognized importance of the early segments of the design process set the priorities for our research, which I will try to reflect in the rest of this paper. I will focus on the patterns we observed designers using to manipulate and use information during the early parts of the design process.

III. THREE VIEWS OF THE DESIGN PROCESS

At various times during the analysis of the protocols, we found it enlightening to think about the early phases of the design process from different points of view. Three of the most useful viewpoints are described briefly below. The first derives from the literature on the psychology of problem solving. The second responds directly to the tangible output we observed from the design process. The third is a more abstract view derived from an information processing model of human cognition.

A. Design as an Ill-defined Construction Task

If we examine design from the viewpoint of the individual designer as a problem solver, we find that two aspects of design tasks have an important impact on the way design problems are solved: design tasks are *construction tasks*, and design tasks are *ill-defined*.

By a construction task we mean a task which has as its main goal the creation, by cumulative action, of a persisting external product—a product which must satisfy a demanding set of criteria before the task is said to have been successfully completed. Industrial design is an example of a construction task because its main goal is the production of an external representation of the solution which must satisfy criteria of utility, clarity, aesthetics, etc. Other examples of construction tasks are computer programming, sculpting, carpentry, and assembling a HEATHKIT amplifier.

Construction tasks vary widely in the demands they place on the problem solver for decisions about the product form. Tasks such as routine carpentry and HEATHKIT amplifier assembly place minimal demands on the problem solver for decisions about the nature of the end result. In contrast, tasks such as sculpting and computer programming impose major demands on the problem solver to contribute to the shaping of the final product. Tasks such as these two are described (Reitman, 1965) as ill-defined because they do not provide the problem solver with sufficient information to specify the final product. Problem solvers must, therefore, provide information through their own decisions which further specify the task and define the form of the final product. There are a number of consequences which follow if a task is an ill-defined construction task:

- The partially completed product becomes part of the task environment. That is, what has been constructed up to now influences the subsequent course of problem solution and the shape of the final product. In design, the designer's current sketches are used as a source of ideas and inferences which shape later design decisions.
- Because the partially completed product is continually changing, the task environment is continually changing. These changes in environment stimulate new ideas and inferences. As a result, invention (finding a new way) is stimulated continuously throughout the course of the solution, right up to the moment when the final drawing is completed.
- Because construction tasks produce a final product which must satisfy external criteria, such tasks necessarily involve a commitment of resources. The more expensive the commitment of resources in producing the final product, the more desirable it will be to plan. All construction tasks benefit from planning because all involve commitment of expensive resources.

B. Design as a Visual Task

As we have seen, design resembles other ill-defined construction tasks. However, there are also some important ways in which design differs from the others. Perhaps the most important difference has to do with the "visual" nature of design. If a designer has several ideas in mind, he typically cannot evaluate the relationship among them until they are made visible. A drawn external representation is a very important aid to the designer in making spatial inferences. Experienced design teachers find that students who claim to have designed wonderful things "in their heads" usually discover disastrous flaws in their designs when they try to put them on paper.

Furthermore, experienced designers mix drawing with other types of external representations. The designer in our in-depth study made use of seven distinct types of representation in the course of designing the *Easybanker*. Each of these appears to serve unique functions in aiding the design process:

- *Procedural Representations* - encoded action scenarios. A designer engages in a brief play. For example, he uses a procedural representation to describe how the user will operate the *Easybanker*.
- *Solid Models* - three-dimensional materials. They help a designer perceive volumetric relations, to orient volumes in space, and to facilitate the arrangement of volumes with respect to one another in space.
- *Matrices* - two-dimensional (usually) grids of information. A designer constructs a matrix to help him to choose which of the alternative parts he wants to work with. The rows of the matrix represent part types

An Experimental View of the Design Process

(keyboards, displays, etc.) and the columns represent alternative models of each type. Information concerning each model is contained in the cells. Thus, keyboard number 1 costs $1.20 and is "big," while keyboard number 2 costs $1.10 and clicks when the keys are pressed.

o *Orthographic Projections* - views perpendicular to the principal Cartesian planes. Many orthographic views were produced, including one with a Polaroid camera. The photograph provided a crude plan view of an arrangement of the solid models. It was used as a record of the locations of the parts with respect to each other.

o *Notations* - words, numbers, letters, and related symbols. The notations may vary in length from a single character to a few lines. Usually, they were no longer than a few words.

o *Perspective Drawings* - graphic conventions which imply a "spatial" view of objects. Many perspective views were produced. The perspective sketches served to record the designer's decisions about the orientation of the parts and about the appearance of the envelope that contained them.

o *Dimensions* - representation which combines symbolic and graphic components. In this representation, meaning is conveyed by both the symbolic and the graphic components. If a designer changes either the numbers or the positions of the arrows, the meaning of the representation changes.

It is evident from the session protocols that a designer uses the various representations episodically, tending to do all the work he can in one representational mode at a time. When that work is done, he switches to another representational mode to do the next block of work. Figure 3 shows some of these episodes graphically.

C. Design as a Series of Information Transactions

At the procedural level, the design process can be explained by the cyclical model shown in Figure 4. The model is based on principles of human cognition (Newell & Simon, 1972) and is useful in describing where information comes from and goes to at any stage of the design process. It accounts for the information held in the designer's head and the information held in the external environment, or held in a computer in the case of an automated design system. Most importantly, the model puts representations in a position of central importance as the medium of exchange between the designer and the computer.

We have observed that designers do not know many details about what they are going to sketch until part of the sketch is made; "knowing" is in the observing of the external sketch. According to the information transaction model, a designer begins sketching (enters the transaction) with only partial information, retrieved from his own memory, about the sketch that will be

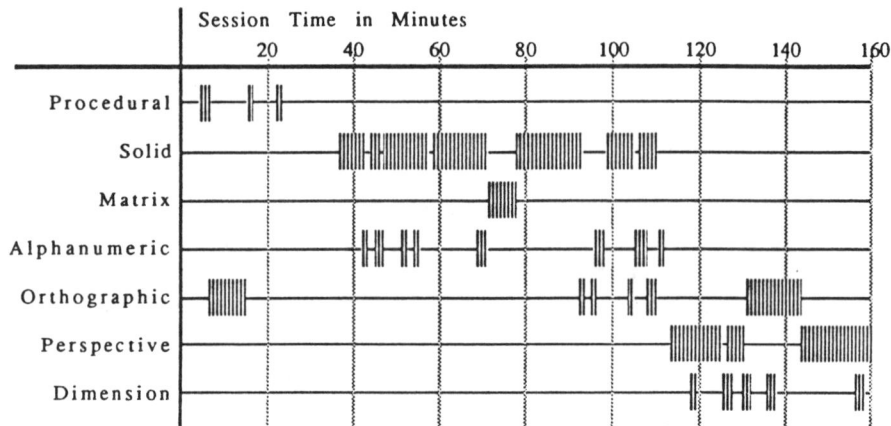

Figure 3. Alternation of Representational Modes. The seven modes of representation were used by a designer during an experimental session that lasted about 160 minutes. The horizontal bands of marks indicate where each model was actively used, with each mark indicating about two minutes of activity. The episodic use of representations can be seen in the clustering of activity in each of the modes. In particular, note the long episodes of solid modeling early in the session and the alternation between orthographic and perspective toward the end of the session. Aside from some time in the first 30 minutes of the session, when the subject was studying the information package, there was nearly continuous activity in one or more modes.

produced. Additional information is retrieved from the task environment, including computers in the environment. A record of the information processes is encoded in a representation, which then becomes a part of the designer's memory and of the task environment. The representations are external and can take a variety of modes, as described above.

If one imagines putting these information transactions end to end, the external artifacts of the design process would appear as a chain of representations, the majority of which are sketches. The large number of sketches produced in the protocols indicates the importance of sketching to the creative design process.

IV. PATTERNS IN DESIGN PROBLEM SOLVING

By using any of our views of the design process, patterns of design problem solving behavior began to emerge. I believe these patterns represent the most interesting and useful insights we gained from this research. We observed idiosyncratic variations in the overall design process depending on the experiences and preferences of the individual designers. Regardless of how the

An Experimental View of the Design Process

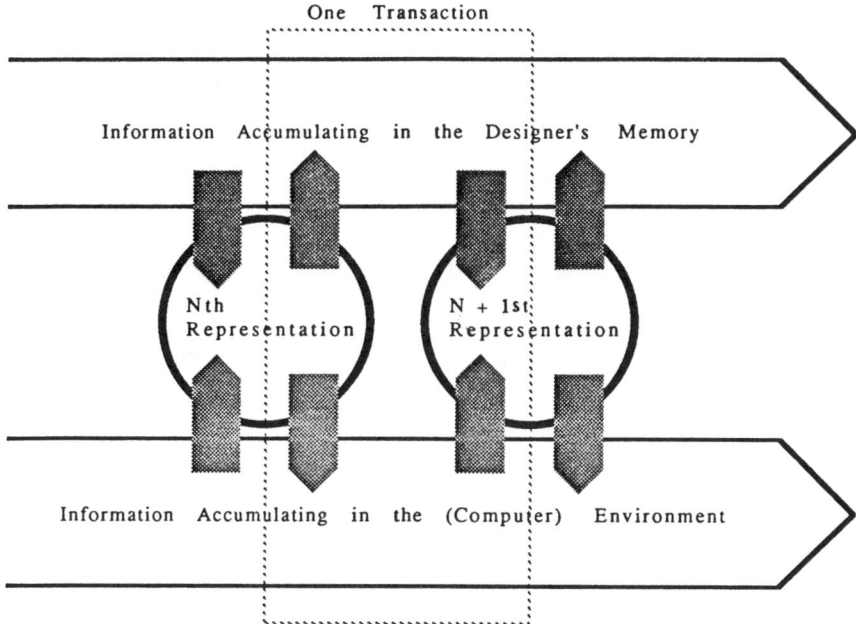

Figure 4. Transaction Model of the Design Process.

process varied in its organization between these basic patterns, there was consistency within the patterns. Figure 5 graphically describes the patterns of three kinds of activity in the overall design process—*Planning, Routine Primary Processes,* and *Workplace Management.*

Figure 5. Overall Design Activity Patterns. These three kinds of design activity constituted the process of a designer during a 160-minute experimental session. The horizontal bands of marks indicate the duration of each activity, with each mark indicating about two minutes of activity. Routines involve the use of well-rehearsed professional skills toward an interim goal. Note how bursts of planning (setting new interim goals) separate the routines. The management activities focus on organizing the designer's work environment to facilitate the routine activities.

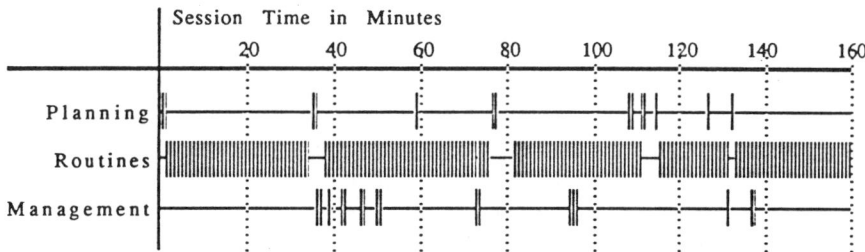

A. Pattern: Alternating Planning and Routine Primary Processes

As shown in Figure 5, the overall design process is divided into large blocks of primary processes which are similar to the process segments of our initial model (e.g., understanding the criteria, producing alternative component arrangements, etc.). These primary processes seem to be well-rehearsed routines. While the routine processes may be interrupted for such purposes as evaluating and editing, these interruptions are germane to the ongoing process. Periods of planning occur between these routines. They are an intense combination of reviewing alternatives, reviewing criteria and constraints, and making major decisions regarding which processes to employ next. The blocks of primary processes are long, typically 20 to 30 minutes each out of a 160-minute session. The planning periods are short, less than a minute each. They represent only about 2% of the total session, but appear to be crucial to the effective management of the primary process blocks.

B. Pattern: Workplace Management

This is the organization and reorganization of the immediate environment in which the designer works. It includes preparation and maintenance of tools; selection of materials; and arrangement of notes, drawings, or other information for comparison and easy reference. While some workplace management is distributed throughout the design process, there are periods of concentrated management following a period of planning. In these, the designer reorganizes in order to carry out the plans he has just made. He responds to the anticipated needs of the next primary process. These periods of concentrated management can exceed 10% of the session when many changes in the design process are called for.

The designer needs to consider a large amount of information in order to create a good design. However, this requirement also means that the designer needs some techniques for organizing or managing this information in his workplace. These techniques help to make the design process more efficient by minimizing the amount of time spent on retrieving, searching, and sorting information. Additionally, these techniques can also minimize errors such as losing or regenerating information. These techniques are discussed in more detail in section D.

C. Pattern: Selective Refinement and the Dimensions of Sketching

We examined the sketch output from our design scenario along with the videotaped protocols of the sequences in which these sketches were produced. As a result of this examination, we identified three variables or dimensions of

An Experimental View of the Design Process 75

sketching which help describe the process by which information is put into sketches. This process is more structured than the traditional view of sketching (i.e., merely fast, inaccurate drawing) and has been given the name "selective refinement."

Most sketches do not start "from scratch." In keeping with the information transaction model, they build on information that is extracted from previous representations. In paper-based systems, it is common to use tracing paper to extract information in an obvious external way. We also have evidence that a transaction happens at the cognitive level, too. In one protocol segment, a subject sketched four different component arrangements without involving any tracing. The first sketch took 13 minutes, the others about 5 minutes each. Clearly some information, probably in the form of partial solutions, was being extracted from the first arrangement and being used in the succeeding arrangements. We call this a decision about inclusion.

Inclusion, coherence, and precision are the three dimensions that describe the ways in which sketches can differ from "finished work."

o   *Inclusion* - the amount of information about a form that is represented in a sketch. It can be thought of as the level of detail or as the "grain size" of the information in the sketch.
o   *Coherence* - the degree to which different pieces of information agree with or support another. It reflects whether the partial solutions to subproblems have been reconciled to one another.
o   *Precision* - the dimensional refinement with which an intended configuration is represented. This is the most commonly understood of the dimensions.

While planning, a designer makes several choices. He chooses a mode of representation and levels of precision, inclusion, and coherence. The choices are made, based on training and experience, to be appropriate to the state of the problem as the designer understands it at that time. The choice of a representation mode is obviously an important decision. Perspectives show information that numerical dimensions do not and vice versa. That is why a completed page of sketches is often a combination of modes. The decisions about inclusion, coherence, and precision are involved in the design process in other ways which will be discussed. The important point is that for experienced designers, these variables are under the designer's control. The looseness of a sketch is a conscious choice, not simply the result of being sloppy or in a hurry. Sketching is not so much the precursor of a drawing as it is the record of a thought. It is a way of controlling the information in the early representations of a design solution. The process can be applied to a wide range of representational modes including sketch perspectives, sketch cross sections, and sketch diagrams.

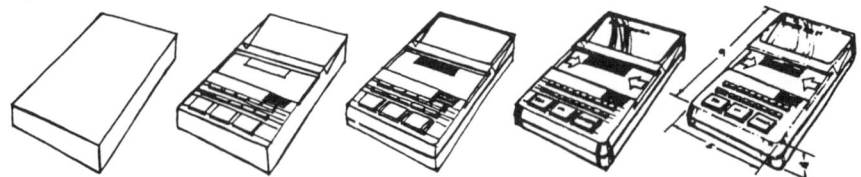

Figure 6. Sequence of Designer's Sketches.

1. Inclusion Pattern

Through inclusion, a sketch can accumulate information as it is worked on. Interestingly, the information in a sketch seems to get added in a consistent sequence that is partly idiosyncratic (the result of a designer's training and experience) and partly a response to the demands of the particular design problem. This implies that computer-supported design might include a personalized expert system that responds by adjusting to specific problem types or to specific users. One designer added information in the following sequence in every orthographic and perspective sketch he produced (see Figure 6):

1. Proportion of envelope (the product housing).
2. Component locations.
3. Mechanical and aesthetic refinements.
4. Material and surface attributes (surface reaction to light).
5. Notations and dimensions.

2. Coherence Pattern

A complete page of sketches often includes several views of an object. Coherence requires that these be in agreement with one another. We have some interesting protocol evidence of a subject dealing with the coherence of a sketch. In reconciling between the orthographic and perspective views, two patterns were observed.

First, as shown in Figure 7, while the subject alternated between orthographic and perspective views, the amount of time spent on each view decreased in a very regular sequence. Less time was spent with each alternation, starting with 15 or 20 minutes on each view and decreasing to a minute or less per view. The typical number of alternation cycles was three or four.

Second, the piece of information, such as a shadow or a detail line, that was the last to be included in one view was consistently first to be included in the next view. For example, while the designer worked on the perspective sketch, the last detail he added was a representation of light reflection to indicate that the digital readout was covered by a transparent window. The first detail he

An Experimental View of the Design Process

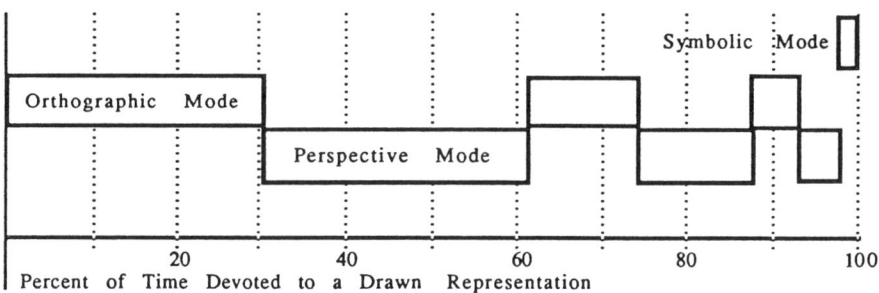

Figure 7. Pattern of Alternating Views.

added upon returning to the orthographic sketches was the same representation of light reflection.

The information processing model of human problem solving indicates a very limited capacity to carry complex information in short-term memory. These patterns of alternation and information carry-over seem to be an effective way of minimizing the amount of information that has to be dealt with at one time. The subject stays in one view and switches from one subproblem to the next until the current view is insufficient for solving the next subproblem, or he stays with one subproblem (representing the window, in the example above) and shifts to another view.

3. Precision Pattern

Choosing a level of precision is clearly related to decisions about economy and available time; a rough sketch is not simply the effect of sloppiness. If that were the case, we could expect a randomness to the precision of line making within a sketch or between sketches. We have made comparative measurements of line precision and the results do not support the sloppiness hypothesis. Rather, the overall precision of an entire sketch changes systematically from low to high as the design process moves toward completion.

Imprecision seems to have a value of its own. In ill-defined construction tasks, we should expect that invention (the substitution or addition of new ideas) will go on throughout the design process. By implication, there is the prediction that we should see final solutions being postponed as long as possible. Both patterns show up in the protocols. Low precision can be seen as evidence of the tendency to postpone final solutions, thus contributing to invention throughout the design process. At any stage of the design process, a designer is more willing to abandon the ideas represented in economical, low precision sketches than those that represent a large commitment of time, thought, or money.

Also, sketches that are low in inclusion, coherence, or precision can provide a dissonant visual field which encourages and sometimes demands invention for the solution of subproblems that were ignored at the beginning of the sketch.

The dissonances that occur may be partly intentional and partly the result of errors in information carried over between views. However these mutations occur, the designer periodically checks for agreement between views and discovers the dissonances. They tend to be located in areas of a sketch which have not been thoroughly solved; thus, the dissonance can be interpreted as representing a range of possible solutions which will require further refinement or the invoking of additional constraints. Alternatively, the dissonance can be interpreted as a spatial paradox or geometric impossibility which will require some invention for its resolution.

D. Pattern: Information Management

A noticeable feature of the design process is the large amount of external information the designer uses in reaching a final solution. In the first 160 minutes of our in-depth study, the designer used a nine-page text, drawings of ten components, foamboard models of seven components, cost analyses of four partial solutions, photographs of four partial solutions, rough perspective drawings of three partial solutions, and partially rendered perspective drawings of three partial solutions.

The large amount of information present in a design task poses the problem of how best to retrieve and store it. We have identified four techniques by which designers manage information: *Partitioning, Grouping, Labeling,* and *Combining*.

1. Partitioning Pattern

The designer maintains three distinct areas for storing information which are ordered by the accessibility of the information contained in them. The most accessible area is the designer's desktop, and this contains the information and representations currently being worked on. The next most accessible area is the wall area around the desk on which the designer pins various drawings of partial solutions. The third is out-of-the-way storage. In it the designer has access to an enormous amount of information that is not kept on the desktop or on the walls; e.g., drawings from previous design projects, tables of standards, etc. This type of information differs from the other two because it is not all contained in any single area. Furthermore, this information is not as accessible as that contained in either of the other areas because most of it is not immediately visible.

The desktop contains most of the information the designer uses for decision-making at any one time. Information needed for making a decision, but not present on the desktop, is typically brought here. For example, the designer would occasionally take a drawing from the wall and trace some aspect of it. Information on the walls is generally used for reference. For example, the designer would take a measurement from a drawing pinned to the wall and use it to confirm some dimension of a drawing on the desktop.

An Experimental View of the Design Process

Partitioning is useful as an information management technique because locating information in separate areas reduces the workload of retrieving information from any single area. Furthermore, the ordering of the areas by their accessibility ensures that there will be minimal interference between the information contained in the different areas.

2. Grouping Pattern

The two characteristics of this technique are that sets of information are grouped together, and the groups are defined by virtue of having different locations. This technique is similar to partitioning, except that the groups are only temporarily assigned to locations. In contrast, the areas used in partitioning are unchanged throughout the design process.

Probably the simplest and most common example of grouping was the designer's tendency to create triads of drawings on a single sheet of paper: a plan view, a side view, and a perspective view. In this way, the designer groups several sets of information (the different views) at a single location (the sheet of paper). Another example occurred at the beginning of the second session when the designer placed all photographs, models, and rough sketches from the first session into separate groups around his desktop while he was reacquainting himself with the design problem.

3. Labeling Pattern

This is simply the act of giving an identifying label or name to a partial solution. The label identifies an object or a collection of objects as being part of a labeled solution. The designer maintained labels for each of four partial solutions he generated. These labels were transferred as each new representation of a partial solution was generated, for example, from photograph to rough perspective to rendered perspective.

During one session, the designer made an error that illustrates the importance of maintaining a labeling scheme. He began to remake, unintentionally, a perspective drawing that he had already completed. Indeed, he used the original perspective to help format the new one on the page. When the designer was asked which partial solution he was working on, he first compared the labels of the new and old perspectives and wondered if he had mislabeled one of them. However, after he had compared the drawings themselves he realized his mistake. This example suggests that the labels provide the designer with a quick method of distinguishing different solutions instead of comparing details in the drawings, which is a relatively slow process.

4. Combining Pattern

In this type of activity, the designer appends information to a representation which did not include that information. For example, the designer always wrote the price of a component onto its foamboard model. Thus, the cost information, something that is not inherently represented in a foamboard model, was appended to the representations of the components. In other instances of combining, the designer wrote the cost of each component in a configuration of models and glued it to the photograph of that configuration. Later, he annotated the rough perspective sketches with approximate dimensions.

Combining information on the various representations appears to free the designer from unnecessary searching. The designer added information which was not inherent in the representation but was useful for making decisions which involved considering information from different representations. One implication of this technique is that the designer already has a good idea of what information will be useful in making decisions.

V. CONCLUSION

A. Observations

1. In sketching, designers are engaging in a set of intentional acts. They make an economic choice about what kind of representation and what level of inclusion, coherence, and precision is most appropriate to explore the problem at hand.
2. The designers perform complex information processing by extracting information out of their own experience and out of previous sketches, adding information and reconciling information conflicts in their sketches.
3. Through *ex*clusion, *in*coherence, and *im*precision, the designers provide their sketches with enough ambiguity so they can take advantage of inventive opportunities right up to the end of the design process.
4. It seems clear that the more kinds of representation a designer can use, the better able he is to work through complex problems. Systems which force a designer to make an early decision about the inclusion, coherence, or precision of information will be counterproductive. They will tend to close down a designer's inventiveness too early in the design process.
5. It is important to recognize how the role and function of the computer changes as the design process progresses. At the beginning, it is a tool for individual problem solving, and sketchy representations are appropriate. Toward the end, it becomes a tool for group management or communication, and complete representations are required. I have been focusing on the earliest phases of the process. For this role, we have identified several qualities that a computer-aided design system should have:

- Users should feel like they are working on the representation, not on the computer.
- In graphics applications, the primary information is the image itself; image manipulation must be direct and intuitive.
- Solid models assist cognitive aspects of spatial problem solving; computer-aided design systems need a surrogate for solid models.
- A model's system of spatial reference should be natural to the task; coordinates are appropriate for specifications but not for ideation.

B. Proposed Studies

There are many areas of incomplete knowledge about the design process. We have proposed three kinds of experiments which we believe would help to bring the requirements for computer-aided design systems into focus:

o *Composition/Decomposition Experiment* - explores the processes by which simple spatial objects are composed or assembled into more complex objects. It also explores the related process of decomposing or analyzing a complex object into simpler parts.

o *Orthographic/Perspective Experiment* - focuses on two conventional systems for representing spatial objects on a simple two-dimensional plane. It is known that experts alternate between these systems when solving problems, but it is not clear how the conventions affect problem solving.

o *Default Representations Experiment* - is a survey of the actual representational parameters used by architects, engineers, and designers in the process of sketching to solve spatial problems. Experts have many available options for such parameters as viewing angle, eye height, hidden lines, and line weights, but in the early stages of the design process, they frequently resort to a small set of default representations.

ACKNOWLEDGMENTS

The research referred to in this paper was supported by contracts with the Kingston, NY, Laboratory of the IBM Corporation. I am grateful for their support. This research required the talents of my colleagues in several departments at Carnegie-Mellon. Thanks are due in particular to Karen Graham of the Department of Design and to John R. Hayes and David Fallside of the Department of Psychology.

REFERENCES

Jones, J.C. (1970). *Design methods.* London: John Wiley & Sons.
Newell, A., & Simon, H.A. (1972). *Human problem solving.* Englewood Cliffs, NJ: Prentice-Hall.
Reitman, W.R., (1965). *Cognition and thought.* New York: Wiley.

Chapter 6

THE TOWER OF BABEL REVISITED: ON CROSS-DISCIPLINARY CHOKEPOINTS IN SYSTEM DESIGN

Kenneth R. Boff

Human Engineering Division
Armstrong Aerospace Medical Research Laboratory
Wright-Patterson Air Force Base, Ohio

> ...the Republic of Learning is breaking up into isolated subcultures with only tenuous lines of communication between them... an assemblage of walled-in hermits, each mumbling to himself words in a private language that only he can understand.
>
> *(Boulding, 1956, p. 198)*

> In science by a fiction as remarkable as any to be found in law, what has once been published, even though it be in the Russian language, is spoken of as *known*, and it is too often forgotten that the rediscovery in the library may be a more difficult and uncertain process than the first discovery in the laboratory.
>
> *(Lord Rayleigh, 1884)*

ABSTRACT

System effectiveness is in large part a function of the cumulative "goodness" of decisions and tradeoffs made in the process of design. Much potentially useful data and technology exist that remain unexploited for purposes of design. A key reason for this is the general difficulty and low perceived value in accessing, interpreting, and using research data from primary sources. This is further confounded by the explosive growth and fragmented organization of technical information in the many disciplines of science and engineering. Hence, given the typical constraints of time, available resources, and training and experience resident with the designer and the design team, it is likely that the technical information factored into design decisions is not the best information objectively available. Eliminating chokepoints in the cross-disciplinary access and utilization of technical data could meaningfully contribute to "design effectiveness."

I. INTRODUCTION

In the biblical story of the Tower of Babel, the Lord confused the builders' tongues, creating a diversity of languages to punish them for their presumption of erecting a tower that could reach heaven. System designers today bear the legacy of their audacious forebears. System design in its simplest embodiment is dependent on a convergence of specialized information from multiple technology domains which contribute integrally to either the function and embodiment of the system or to the processes needed to manufacture it. It is unlikely that a single individual today can be a master integrator of many technologies, let alone competently keep abreast of the developments in a single field. Nowhere is this problem more evident than in the transition of new data and technologies to application. As a result, I believe it is frequently the case that the technical information factored into design decisions is *not* the "best" information objectively available. Indeed, in these instances, the tendency for designers to "satisfice," in the sense of not being willing to invest the effort required to acquire the "best" information, could have negative consequences for design effectiveness.

In the sections which follow, I critically examine some of the bottlenecks in the multidisciplinary access and utilization of technical data and speculate on several conceptual schemes for improving the flow of new technologies into the process of system design.

II. DESIGN EFFECTIVENESS

System design may be characterized as a creative integration and/or skillful blending of technologies counterbalanced to accomplish a predefined function within material, cost, and schedule constraints. It is an intellectual attempt on the part of the designer to meet specific design goals in the best possible manner. *Design effectiveness* is the degree to which system function meets design requirements within these constraints. It therefore follows that design effectiveness is a function of the cumulative "goodness" of design decisions and tradeoffs, which are in turn dependent on the information factored into these decisions. Conversely, design decisions made without consideration of potentially leveraging information may be suboptimal and may collectively, depending on their impact on system function, undermine design effectiveness.

The development of a relatively simple system, for instance, a ball point pen, is dependent at some stage in its design on consideration of technical data regarding inks, plastics, molding and manufacturing techniques, mechanical principles, ergonomics, target market demography and user expectations, regulatory constraints, etc. Regardless of whether it is an "adaptive" design, a "variant" design, or an "original" design (see Pahl & Beitz, 1984, for a discussion on each of these), the search for pertinent technical information will begin with a review of related system designs. In both adaptive design (i.e., adapting a known design solution to a changed task) and variant design (i.e., varying parameters such as size, arrangement, or timing of a known design solution without changing the basic design), a high percentage of the variance in the design's effectiveness may already be accounted for by the selected baseline. Remaining design decisions are determined after consideration of those aspects of the baseline which are either not responsive to current system requirements or are not "effective" in meeting these requirements. Thus, effective baselining of a system design—identifying system designs analogous to the one under development—will reduce risk to design effectiveness. Original design is far more risky to the extent that it depends on untested approaches or technology.

III. THE FLOW OF TECHNICAL INFORMATION IN SYSTEM DESIGN

An integrated system design which meets required performance and cost goals is typically the product of a multitude of design tradeoffs and decisions which are dependent on a reliable and manageable flow of technical information. The flow of information into and through the process of design to its implementation in design decision making is both motivated and constrained by a variety of factors. These may include existing organizational biases for implementing or employing a particular technology or method without consideration of alternatives; manufacturing and marketing trends; mandatory standards, specific requirements, and/or regulatory constraints; etc. These factors have a relatively well-defined impact on the use of information resources. Less obvious (and of greater interest to me) is the influence of designer biases on the flow of information and technology from fundamental research, particularly from technical domains outside the expertise of the designer/design team.

Depending on the choice of baseline system and the degree to which it satisfies new design goals and constraints, the designer is faced with design tradeoffs and decisions to be made. Boff and Martin (1980) characterized this design decision process as a subjective integration of "valued" information within a typically limited time and resources window. If some data are unavailable and not deemed critical, then a "judgment call" using the "best"

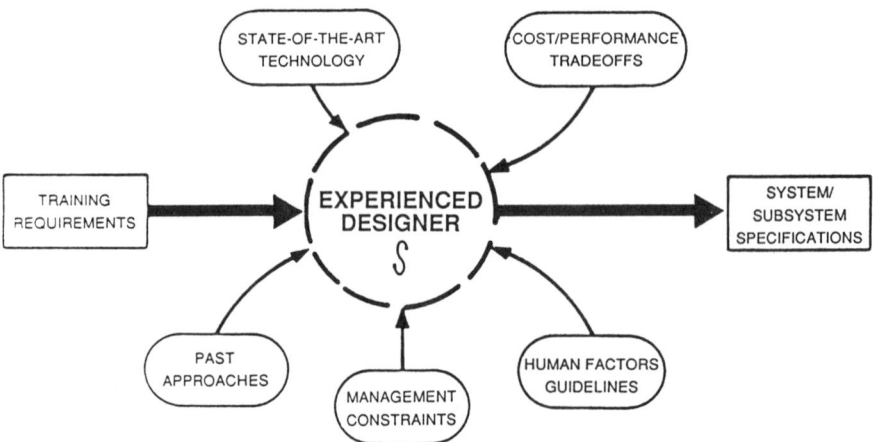

Figure 1. Design Decision Process. Design requirements and specifications are determined by the subjective integration of a range of variables. (From Boff & Martin, 1980).

information available will "satisfice." This fundamental relationship, schematically illustrated in Figure 1, is iterative throughout the design process, possibly by different individuals in the role of designer, and may indeed account for a large uncontrolled source of variance in system design. The extent to which the criteria for the "best" information to be factored into design decisions are driven by the factors associated with "perceived value" or "availability" could, under appropriate circumstances, have profound implications for design effectiveness, future retrofits/redesign, or system failure. Numerous examples of this exist in the design of human-system interfaces (e.g., nuclear power plant control displays and warning systems at Three-Mile Island), though it is my guess that "bad" designs often occur involving other technologies and other domains of design for these same reasons.

At the system level, at least, multiple designers/design teams using equivalent requirement inputs would normally be expected to produce system specifications which vary within a predictable range. Discrimination and selection of the most "effective" design alternative is a critical juncture in the design process where the flow of technical information is also vital. Typically, in the case of government source selection, design proposals are matched against functional specifications and requirements using published selection criteria. Usually, unambiguously bad or good design proposals with respect to system requirements can be discriminated on this basis. Finer discriminations on an *objective* basis other than cost may be impossible. Indeed, source selection evaluators have little recourse but to evaluate technical details of the proposal (i.e., a "judgment call") using the "best" of *their* prior training and experience. Indeed, in the procurement of some complex systems, proposals per se may be funded up to the prototype stage in order to provide some visible performance basis for evaluations. This poses interesting possibilities for design support systems that can also be used during source selection assessments to provide low cost (i.e., in terms of time, effort, and money) access to the information audit trail for major design decisions and tradeoffs. Enhancing visibility of the basis for design decisions might also have implications for the perceived value and use of particular information resources at the individual designer as well as the corporate level.

A. Factors Affecting Cross-disciplinary Search and Utilization of Information and Technology

If information needs with high perceived value to a design effort are identified but cannot be satisfied by the knowledge base at hand, information seeking behavior by the designer/design team will be motivated. According to Allen (1977), and reaffirmed in an unreported study by Boff and Carr (1984) using simulator engineers in DoD and industry settings, information seeking will include consultation with subject matter experts and search through the extant literature. Further, depending on the stage of design and the availability of reliable information, empirical means may be employed such as experimentation, simulation/modeling, or prototyping. This problem is exacerbated when the information needed falls outside the disciplinary boundaries in which the searcher is conversant. In the section which follows, I outline what I believe to be the principal chokepoints to cross-disciplinary search and acquisition of existing information and technology for system design.

1. Knowledge or Ignorance of the Existence of Useful Technical Data

To initiate a search for information, it is necessary to have a motive for and an object of search. For a designer, it may be awareness or belief that some technology area could influence the function or embodiment of the system under design. The main point is that technical information of which the designer and/or design team are ignorant will *not* be sought after, and will, as a result, *not* be factored into applicable design decisions. Indeed, it is the thesis of this paper that much potentially useful technical data exist which remain unexploited for design purposes.

2. Understanding the Value of Specific Technical Information

The role of experiential and training biases in assigning a value and cost to information factored into design decision making seems understated in the literature. Rouse (1986) suggests that for information to be perceived as valuable, it must "have the potential to reduce uncertainty...be relevant to near-term stages/tasks and to have an appropriate form." Dealing with technical bias by designers with traditional engineering backgrounds is a familiar problem to human factors practitioners. Domains of technical information *believed* not to be of substantive value will *not* be accessed no matter what the real value of the information.

3. The Nature of Design and the Cognitive State of the Designer

The essence of this constraint hinges on the form of reasoning involved in design. Depending on the design problem and the stage of design, the design process may be characterized as investigative, creative, or rational, which in turn will govern whether the designer draws on analytic, recognitional, or other forms of reasoning. The search for information, both in terms of strategy and with respect to the kinds of information that might be considered useful, is a function of the cognitive state of the designer and the demands of a specific design at any given stage of the process. While recognitional skills are needed for the baselining of systems, which may indeed account for a large percentage of the design variance, it is also evident that specialized institutional knowledge and expertise are needed for detailed specifications that depart far from intuition, experience, and tradition, particularly in the design of complex systems. The degree to which the cognitive state of the designer is *matched* to the needs of the process at any given stage of design will influence the perceived value and cost of acquiring technical information and technology.

4. Linking Methodology for Decomposing and Relating Design Requirements to Informational Needs

Design may be viewed as a form of problem solving. As shown in Figure 1, the designer's task is to translate requirement inputs into design specifications. The decision space (i.e., the acceptable zone for "satisficing") that the designer will have varies directly with the specificity of input requirements. The task of converting ambiguously stated functional requirements into detailed system specifications usually boils down, at best, to educated guesswork by the designer/design team, particularly if there are few good working baseline systems on which to draw. Analytic approaches can support reduction of system requirements, but are ultimately dependent on the personal skills and knowledge of the designer in dealing with the gap between the best reduction of the design problem and best resolution that acquired data resources are able to provide. Technical data that are not readily understood (i.e., do not share a framework in terms of common goals, approaches, and criteria; see Rouse, 1985), with respect to the design problem or the products of design problem reduction, are likely to be ignored. This is the fate of much fundamental research technology that does not succeed in transitioning to applications. The problems typical for design of human-system interaction are a good example of where requirements tend to be ill-defined and solutions are often trial and error, even in instances where potentially useful information is available.

5. *Organization and Distribution of Pertinent Technical Knowledge*

The successful acquisition and interpretation of relevant information from primary source literature can be a formidable task. In addition to the difficulties in communicating across disciplinary boundaries, this problem is exacerbated by the staggering volume of existing data and the manner in which it is organized topically and distributed physically over a wide number of journals and report media. The contextual and theoretical frameworks within which researchers typically generate, disseminate, and otherwise organize technical information are not necessarily common with the logical framework or the needs of the practitioner. Practically speaking, the designer may not readily find needed information where it is expected to be located. Indeed, the degree of fragmentation of individual domains into separable research subcultures places a considerable burden on searches *within* the domain as well. In a recent project, Boff, Calhoun, and Lincoln (1984) were easily able to subdivide the human perception and performance literature into over forty subtopic domains of varying degrees of independence. It is, therefore, easy to see why the unaided designer searching for information outside his domain of expertise is likely to experience costly difficulties while attempting to discriminate hits from false alarms among the products of a literature search.

6. *Context and Packaging of Technical Information Resources*

Primary research reports in the sciences are typically prepared in accordance with traditions of reporting scientific data which were established with the objective of communicating the evidence for scientific findings in an unambiguous and verifiable manner among members of the same scientific community or subculture. While it is questionable in my mind as to whether the current generation of scientific journals has, indeed, adhered to this tradition, there is no doubt that packaging data for purposes of scientific communication adds considerable overhead, and in fact may be a barrier for the practitioner. The human factors discipline has been particularly guilty of not "human factoring" the presentation of human factors data for practitioners (see Boff et al., 1984, for further discussion of this issue).

7. *Communicability of Specialized Knowledge*

While precise communication among specialists within a given technology can be a tedious process, it pales by comparison with the difficulties involved in attempting precise communication among specialists from across multiple disciplines. Specialized technical domains typically evolve specialized languages in order to communicate ideas and data unambiguously within a given frame of reference among specialists. Hence, in order for specialists to communicate effectively across domains, they must find a common frame of reference. The larger the number of individuals involved or the greater the differences in domains of expertise, the more difficult it will be to develop a common frame of reference and the more likely the potential for misunderstanding in communications. This is probably why a single knowledgeable mind is a more efficient integrator and innovator (within bounds) than is a collection of minds (e.g., a team) that possess an equal total sum of information. With the existing emphasis on specialization among data producers and practitioners in our culture, there is a high potential for technical misunderstanding. When potentially useful information or technology cannot be understood with respect to the designer's problem, it *will not* be used. Hence, frames of reference provide an important enabling ground for producing and communicating technical information in a given domain—but the same frames of reference also may serve as barriers or chokepoints in the transfer of this domain knowledge to the system designer.

IV. FACILITATION OF TECHNICAL INFORMATION TRANSFER

Applying knowledge from one specialized domain to another requires, as a minimum, an adjustment of the frame of reference and context in which the information was gathered or produced. While this does pose potential difficulties and pitfalls, such linkage is accomplished and is, in fact, crucial to feeding informational needs in a complex technological society. Mechanisms and necessary conditions for linkage and technology transfer (Havelock, 1969), the role of "technological gatekeepers" (Allen, 1977), the linker role in technology transfer (Essoglou, 1975), and many others are discussed and modeled in the literature. The most basic embodiment of the linker is as a "conveyor" who acquires information from expert sources and passes it on to nonexpert potential users. Conveyors may exist as salesmen, advertising agents, technical writers, or as technological gatekeepers whom Allen (1977) suggests are critical to the flow of technology in R&D laboratories by translating material from referred journals for use by the average technician. As such, conveyors are an important driving force in the transfer and utilization of technical information.

Linkage in one form or another is quite common in daily discourse. Second generation conveyors abound in the form of satisfied consumers or successful baseline designs. Also, while domains of knowledge are widely distributed among individuals who are keepers of specialized knowledge, we all know of experts in particular technical domains without having specific knowledge of the particulars of their expertise. The immediate implication of this is that if you do not know the answer to something, you might know who to ask or where to start looking. Indeed, if such a contact is available, using it should have the lowest immediate cost to the searcher. Interestingly enough, however, this prediction is contradicted by data from Allen (1977), which shows that when beginning a search, searchers in R&D organizations try accessing the literature before using personal contacts.

A related issue where linkage may be important is the transitioning of basic research findings to application. There is an active controversy in the human factors literature regarding the transferability and/or inherent usefulness of basic research in human performance for purposes of application. Many believe that the differences in goals, approaches, and criteria between basic researchers and practitioners cannot be linked by a common framework and consider attempts to do so as futile. Meister (1985) appropriately extends this controversy to include human factors applied research and proposes that resolution of this problem is dependent on a shift in conceptual framework by researchers toward that of practitioners.

All of this begs the question of "usefulness"; i.e., what is "useful" to designers as consumers of research data? It seems to me that the usefulness of basic data for design applications is a function of the statistical properties of the data, the situation to which it is to be applied, and the user's ability both to generalize from the data and to reliably decompose the design problem. I do not intend to suggest here that a case cannot be made that certain classes of basic data or research are useless for applications, but rather that the usefulness of existing data is also dependent on the user's ability to reduce or constrain the statement of the problem to terms compatible with the predictive power of the data. For example, the design of a human-machine interface optimized for information transfer should depend, at some level of its development, on specific data regarding the variables which influence the human's ability to acquire, process, and make effective control decisions. This assumes that the designer can systematically decompose the interface design to equipment requirements and options (e.g., quality/tolerances for display and control subsystems, and definition of requirements for information content, format, and portrayal). These can, in turn, be considered for tradeoff against human performance requirements as well as other system performance and manufacturability factors. While a working knowledge of the variables which may influence the human operator's performance within a prospective system environment is somewhat less than intuitive, recourse to search and evaluate the literature is both risky and costly in terms of time, money, and effort. It is likely that a given literature search will turn up *either* applied studies involving complex system variables that are difficult to relate to the design at hand, *or* basic research in which the variables of interest have been studied within a constrained framework from which it is difficult to generalize. In either case, the usefulness of the data ultimately depends on the designer's ability to relate the system design requirements and options to the human performance data, as well as use good judgment regarding the reliability of any assertions and assumptions based on generalizations from these data. Good judgment, in these instances, may dictate the necessity for verifying the data through further literature search or some empirical data collection process.

This process of fitting data to design needs is not unlike the baselining (i.e., search for analogous designs) accomplished at a macro system level early in design. Hence, it would seem that the real value of research data is in their "implications" for design tradeoffs and decisions rather than in their direct use in cookbook fashion. Historically, linkages between basic human performance data and design have been successfully established for problems of control/display quality and show considerable promise for current DoD problems concerned with information portrayal. Future design support systems may also succeed in achieving linkage between other basic research areas and design applications.

V. IMPLICATIONS FOR DESIGN SUPPORT SYSTEMS

A. A Designer's Associate Concept

To the extent that "design effectiveness" is dependent on the flow of information, the design of design support systems must take into account the factors which inhibit and/or facilitate the access and utilization of pertinent technical information. Hence, in my view, an *ideal* design support system would function as a Designer's Associate or collaborator, with a principal role as conveyor or technological gatekeeper. To meet the information demands of the designer in a timely and cost-effective manner, this hypothetical support system must be able to overcome critical "chokepoints" in the flow of information. The following system characteristics for a Designer's Associate parallel the "chokepoints" discussed earlier:

o   The system must be able to help identify as well as access information as an object of search. In other words, it must help the designer ask questions that he may not know to ask.

o   The system must be able to enhance the perceived value and reduce the perceived cost of acquiring and using technical data in design decision making.

o   The system's selection and portrayal of information must be adaptively matched to the cognitive state of the designer and the requirements of the design process as a function of the stage of design and the demands of the design problem.

o   The system must include a basis for aiding the designer to decompose

The Tower of Babel Revisited

the design problem to a level of resolution from which a) reliable targets for an information search may be derived, and b) there is some likelihood of matching up to the "best" resolution that data in a particular area has to offer.

o The system must encompass multiple domains of knowledge that are relationally organized and heuristically managed.

o The system must be able to integrate data files across technology domains in an adaptive manner that is consistent with the needs of the designer and the constraints of the technology domain.

o The system must be able to encompass and manage multiple frames of reference in communicating with the designer/design team.

Development of an advanced design support system that possesses these characteristics poses some intriguing demands on the information management technology needed to support it.

B. Technology Chokepoints for the Foreseeable Future

In a recent evaluation of automated information management technologies, Reitman, Weischedel, Boff, Jones, and Martino (1985) attempted to identify information management chokepoints and demands on technology for developing a hypothetical Designer's Associate system. Principal among the areas defined as technology chokepoints was the development of reliable knowledge engineering techniques for identifying, extracting, representing, maintaining, and communicating expert knowledge. Predictably, a serious weakness in current and projected technologies is the facilitation of interdisciplinary communication of expertise. Table 1, from the summary of the findings of this report, provides estimates of the probability of success for overcoming these technology chokepoints and estimates over time the level of effort required to achieve needed breakthroughs. In my view, these estimates are somewhat optimistic given that the number of competent researchers needed over time exceeds the extant population of investigators focusing on these problems. However, it may be inferred from these estimates that design of a Designer's Associate is a feasible undertaking, given current directions and progress in machine intelligence research.

VI. CONCLUSION

Constrained by factors such as compulsory regulations, limited resources and available time, and the expertise resident with the design team, it is likely that the best information considered in design decisions is *not* necessarily the best

Table 1. Forecasts for Overcoming Technology Chokepoints (Adapted from Reitman, Weischedel, Boff, Jones, and Martino, 1985).

AI AREA	PROBABILITY OF SUCCESS	MILESTONE/TECHNOLOGY DEMAND	TOTAL ESTIMATED LEVEL OF EFFORT (PERSON-YEARS)*	PERIOD (YEARS)
Expert Systems	.9	A Full-Scale Explanatory Capability	99	15
		Reasoning Across Domains for Integrated Knowledge Sources	75-85	10-15
Natural Language Processing	.8-.9	A Well-Scoped, Practical Domain	--	--
		Understanding User Intention	40	4-8
		Unifying Natural Language Generation and Understanding	42-48	6-8
		Explaining and Paraphrasing	99	17
		Natural Language in a Broad Domain or Across Narrowly Scoped Domains	364	15-20
Knowledge Representation	.8-.9	Data Base Content and Purposes	10-15	3-5
		Analogical Reasoning	42-46	4-10
Planning	.6-.7	Common Sense System	104	8-13
AI Tools and Environment	.5-.9	Highly Parallel Programming	64	10-15
Speech	.8	Continuous Speech Understanding in a Broad Domain	230	20

*These estimates are for additional effort over the current funded baseline.

information available. This problem is confounded by instances where the perceived value of information is not correlated with its intrinsic worth, but rather with ease of access and utilization. In this paper, I have discussed a variety of chokepoints to the transfer of data and technology to system design. Within the constraints of the design environment, these chokepoints may act to mask the potential "usefulness" of data by raising the perceived cost and risk to obtaining it. This is consistent with Allen's (1977) finding that 92% of the technical information used by engineers is already resident in personal files or with colleagues at the time that the information is needed, and with the notion that designers are likely to seek technical data for "satisficing" rather than optimizing design decisions. Hence, the factors driving utilization of technical information may not be positively correlated with the factors which contribute to design effectiveness. An optimal design support system should provide its user with low overhead access and expert guidance to a wide range of technologies germane to design information needs. Eliminating the "Babel" from the cross-disciplinary flow of information must be made an explicit objective for the design of design support systems.

VII. ACKNOWLEDGMENTS

I am indebted to Robert G. Eggleston and Edward A. Martin for helpful suggestions in revising the draft of this manuscript. I am also grateful for the support provided by Tanya Ellifritt in preparing this manuscript.

REFERENCES

Allen, T.J. (1977). *Managing the flow of technology: Technology transfer and the dissemination of technological information within the R&D organization.* Cambridge, MA: The MIT Press.

Boff, K.R., Calhoun, G.L., & Lincoln, J. (1984). Making perceptual and human performance data an effective resource for designers. *Proceedings of the NATO DRG Workshop* (Panel IV). Royal College of Science. Shrivenham, England.

Boff, K.R., & Carr, T. (1984). [Simulator engineers in DoD and industry settings]. Unpublished data.

Boff, K.R., & Martin, E.A. (1980). Aircrew information requirements in simulator display design: The integrated cuing requirements study. *Proceedings of the Second Interservice/Industry Training Equipment Conference,* 355-362.

Boulding, K.E. (1956). General systems theory - The skeleton of science. *Management of Science, 2,* 197-208.

Essoglou, M.E. (1975). The linker role in the technology transfer process. In

J.A. Jolly and J.W. Creighton's (Eds.), *Technology Transfer in Research and Development*. Monterey, CA: Naval Postgraduate School.

Havelock, R.G. (1969). *Planning for innovation - A comparative study of the dissemination and utilization of scientific knowledge* (HEW Contract No. OEC-3-7-070028-2143). University of Michigan: Center for Research on Utilization of Scientific Knowledge.

Meister, D. (1985). The two worlds of human factors. In R.E. Eberts and C.G. Eberts (Eds.), *Trends in Ergonomics/Human Factors II*. Elsevier Science Publishers.

Pahl, G., & Beitz, W. (1984). *Engineering design*. New York: Springer-Verlag.

Reitman, W., Weischedel, R.M., Boff, K.R., Jones, M.E., & Martino, J.P. (1985). *Automated information management technology (AIM-tech): Considerations for a technology investment strategy* (AFAMRL-TR-85-042).

Rouse, W.B. (1985). On better mousetraps and basic research: Getting the applied world to the laboratory door. *IEEE Transactions on Systems, Man, and Cybernetics, SMC-15*(1), 2-8.

Rouse, W.B. (1986). On the value of information in system design: A framework for understanding and aiding designers. *Information Processing and Management, 22*, 217-228.

Chapter 7

PSYCHOLOGY OR REALITY

Paul R. Chatelier

Perceptronics, Inc.
Arlington, Virginia
formerly of the
Office of Undersecretary of Defense
for Research and Engineering
Washington, DC

ABSTRACT

System design and the application of human factors techniques is a favorite topic of those wishing to improve operator-hardware design. However, even after years of research and development of new, modified, and/or warmed-over approaches, systems are still designed without adequate attention to the people who must use and maintain them. The reality of the situation is that human factors professionals must develop techniques and methods that are understandable to the design engineers. Developing the "right approach" from a human factors point of view does not guarantee the system's designers, managers, or bureaucrats will apply it if the human factors solution slows down the schedule, costs more, or requires an approach that is different from the current design. Reality of design for human factors today means that these professionals must understand the established procedures used by system designers, acquisition specialists, and program managers, and they must become an integral part of the accepted process from the beginning of the concept to the creation of the product.

I. INTRODUCTION

As I began this position paper, I was in a bit of a quandary about exactly what was expected from a workshop entitled, "The Psychology of System Design." After some thought about the overall professional sophistication of the gathering, I decided to assume that, in this case, psychology was to mean "real world." Consequently, I will attempt to explain what "really" happens in the governmental process of system design that is currently attempting to make use of human factors.

With that goal in mind, let me suggest that the "psychology of system design" is perhaps in the eyes of the beholder. More importantly, at this stage, it is dependent on which discipline, personal interest, or organizational position one wishes to emphasize as being important to the design process. It is this direction of purpose that I believe is the reality of the psychology of system design today. Furthermore, I suggest that there are no easy or even predictable micropaths to developing an effective design, but there might be a macropath. That is to say, it is possible that an overall management-oriented approach, responsive to both engineering and psychological principles, might be workable. However, I would like to provide my view of how designs are initiated. First of all, it is a complex, interrelated matrix of politics, policy, buyer goals, corporate goals, designers, and users. Unfortunately, or perhaps not so unfortunately, there is neither time nor space to deal with all these aspects in detail. Thus, I have chosen to discuss only a few that are the most significant from my organizational perspective.

II. DESIGN OF MILITARY SYSTEMS

Given my background, this information about system design will relate primarily to the design of military systems. The first reality of design is that it is very strongly affected by political influences. In fact, the influence of politics is the basis for many policy decisions that affect what can and cannot be purchased as well as how one goes about the business of acquisition. The policy now in use began in 1969 when Congress chartered the Congressional Commission on Government Procurement with Senator Lawton Childs as its chairman. Based on its review of procurement case studies, the Commission believed that competition and the open viewing of contracts were necessary to save money and to provide the best possible design of systems. Therefore, in its 1971 report to Congress, the Commission stated that the Federal Government had to establish "top-side reforms to the procurement system which would permeate throughout the entire government." In 1974, President Ford signed the law that established the Office of Federal Procurement Policy (OFPP). This office was placed under the Office of Management and Budget for administrative supervision.

OFPP soon produced rules and regulations that the top management levels considered the way to gain control of the system development and procurement process. However, as soon as those responsible for explaining the policy and implementing its interpretations received their copy, the highly touted process seemed to be viewed as something less than perfect. This happened because middle management's preceptions of policy are based on past experience and not necessarily on data. As usual with a new policy, the first efforts are to establish new layers of management, each with its own charters, steering committees, advocates, and potentially available bypass channels. It is significant that OFPP did not help the workers design systems. Instead, these policy makers said that if one used a front-end analytical technique, determined alternatives, competed the functional requirements, and did a good functional evaluation, the government would save money and receive a superior product. It was supposed to work for everyone in an efficient and effective manner because it was "good and right," and as I was told, it was the Jeffersonian way. In other words, it was an upper-level management overview with a philosophical direction. It was not an engineering guideline to be used in the day-to-day development process.

Nevertheless, this policy activity did begin to permeate through both the government and the corporate structures. Everyone pursued his interpretation of the policy, and human factors engineers, recognizing the opportunity to be involved earlier in system design, jumped into the issues of front-end analysis. Consequently, today's literature includes an abundance of papers on the front-end analytic process written during this time describing how the process would allow designers to build user-friendly or at least human-compatible systems. Unfortunately, all of this activity slowed down when industry began to place a price tag on front-end analysis. Furthermore, when requests for proposals were sent out by the Government, human factors was not a line item to be evaluated or paid for during proposal evaluation or early system definition. Thus, we found ourselves with a lot of policy and management involvement in reforming the development and procurement system, but it was only business as usual at the bench of the designers, especially as it concerned human factors.

Although this happened during the middle 1970s, it is worth pointing out that we are currently experiencing some deja vu. On February 28, 1986, at the request of the President, the Blue Ribbon Commission on Defense Management published its report on the subject of procurement. It was known as "the blueprint for change," and it was driven by many horror stories concerning the high cost of developing, maintaining, and purchasing for systems or parts of systems. Among the commission's recommendations were those meant to improve the control and supervision of the entire acquisition system. However, as often is the case with such high-level studies, the solution ended up being a management approach not unlike the previous solution. The difference is that the earlier solution improved acquisition through a more detailed and programmatic approach. This current solution reduces the details that are considered too complex and, therefore, too costly. However, neither approach looks at the design procedure as a systematic procedure or attempts to improve the process so the buyer knows that hardware, software, and people are factored into the design. The smart buyer still does not know specifically what can be expected of the product before it is delivered and used.

A. Other Factors Influencing Designs

If the program requires extensive design efforts, there are usually several influencing factors besides politics affecting its development. I suggest those factors are money, time, and acceptance of results by management. (Please note that last point because I did not say the user(s) but *managers* who are in charge.) Furthermore, if there are several problems to be solved, it is possible that the process can be made more complicated by sincere efforts to find an engineering solution. For example, when a person operating a prototype of a sensor system cannot find the target, whose problem is it? First, the sensor designer gets the responsibility. Perhaps next, the engineer believes that the sensor could not resolve the target. Following a chain of events, one eventually gets to the processor or computer in the system that is considered to be not doing its job. Eventually, the problem is thought to be with the display because it cannot provide the "good" information available from both the sensor and the processor, or the display is incapable of displaying what the sensor and the processor can handle. Thus, it can go on and on without much being done to solve a system's problem through specific engineering design efforts.

At this point, it is appropriate to ask, can a human factors engineer solve the controversy? The answer is he probably cannot at this time, but only because he has no way of knowing that the question, problem, or controversy exists or the ramifications of the issue. Furthermore, to be effective in such a situation, the human factors effort must be part of a team that is multidisciplinary and multifaceted. That is to say, there must be human factors personnel who understand the politics of design, the policy of design and development, and the techniques and science of man-machine design. It takes a total awareness of the problem and teamwork to know how to focus on a workable solution.

B. Human Factors in Design

My experience shows that during system development, the design process is dominated by a large amount of nonstandard, common sense human factors. There is a bit of art, a bit of science, and a lot of what I would like to call "role playing." By this I mean a person involved in design becomes the manager, the layout engineer, the system integrator, etc. However, it is interesting that seldomly does this person take the role of the logistician or trainer. As a result, one can find at least two different task analyses and many different assumptions regarding use, maintenance, personnel, etc.

In spite of the best intentions of those improving the acquisition policies and procedures, we still do not "plan" to perform the "design procedure" any better. I believe the varied background and experience level of the human factors engineer make him an efficient design integrator if the front-end planning allows for a truly integrated process.

III. WHO ARE THE SYSTEM DESIGNERS?

It is my view that systems are actually designed by three groups within an organization. The first are those experienced, well-established engineers who are always ready to produce a new design that will reflect well on their company. This orientation allows a design to look something like the last one but with some advances from the new, exciting technologies recently made available. This kind of professional will take risks only when the options are well understood.

The second group that designs the systems are the young tigers who are learning the profession. They want to do well enough to eventually be in top management like their bosses are now. They rely on handbooks, specifications, standards, and the guidance they are given. They are free to roam the plant and think of solutions as long as they do not take up too much time or require the use of risky or costly items.

The third and last group working on the system design concepts are those who have been there a long time. They are happy to do a specific job on the design team. These people have the ability to design a communications panel or write the integrated logistics support requirements for a landing gear while they sleep. They are not really interested in developing anything new or innovative. Unfortunately, the result of these methods is a system where policy and organization can limit the freedom of expression in systems design.

Without top-level policy that states the requirement to pay for effective human-system design throughout the design process, such improvements will not happen. Design must be paid for at each and every phase of the acquisition process. That means it has to be on the "must do" list during proposal evaluation. There must be professional program monitors and progress reports at critical milestones. There must also be high-level executives in the procurement organization asking about the human-system design.

As good as it would be for procurement, what about the systems design developer? I submit that the developer has a more difficult time. He must have an organizational structure that allows for all forms of technological cross-feed and open information flow. Some designers must be artists whose free expression can be shared with those who know the rules but also know how to bend them to produce a better product. The recent addition of Computer-Aided Design (CAD) has assisted the transmission of design data to all those who want it or should have some knowledge of it. However, I have not heard of a system being designed where the entire team, including human factors professionals, were on a total CAD network. Such a teaming network would allow "off-line developments" and might even increase the output of inventive solutions from those designers who have not developed their interpersonal communication skills. I hope that it is possible to demonstrate this form of networking in the future.

IV. CONCLUSION

If we can encourage top-level policy to include human-system design in all stages of design development, we will have reached the point where we can start to do our job the way it was intended to be done. This does require that human factors professionals understand policy as well as design changes. We are not there yet, but we can see the goal. It is my hope that all who believe in the systematic development of human-machine (and software) systems will continue to pursue a goal-oriented, not rule-oriented, process.

This new design team consisting of human factors and engineering specialists must be of varied backgrounds to make the development and design process work. This team must work with conviction, determination, and optimism to make a worthwhile difference.

Chapter 8

SOME INTELLECTUAL REQUIREMENTS FOR SYSTEM DESIGN

Anthony Debons

School of Library and Information Science
University of Pittsburgh
Pittsburgh, Pennsylvania

ABSTRACT

This essay examines two propositions on system design. These propositions are neither new nor mutually exclusive. The propositions suggest that the designer's education and experience, together with a conceptual model of a system, are considered fundamental to the design process. Because the current academic institutions emphasize discipline-oriented educational programs, the availability of educators to provide interdisciplinary perspective to system design is limited. Governmental support of those institutions that emphasize interdisciplinarity as part of their academic philosophy could aid in providing the kind of professional competence that system design requires. The system design process is also aided by the development of system concept, which identifies the necessary and sufficient components of a system to accomplish its objectives.

The thesis of this essay is that the conceptual model should recognize the distinction between the design of the overall "total" system and the conceiving of the overall or "governed" system as the component of such a system. The EATPUT model generated from considerations pertinent to C^3 systems attempts to provide the theoretical and operational foundations of such systems.

I. EDUCATION AND TRAINING

There are two realities that need to be reconciled. First, the design of systems is a practical, operational activity. The activity requires multidisciplinary involvement, inviting the skills of individuals who, by training, can bring together the knowledge, insights, and experiences of a number of different disciplines. Second, the present ability of academic programs to develop interdisciplinary synthesis from multidisciplinary exposure is inadequate.

My claim is that although colleges or universities provide individuals the opportunity for a multidisciplinary experience, they fall short in providing individuals *with* an interdisciplinary experience.

Multidisciplinarity - A variety of disciplines, offered simultaneously, but without making explicit possible relationships between them. ("Interdisciplinarity," 1972, p. 106.)

Interdisciplinarity - A common axiomatic for a group of related disciplines is defined at the next higher hierarchical level or sublevel, thereby introducing a sense of purpose. ("Interdisciplinarity," 1972, p.106.)

Organizational and administrative philosophies of academic institutions do not provide for the support that interdisciplinary education requires. Departments are inclined to focus on safeguarding existing paradigms. As Churchman (1979) has indicated, disciplines are "political entities, not scientific." In this climate, interdisciplinary thought may not be welcomed. For example, in some colleges and universities it is difficult, if not impossible, for a faculty member whose research and professional interests cut across several disciplines to obtain tenure. Faculty of specific disciplines are less tolerant in permitting their teaching and research resources to be consumed by other departments, particularly when the compensating costs are borne by their department.

As indicated, the second reality concerns the intellectual process that defines interdisciplinarity—namely synthesis. When we refer to systems and the systems approach, we are actually inferring the business of synthesis (Bloom, 1982):

Synthesis - putting together of elements and parts to form a whole...working with elements...and combining them in such a way as to constitute a pattern or structure not clearly there before.

This concept of synthesis is implied in the definitions attributed to system and related terms (Sippl, 1976):

System - An assembly of components *united* by some form of regulated *interaction* to form an *organized* whole.

Design - Development of a plan; the *ordering* of components.

System Design - The specification of the working *relations* between all the parts of a system in terms of their characteristic action.

There are two views that can be discussed regarding acquisition of synthesizing skills. The first view is that the capacity to synthesize results from the assimilation of a diverse set of field experiences, and each experience requires a multidisciplinary perspective. The second view is that synthesis is the result of the intellectual action by the receiver in response to the input by the sender. Applied to the academic situation, the faculty member, who can be identified as the synthesizer (the sender), is an individual who, through field experience and other interests, imparts this synthesized knowledge to the student. The student in this case is the receiver of the synthesis. In the second case, the student is the synthesizer of the content given to the student.

Other views I am sure can be entertained about the synthesizing function, one of which could be the merging of the two views discussed above. The point I wish to make, however, is that academic institutions are often unable to provide the climate and/or resources to engage in and sustain interdisciplinarity, which the intellectual function of synthesis, so important to design, requires. We find that those individuals whose ambitions and predilections are toward working with a number of disciplines (bringing together varying perspectives) are often not accepted, recognized, or rewarded by the institution, administration or, for that matter, their peers—many of whom are oriented psychologically and professionally to the discipline in which they received their formal education.

Colleges and universities are often unable to employ and maintain interdisciplinarians—synthesizers. In the professional marketplace, interdisciplinarians are at a premium—as such, they can command renumeration significantly beyond that available in academia. In the absence of such individuals, the discipline-oriented instruction is perpetuated. The significance of this to the system design process, which as discussed previously is a synthesizing function, can be fully appreciated. There is one further point regarding impact which I would like to include at this point. Although design activity can be argued to be largely multidisciplinary, the management of system design can be seen as essentially interdisciplinary. Thus, the lack of an interdisciplinary professional resource to support system design activity actually impacts most at the critical level—namely, the planning, operating, and control of systems at the management level.

What can be done concerning this state of affairs, if anything? How can interdisciplinary perspectives be instilled in our academic institutions? It seems that although colleges and universities fully recognize the importance and value of interdisciplinary education, quite often the administrative problems are difficult to surmount, given the present strength and political power of disciplinary departments. A model of college/university interdisciplinarity is needed. Such models do exist. These colleges and universities need to be recognized and rewarded. They should be the focus of support by private and governmental funding agencies.

Not much is known about interdisciplinary education ("Interdisciplinarity," 1972). Research is required on pedagogical factors that contribute to the synthesizing functions. Again, there is a basis for this research. Interdisciplinary-multidisciplinary programs in different areas (information science, communication science, social engineering, etc.) are available from which data can be collected on the development and assessment of curricula correlated to post-graduate employment effectiveness.

II. SYSTEM APPROACH AND THE DESIGN FUNCTION

Hand in hand with the personnel resource problem that has been discussed is the absence of a conceptual model of a system on which the design can be predicated. This state of affairs is well illustrated in the prevailing activities and practices that are relegated to command-control-communication systems (C^3). The problem with C^3 systems (information systems) is that they are conceptually transparent. They include much more than "meets the eye." Yet, design demands concreteness by definition. This state of affairs seems to support the need and importance of a conceptual model.

At the outset, it should be recognized that C^3 systems are part of a class of systems referred to as information systems. What are information systems? All organisms are information systems to the extent that the various organs and functions of organisms (sensors, brain, central nervous system, etc.) are biologically available to enable the organism to develop, respond, and survive in a physical/social environment (Miller, 1978). When reference is made to "information systems," the phrase refers to an environment of people, technology, and procedures that provides the human with the capability of augmenting his biological endowment in solving problems and making decisions. Licklider (1964) refers to this phenomenon as human-machine symbiosis.

The purpose and objective of the C^3 system is to aid the military in solving problems and making decisions. Another way of stating this is that such systems augment the human functions that relate directly to problem solving and decision making. Decision support systems (DSS) and management information systems (MIS) are designations of systems with more specialized orientations (planning operating, control operations, etc.). DSS and MIS are commercial designs of military C^3 systems.

What are the basic aspects of such systems? This question seems to bemuse those who *are* in the business of designing such systems. There are a variety of models which presumably attempt to identify the fundamentals of such systems (Debons & Larson, 1983). Despite the fact that many such models have been privileged to be considered as "generalized models," many remain essentially

Some Intellectual Requirements for System Design

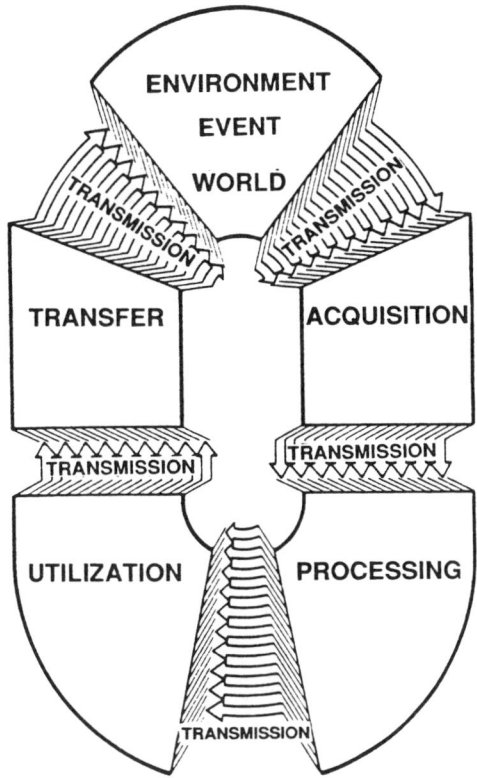

Figure 1. EATPUT Model 4.5: Debonian Information System Model. (From Morton, May 1985, with permission from the Journal of Information and Image Management.)

component-oriented. For example, many of these systems orient the total design of C^3 to the data processing or display function rather than placing these elements in a larger frame of reference—a frame of reference that embraces the total system. Consequently, design activity is oriented toward selecting *technology* to meet *component* requirements that are function-specific and cost-efficient. An overall system orientation would consider relevant human technology interfaces within a broader concept of efficiency and effectiveness.

The C^3 (information) system reflects the six necessary and sufficient components that enable the transformation of data to information. The system provides the decision makers and problem solvers with the ability to deal with a wide spectrum of contingencies in everyday life encounters (see Figure 1).

Event World - consists of occurrences both in space and time (environment) which the system is intended to capture. The representation of these events is included in the data that provide the bases for decisions made by humans. Event world analysis is influenced by the symbols and signs we use to identify occurrences, by the manner in which we order the characteristics of the event (categorization and classification), and by the tools (measures) we use to differentiate one aspect from another.

Acquisition - refers to the capturing of events in space and time. The capturing is achieved through technology (radars and sensors) or through human sensory (light, sound, movement, etc.) capacity.

Transmission - refers to the movement of signals that are captured from the event world to the sensors. Transmission is equivalent metaphorically to the human circulatory systems through which the basic ingredients of life are transported from one organ to another through the living systems. Transmission for information systems is basically an electronic phenomenon which interlinks the various components of the total information systems.

Processing - the "brain" of the system where data received from acquisition through transmission is subject to collection and is stored for possible retrieval at some point in time for the purposes dictated at that time. The processing includes consideration for the development of rules which allow for the ordering of the data on an electronic display or paper hard copy (both transmission media) based on the demands of a user (decision maker or problem solver). This component enjoys considerable attention by current information system designers.

Utilization - consists of all the formal rules and concepts which can guide the human interpretative functions relating to available action options. The use of the rules in formulating decisions or in the process of solving problems *precede action.*

Transfer - the communicative aspect of the system. It represents all the considerations relating to the act of effecting the output from interpretation of data about an event and the related consequences. It includes the numerous aspects of language, including its structure and its role in social interaction.

If the system approach is to be applied in the design of such systems, the laws, theories, and principles relevant to each component and the relationships among the various components are to be taken into account. Table 1 provides a matrix of such laws, theories, and principles governing the component and the relationships among the components.

To illustrate part of the complexity that is involved, we can examine briefly a cross-section of the variables that are implied in the matrix. In matching human and technological resources for capturing events (i.e., E horizontal to A vertical in the matrix), a number of factors (t, f, etc.) are involved. For example,

Table 1. Information System Components: Mapping Concepts on Information System (EATPUT) Component Interaction. (Completeness of the concept inclusion cannot be assumed).

	E	A	T	P	U	Tr
E	c; t; a_2; c_3; l; s_4	t_1; f; r; k; s_1; e_1; e_2; h_1; p; p_3	t_1; d_2; n; f_1; a_2; h_1; c_3	c; d_3; d_4; k; f; h_1; s_2; l	f; p; h_1 l	l; f
A	e_1; e_2; f; s_1	e; e_1; t_1; e_2	f; d_2; s_4	f; d_2; p_2; k; p_3; s_2	f_1; h_1; p; s	f; d_2; c_2
T	i; f; n; t_1; a_2	d_2; n; a_2; s_4	i; e; n; e_1; e_2; n_1; a_2; c_3	p_2; d_2; s_2	f; h_1; t_1; a_2 s_4	f_1; c_1; c_2
P	f; c_1	f; p_2; d_2	d_2; p_2	a; a_1; d_3; d_4; e; e_1; e_2; s_2; h; l; x	h; h_1; a; d_2; d_4; s_2; r	f; l_2; l
U	f; c	f; d_2; h_1	i; i_1	s_2; a; d_2; l_2; s_2	d; d_1; c_1; x; h; h_1; p_1; m; s_1; d_2	f; l_2; s_3 l_3; s_1
Tr	f; a_2; g_1 c_1	f; c_1	l; n; c_1	h; l; s_2	f; c_1; s_3	l; c_1; c_2; s_4; e; n_1; h_1; s_1; m; o; l_3

Legend
- a Artificial intelligence
  - a_1 Automata theory
  - a_2 Ambience
- c Classification theory
  Categorization theory
  - c_1 Cybernetics (control)
  - c_2 Communication theory
  - c_3 Code theory
- d Decision theory (decision support)
  - d_1 Problem solving
  - d_2 Display capabilities
  - d_3 Data base structure
  - d_4 Data management
- e Hardware standardization
  - e_1 Hardware thresholds
  - e_2 Hardware calibration
- f Feedback (control)
- g Change dynamics
- h Human information processing
  - h_1 Human factors application
- i Information theory (mathematical)
  - i_1 Information theory (communications)
- j
- k Knowledge representation
- l Linguistics
  - l_1 Learning theory
  - l_2 Literary science
- m Management theory
- n Noise
  - n_1 Network
- o Operations analysis
- p Perception
  - p_1 Personality style dynamics
  - p_2 Signal prioritization
  - p_3 Human physiology (sensory, neurological)
- q
- r Reaction time
  - r_1 Reliability measurement
- s Signal detection theory
  - s_1 Sociological parameters
  - s_2 Software design
  - s_3 Semantics
  - s_4 Semiotics
- t Time space measurement
  - t_1 Threshold (physical-sensory)
- x Expert system formulations

E = Event World
A = Acquisition
T = Transmission
P = Processing
U = Utilization
Tr = Transfer

if the designer were to choose a technology to capture an event, he would need to account for the capability (threshold(s)) of the hardware to accomplish its objective. With this consideration in mind, assumptions about time and space measurements should also be taken into account. In another case, let us consider "P" on the horizontal axis and "P" on the vertical axis. This cross-section refers to the principles that would govern the matching of one processing resource (human) with another processing resource (technology), or two technological or human resources. In this case, the relevant area of interest would include theories, concepts, and principles that would be found in artificial intelligence (a) and, in particular, automata theory (a_1). It could include decision theory and data base structuring (d_3) or data management (d_4). These illustrations can be applied to all six components shown in Figure 1 as well.

III. CONCLUSION

The system design process is only as good as the professional resources that can be applied to it. Interdisciplinary education and training is a fundamental resource requirement for system design activity. The lack of academic education to provide professional resources oriented to such a requirement is a matter that deserves careful study and resolution. The policy of government to encourage research programs that involve academic-industrial arrangements is one approach to the problem.

System design requires a conceptual basis upon which the structure of the system to be designed is predicated. The EATPUT model can provide a conceptual base for both system analysis and design. Isolating or defining the total information system in terms of a single component (i.e., computer, display, or radar) does not comply with the spirit of the systems approach and can seriously impact the system design process.

REFERENCES

Bloom, B.S. (Ed.). (1958). *Taxonomy of educational objectives: Book 1 cognitive domain* (p. 162). New York: Longman.

Churchman, C.W. (1979). *The systems approach and its enemies.* New York: Basic Books.

Debons, A., & Larson, A. (Eds.). (1983). *Information science in action: System design process* (pp. 10-61). The Hague: Martinus Nijhoff Publishers.

Licklider, J.C.R. (1964). Artificial intelligence, military intelligence, and command and control. In E. Bennett, J. Degan, & J. Speigel (Eds.), *Military information systems: The design of computer-aided systems for command.* New York: Frederick A. Praeger Publisher.

Miller, J.G. (1978). *Living systems.* New York: McGraw-Hill Book Company.

Morton, P. (1985). A model for planning effective information systems. *Journal of Information and Image Management, 18*, 11.

Organization for Economic Co-operation and Development. (1972). *Interdisciplinarity: Problems of teaching and research in universities.* Centre for Educational Research and Organization.

Sippl, C.J. (1976). *Data communications dictionary.* New York: Van Nostrand Reinhold Company.

Chapter 9

THE CHANGING NATURE OF THE HUMAN-MACHINE DESIGN PROBLEM: IMPLICATIONS FOR SYSTEM DESIGN AND DEVELOPMENT

Robert G. Eggleston

Human Engineering Division
Armstrong Aerospace Medical Research Laboratory
Wright-Patterson Air Force Base, Ohio

ABSTRACT

Developments in machine intelligence are altering the basic nature of the human-machine system design problem. This issue is discussed from the perspective of the human-machine interface for intelligent systems. It is shown that the interface design for such systems is vastly more complex and different in ways that call into question some current design practices. It is proposed that the designer must consider the desired form of the human-machine relationship as a design problem and treat the cognitive, motivational/emotional, and social aspects of a relationship as design issues. Current conceptual models and other aids used to guide the design of human-machine systems are critiqued in terms of their ability to support design from the relationship perspective. An alternative conceptual model more in tune with a relationship design approach is presented.

I. INTRODUCTION

Human-machine systems are in the midst of a major change. Networks of digital electronics containing small, flexible, and powerful microprocessors are beginning to populate the core of systems ranging in complexity from advanced airborne weapon systems to home appliances like dishwashers, televisions, and

VCRs. Because systems containing components of this nature can perform complex computations and symbolic manipulations, they can be made to emulate, at some level, the human capabilities of sensing, thinking, and acting. Indeed, for some specific tasks, machines thus endowed can respond to input signals that a human cannot sense, process immense amounts of data and discern patterns and relationships that are undetectable by a human, and solve complex logical reasoning tasks that can overwhelm the human intellect. Given the potential to make "intelligent" machines that can sense features of the external environment, make decisions, and potentially act on those decisions, the following questions arise for the designer: How can effective human-intelligent machine systems be designed? Is there any fundamental change in the nature of the design problem when intelligent machine entities are incorporated in the system?

One place where change is likely to occur is in the type of relationship that must be supported by the human-machine interface. Historically, machines have been viewed as tools to be applied directly by people to perform tasks. This has generally led to the view, often expressed only implicitly, that the tool is an extension of the human and is under the human's control. Thus, the human-machine interface has been formed around a control relationship. This view may not be appropriate, however, when the machine in a human-machine system can reason and make decisions. Intelligent machines can extend human capabilities in other ways. For instance, an intelligent machine could aid a person by suggesting ways to perform a task, explicate features of a problem, generate approaches (options) to problem solution, and/or evaluate alternative solutions to a problem. *Aiding* of this type is clearly different from *using* a machine to directly manipulate something physically present in the environment. Thus, for intelligent systems with abilities like these, the human-machine interface will have to support two-way cognitive transactions. This implies that a different kind of human-machine relationship is needed. The specification of a human-machine relationship, therefore, becomes a central design issue. Clearly, it will affect both the form and function of the human-machine interface, and it is also likely to have far-reaching implications for the design of the entire system.

Unfortunately, the current design process for the development of complex human-machine systems does not explicitly treat the nature or form of the human-machine relationship as an issue to be considered by the design team. Indeed, issues of the human-machine interface are considered during system design and development, but the fundamental nature of how the machine and human are going to relate to each other is never seriously addressed. Generally, the nature of the relationship is assumed to be of one form and never challenged or manipulated as a design variable.

It is argued here that the human-machine relationship must be treated as a design variable during the early stages of conceptual design. This is especially true for intelligent systems since the relevant dimensions of the relationship increase in new ways when the human-machine interface takes on a more cognitive flavor.

II. HUMAN-MACHINE RELATIONSHIPS

The relationship between human and machine components in any system is largely determined by the roles each is expected to play in the system. For conventional systems (i.e., ones without the benefit of machine intelligence), the roles of the machine are limited mainly to direct sensing of an environmental state and/or direct manipulation (e.g., moving, attaching, bending, cutting, breaking, etc.) of something in the environment. As stated earlier, the human's role in conventional systems has often been viewed mainly in terms of control. For simple systems, like a person riding and guiding a bicycle, the human is generally treated as the operator of the device, using it in a controlled manner to achieve some desired end. More complex systems, ones that utilize automation features, permit the human to relax direct control under normal operation and to indirectly control the system through the management of available resources. That is, the human becomes a "system manager."

A limited form of management sometimes used to express the relationship between a human and the machine is called supervisory control. From this viewpoint, human control of the system is still emphasized, but now the human controls task performance normally by his knowledge of the operation and through the assessment of system state parameters presented on a set of status displays. The "manager" takes direct control only when necessary.

Although other roles could be used to define the human-machine relationship for conventional systems, the range is small when compared to the possibilities available to intelligent systems. Since intelligent machines can process information and form and manipulate abstract, semantically rich knowledge, they have the ability to serve many roles when coupled with a human in a system. They can interact directly with a human at a cognitive level. Hence, intelligent machines could serve in the role of 1) a coach to aid a novice system user; 2) as the system's "operations officer" to carry out the directions (policies and general top-level instructions) of the (human) "chief executive officer"; 3) as a tactician or strategist proposing and explaining courses of action to the human; or 4) in virtually any other role that could be performed by a human with similar sensory, mental, and motor ability.

The machine roles indicated here suggest that, under some circumstances, the relationship between the human and intelligent machine is perhaps best viewed as a partnership rather than as the master-slave relationship that emerges from a control viewpoint. A partnership stance would seem most appropriate when both the human and machine can engage in *independent* decision-making and action-taking activities. In order for the system to be effective under these circumstances, all critical independent activities must be properly coordinated across the human and machine entities. Furthermore, the partners must be able, at times, to direct each other; expect certain actions will be taken by the other; and, in general, work together yet individually to achieve the same desired end. This is especially important when neither the machine nor the human can accomplish the task alone.

The payoff of a good partnership is increased system performance beyond that achievable by a human supported by a machine totally under the human's control. Elements of a good partnership are illustrated by the actions of two players on a basketball team as they coordinate their *moves and thought patterns* to overcome the defenders and score a basket. For example, one player, a guard, steals the ball and begins dribbling down the court toward the team's basket. In mid-court, a defender drifts over to block the path to the basket. Seeing this situation unfold and anticipating the defender's move, a teammate breaks wide left and then begins to cut back across the key near the foul line. The guard, having inferred his teammate would follow this route, allows the defender to force him wide right, and then suddenly looks directly left momentarily, but quickly throws a lead pass straight ahead in front of his teammate, now racing toward the basket for an easy lay-up. Without cooperation and coordinated sensing, thinking, and action taking on the part of both players, the goal would not have been achieved. They had to act as a single purposeful unit—as partners with a common objective.

Clearly, a partnership is a two-sided relationship that can take on any number of forms and that can vary in the amount of cognitive involvement required from each side. The desired form of the relationship must be thoroughly considered so that, to the extent possible, it can be designed into the system. It seems likely that new design aids, such as conceptual models and other representation methods sensitive to this and other related issues, may be needed to support the process of developing intelligent human-machine systems.

III. CONCEPTUAL MODELS OF THE HUMAN-MACHINE RELATIONSHIP

It is customary to use models and other representational devices initially to conceptually characterize and analyze the properties of the design of a complex human-machine system. Some of the available models and modeling languages allow human performance to be assessed as part of the system (e.g., Pritsker, Wortman, Seum, Chubb, & Seifert, 1974), but others are restricted to representing the functional form or information flow within a system. Unfortunately, the human-machine relationship is not treated explicitly as a design variable in any available model. System models cast in the form of manual control, for example, treat the total system, including the human, from the viewpoint of system control. They attempt to use the same constructs (e.g., gain, phase angle, lead, lag, transfer function, etc.) to model both the human and the machine elements in the system. While this may be valuable when a control relationship between the human and machine elements is appropriate, the model cannot consider other types of relationships or address control when it is exercised at a more cerebral or cognitive level across the interface.

A Supervisory Control Model (Sheridan & Ferrell, 1967) has been proposed to guide the design of human-machine systems that have a long delay time between input and system response. It has been suggested, for example, as a means to model the human and machine involvement in the generation of power in a nuclear power plant. Unfortunately, this model does not consider the human-machine relationship from other than a single perspective, and thus is equally limited. It does serve to illustrate, however, that new design issues emerge when there is a shift in thinking about the human-machine relationship. According to the Supervisory Control Model, the human's main function is to monitor the system and to detect, diagnose, and correct faulty system performance. These new functions, in turn, create new design issues: 1) How will the system maintain operator alertness over a designated watch period? 2) How will signal malfunctions and their causes be communicated to the supervisor? 3) How should the human be brought back into the control loop in case of an emergency? Clearly, these, and other important issues made evident by the supervisory control stance, are of no consequence and *never surface* when the system design proceeds from a manual control view. Thus, whenever the form of a human-machine relationship is changed, it will have a significant effect on the functionality of a system.

A baseline system concept is often formulated as another means of modeling a system at the beginning of development. Frequently, the baseline itself is not thought of as "being designed," rather it is generally viewed as providing a concrete point from which to begin the "actual" design activities. Because the baseline concept reflects a particular type and form of solution to the design problem, usually based on past experience, it will contain within it some form of the human-machine relationship. Generally, the nature of the relationship will be tacitly accepted without inspection, and attention will be focused on how improvements can be made in achieving certain functions made evident by the baseline. In other words, little thought will be given to departing significantly from the inherent functionality and/or human-machine relationship suggested by the baseline. Thus, the use of a baseline to guide system design is also weak in its ability to highlight the human-machine relationship as a design variable.

An alternative framework for conceptually representing a design is presented below. This framework, called the Symbiosis Model, regards both the human and machine system components as comprised of a network of elemental abilities. A significant portion of the functionality of the system emerges as part of the network developed from interconnecting sets of basic elements. This includes the functionality associated with the human-machine relationship. Thus, since the designer must create the network structure to support a desired relationship, the human-machine relationship must be considered by the designer when the Symbiosis Model is used.

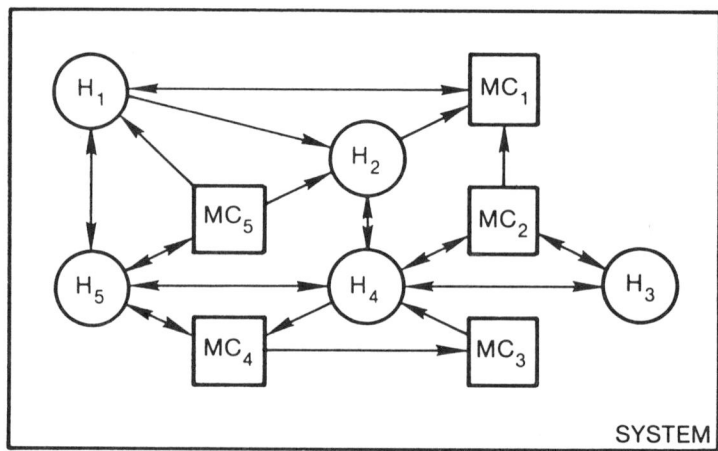

Figure 1. Schematic Representation of a Human-Machine System from a Symbiosis Model Perspective (H=Human Element; MC=Machine Element).

A schematic representation of the Symbiosis Model is shown in Figure 1. The machine part of a system represented by the model is treated in terms of primitive elements labeled MC_X. Each of these is capable of performing one of three different types of elemental processes: sensing, thinking, or acting. Sensing refers to the ability to detect the presence or absence of a state or a change in state (e.g., sensing the position of an object relative to a set point). Thinking refers to the set of processes involved in recognizing a pattern; comparing object values; selecting and evaluating optional courses of action; and recommending an action, step, or sequence. Acting refers to the manipulation of an effector upon a command received from a thinking element. Under some circumstances, acting may involve only issuing instructions to others who, in turn, actually operate an effector.

Human contributions to the system are shown as a set of elements labeled H_X. The set of H_X's are used to suggest that the human contributes to the system in several ways based on the human's ability to sense, think, and act, whether the object of focus is a concrete detail (e.g., is the needle pointed at the correct number?) or a higher, more abstract system-level issue (e.g., is the system operating normally?). Note that the connections between the human elements and the machine elements are not organized hierarchically and that there may be multiple paths from one element to another. Thus, either a human or machine entity could be involved in a particular system function, depending on the availability and capability of each element.

The Symbiosis Model attempts to convey several ideas about the human-machine relationship in a system. As Figure 1 suggests, the human has a wide-ranging and extensive capability that can interact with machine elements in a vast number of ways. The same is true of the machine elements. As a consequence, there is no easily recognizable human-machine interface. The essential idea is that, through proper design, both the human and the machine can aggregate their elemental processing abilities of sensing, thinking, and acting in various ways to form "sources of power" (Lenat, 1984) that can perform different functions. Furthermore, the sources of power can be interconnected in ways to process incoming data such that the system performs unique functions needed to meet higher-level design requirements. These requirements, or the functions performed to meet them, are called system-level attributes since they can be satisfied only at the level of the entire system.

In this model view of a system, the human is not considered to be an "operator." Rather, the human brings to the system a set of abilities which, for some functions under some circumstances, may be used to operate machine elements, but for other functions, may interact with the machine elements in a different way. Generally, many machine and human elements work independently but in a coordinated manner based on a shared "understanding" of the task. In short, the Symbiosis Model maintains that the human and the machine are united in many local ways which yield a global partnership focused on achieving a common goal.

The total system concept takes on major importance when the design problem is conceptualized according to the Symbiosis Model. This is reflected in Figure 1 by the use of a solid line to represent the total system. The system as a whole is emphasized since much of its functionality is satisfied by the form of the entire network structure of sources of power. That is, the higher-order system-level attributes *emerge* as a derived ability from the way elemental units are organized into a network.

The Symbiosis Model draws heavily on the concept of sources of power. This concept was introduced by Douglas Lenat (1984) in an article on machine intelligence to express generically the variety of methods a human appears to have available to solve complex problems. Lenat identified nine human power sources including heuristic reasoning, representation (i.e., use of different ways to characterize a problem), and analogy. I use the concept here in a different but related way. In the Symbiosis Model, "sources of power" refers to root functional abilities that, through proper design, emerge from unique sensing, thinking, and acting processing structures. These structures can yield all of the sources of power Lenat identifies, but they are not limited to this set.

Sources of power available to the two intelligent entities who formed an effective system in the basketball illustration presented earlier, for example, would be such things as their abilities to perform the basic functions of dribbling, passing, blocking, shooting, etc. Each of these functions is achieved through the unique and focused use of sensing, thinking, and acting processes. The total set of intelligent entities derived from a unique network of sources of power elements comprising a basketball team are distributed spatially, with some active and others inactive (e.g., riding the branch) at any moment in time.

Reserve elements can be substituted for currently active ones; however, this will usually result in a loss of capability. The same is true for a human-machine system designed according to the general concept reflected in the Symbiosis Model. A source of power, like the ability to derive an estimate of altitude in an airborne weapon system, can be resident in the abilities of several elements. Each element, perhaps, offers a different degree of accuracy or time required to produce the estimate. The designer must be able to link these elemental sources effectively to several different system-level attributes that rely, in part, on altitude information to perform their functions.

The basketball team also provides a glimpse of at least three system-level attributes. Both players achieved a common understanding of the situation independently based on their view of the problem and their stored knowledge of the game and of each other's style of play. That is, they used sensed data along with the activation of stored knowledge to form independent, but related, "mental models" of the situation. Furthermore, they used this system-level attribute (i.e., an internal model of the situation) together with other sources of power to achieve yet another higher-order attribute—that of mutual support (i.e., formed another internal model or concept of what one's partner needs to do and how to complement it). Finally, to receive the rewards of mutual support, the guard had to *trust* that the teammate would execute the (implicit) plan believed to be unfolding and, therefore, follow through with it instead of activating another plan, perhaps one the guard could execute alone.

System-level attributes or features, such as situation understanding, mutual support, and trust, are just as important for the performance of intelligent human-machine systems as they are for the performance of intelligent human-human ones. They demonstrate that the interface for an intelligent system must support a more abstract, cognitively oriented form of communication.

The view that trust is rightly considered to be a necessary system attribute also implies that the human-machine interface might have to account for some social aspects of a relationship. Attitudes like trust are formed largely on the basis of the general quality of an interpersonal relationship. It is determined, in part, by the degree of sensitivity each person shows toward the needs of the other over time, and, in part, by the general knowledge, gained through experience, one has regarding how the target person behaves in different situations. System-level attributes of this nature, therefore, are not likely to be supported by an interface in an intelligent human-machine system unless the design is constructed in a way to support the formation and maintenance of socially based attitudes.

The Symbiosis Model is a useful conceptual tool for aiding the design of intelligent human-machine systems. First, in an indirect way, it compels the designer to explicitly address the type of human-machine relationship to be included in the system. This is in contrast to either the manual or supervisory control models which lock in a specific human-machine relationship. The Symbiosis Model treats this issue through the fact that no relationship is contained as part of the provided model features. Consequently, the designer must define the desired relationship, determine the functions needed to support it, and formulate ways to achieve these features directly with individual sources of power or indirectly through the internetting of available power sources.

For intelligent machines, it is likely that the relationship will not be simple. To sustain coordination based on a cognitive interchange, issues of trust, pride, and other similar types of qualities need to be supported by the design. Lack of attention to issues of this nature could well result in a breakdown of the interface for intelligent systems. If the decisions made by the system are not understood and trusted by the human, the human will simply "pull the plug." This will be true even if the interface provides for the easy passage of detailed, substantive (e.g., factual) information about the task.

The Symbiosis Model also has the flexibility to support a broad range of human-machine relationships. A restrictive master-slave type of relationship could be represented, or a democratically organized federation of equal power sources could be formed. Since the designer must choose the desired relationship, and the model does not limit the possibilities it can take, the human-machine relationship must be regarded as a design issue.

Once the designer begins to wrestle with the subtle and often abstract cognitive nature of the human-machine relationship, several diffuse functions that need to be achieved by the system as a whole will become evident. These functions are the system-level attributes needed to form and sustain a particular relationship. Here, too, the Symbiosis Model subtly encourages the designer to address these difficult issues since a collection of general-purpose sources of power cannot be formed into a viable system without a proper network structure. The system-level attributes provide the goals for specifying a network structure and, thus, must be established by the designer before a particular network can be designed.

IV. IMPLICATIONS AND CONCLUSIONS

As machines become endowed with more human-like abilities, the nature of the human-machine relationship must be reexamined. It seems clear that this relationship will be more cognitive in nature. Intelligent machines have the ability to use semantic knowledge and sophisticated methods of reasoning to draw inferences from which a course of action can be formulated. The human element of the system will, at times, be engaged in similar activities, but perhaps the object of focus will be a different aspect of the problem confronting the system. At times, each of these entities (human and machine) will have to infer what the other is going to do so that they can complement each other's actions. The human-machine interface for such a system, therefore, will have to accommodate the formation and maintenance of a relationship founded on mutual cooperation. When two entities are involved, a cooperative relationship is typified by a partnership.

A. Design of "Cooperative" Systems

Cooperation is at its highest when the parties involved have a shared and compatible understanding of the task and a strong interpersonal bond. When this is the case, the system will perform in a way that is in the best interest of the group. To succeed in the development of a system of this nature, the designer will have to construct a human-machine interface so that it will engender feelings of trust, confidence, and perhaps even a willingness for an entity to make a sacrifice for the good of the team. These concerns create hardware and software design issues that most designers have not had to deal with in the past. To be sure, addressing issues of this nature seems to be more in the domain of the psychologist and sociologist than in that of the design engineer. I believe the designer cannot avoid them, however, if highly capable and effective intelligent human-machine systems are to be created.

The ability to achieve a "shared understanding" needed for effective cooperation is further complicated by the fact that the full extent of detailed information for many tasks cannot be passed explicitly across the interface. Through the processing of vast arrays of data, each intelligent entity becomes aware of the task in many different ways. For example, an entity understands the physical situation from one or more perspectives, it understands the tactical situation, it formulates and understands a set of immediate and long-range goals, etc. If detailed information from all of these and other perspectives had to be passed between the human and the machine, then clearly the interface would quickly become overloaded. The only alternative is to synthesize the relevant information and pass a single global characterization of the problem. "Shared understanding," therefore, has to somehow emerge from the implicit information embedded in the synthesized carrier. This will require the designer to consider the interface from yet another perspective; that is, how to provide ready access to *implicit information* needed to support cooperative, yet independent, action taking. The interface, therefore, will have to be designed such that it can assist the intelligent entities in answering the following type of questions: 1) How do I (human element in the system) know if the machine correctly understands the problem and my goals and objectives? 2) How do I know it is coordinating its activities to mine and taking care of important tasks that I am not doing? 3) How do I (human or machine) know my partner has redefined the problem or is now working on a new one?

B. Designing "Mental" Models

One possible approach that a designer could use to build a synthesized, global "understanding" into the machine elements is to cast the design problem in terms of the construction of an internal (mental) model. The communication channel between the human and machine would then have to support the passage of appropriate model parameters to induce the desired understanding of a partner's intentions and actions. For dynamic tasks, which are the most interesting and difficult to handle, the set of mental models in the system would have to be updated at designated intervals. Information used to form, maintain, and update these models would be provided by the processing performed by a network of sources of power resident in the system. Therefore, in my view, internal models of this type probably will not be localized entities. There will not be any black box whose job it is to construct a mental model. This is unlikely because some of the features of a mental model are qualities that can emerge only through the interactions of system (both human and machine) elements. It is more likely, therefore, that mental models will be a diffuse collection of perceptually and semantically based knowledge that resides in the system as properties of some portions of a system's network structure. In other words, internal models will be one type of emergent system-level attribute. Their creation, as well as that of other system-level attributes mentioned throughout this paper, will pose a significant challenge to the designer.

C. Design Practice: Integration Versus Holistic Design Stance

Current design practice for large-scale systems generally follows an integrationist design stance. This stance places emphasis on early definition of the desired system functions, and usually employs the use of a baseline system concept to guide the identification of functions. Design begins with an analysis of the problem in an effort to formulate the set of functions that the to-be-designed system must achieve. Once the functions are identified, it is customary for them to be decomposed and grouped in such a way that separate design teams can focus on smaller and more narrowly and completely defined design problems. The resultant components and subsystems are then pulled together in a final "system integration" design step.

An integrationist design stance may not be well-suited to the design of intelligent systems. Some important aspects of the system are the abstract qualities such as mental models, mutual support, trust, and other system-level attributes. An integrationist would attempt to decompose these aspects of the system into component parts. But, as I have indicated, these attributes probably can exist only as emergent properties of a network structure developed around a set of elemental system components. Therefore, by nature, they cannot be decomposed and will not automatically appear when a set of components are integrated. Thus, they will not be recoverable through integration unless prior thought is given to the design of a network structure and what requirements it imposes on the design of each component.

An alternative to the integrationist design stance is a holistic design stance. This stance focuses on the development of an early synthesis of the entire system concept. Once the functional capabilities of the system are specified conceptually, attention shifts to the identification and design of a set of primitive elements. (An example of a set of primitives is the instruction set, e.g., increment, decrement, branch if equal, etc., programmed into a microprocessor.) these primitives are individually characterized and reviewed as a set in an effort to determine the reasonableness of using them to construct *all* of the system features. Consequently, design from this stance immediately involves the designer with issues such as 1) the definition of system-level attributes, 2) the definition of the human-machine relationship desired for the system, 3) the functional requirements of the system, 4) establishing the admissible form of primitive components, and 5) the definition of the set of rules and standards to be used in the design of the network structure. This type of approach to system design is precisely what is needed to create systems comprised of a large set of distributed processors that can be linked to form desirable sources of power, and, hence, is well-suited to the design of intelligent systems.

D. Use of a Baseline Concept

The question of when and how to use a baseline concept in the design process must be reexamined. The formation of a baseline to guide design in the conceptual stage of system development appears to follow from an integrationist design stance. A baseline has two roles. First, it is used to characterize the level of performance available in an existing system to determine what enhancements one would need to add to it to achieve the new requirements. Second, a baseline is used later as a benchmark against which expected performance from a "synthesized" new system can be compared. (The synthesized system often consists of reshaping the baseline on the basis of new technology known to be under development.) The "add to" approach followed in the use of a baseline naturally leads to the need for integrating these new capabilities into existing ones.

From a holistic design stance, the system architecture is central to the rest of the design. Its design leads, rather than follows, the detailed design of individual components and subsystems. For a baseline concept to be useful early in design from this stance, it would have to be viewed from an architectural perspective. That is, the requirements for a new system would be viewed in terms of how they might impact (e.g., require modification in) the architecture of a baseline system. It is not clear that a baseline can be used effectively in this manner.

The use of a baseline as a benchmark for comparative purposes later in design, however, would still be valuable when a holistic design stance is taken. The designer will always need a criteria against which to evaluate the performance of a new system.

E. Need for New Design Tools

It seems evident that new models, methods, tools, and other design aids will be needed to support the creation of intelligent systems of the type discussed here. The Symbiosis Model offers a crude framework for thinking about the design of cognitively oriented systems. Certainly other tools will be needed to more directly help the designer identify alternative interface concepts, to determine what system-level attributes are desired, and to form and test various network structures that can induce necessary emergent qualities. Tools may also be needed that are sensitive to the design considerations needed to support the socialization process of the human and machine "getting to know one another." These cognitive, emotional, and sociological aspects of systems will take on more meaning as the decision-making capabilities of machines increase. Now is the time to equip designers with the knowledge and tools needed to handle these new requirements.

The potential offered by intelligent human-machine systems is far-reaching. The task confronting the people who will design these systems is made difficult by the abstract nature of some of the features that must be included in a system, as well as by the fact that many aspects of the design will be properties that are embedded in the total system network structure. It will, no doubt, take time and a great deal of creative energy to learn how to design well-formed intelligent systems, but the fruits of this effort should be something to behold!

REFERENCES

Lenat, D.B. (1984). Computer software for intelligent systems. *Scientific American, 251*(3).

Pritsker, A.A.B., Wortman, D.B., Seum, C.S., Chubb, G.P., & Seifert, D.J. (1974). *SAINT: Volume I. System analysis of integrated networks of tasks* (AMRL-TR-73-126 [AD A-014843]). Air Force Aerospace Medical Research Laboratory.

Sheridan, T.B., & Ferrell, W.R. (1967). Supervisory control of manipulation (NASA SP-144). *Proceedings of the 3rd Annual Conference on Manual Control.*

Chapter 10

DESIGNING IN VIRTUAL SPACE

Thomas A. Furness III

Armstrong Aerospace Medical Research Laboratory
Wright-Patterson Air Force Base, Ohio

ABSTRACT

Design is described as consisting of visual thinking processes which modern computer-aided design interfaces fail to exploit. An ideal design interface is postulated, leading to the concept that virtual space can be used as a design medium. Recent advances in technologies to synthesize virtual worlds are discussed. The concept of a virtual terminal is presented with speculations regarding its application to design and other uses.

I. INTRODUCTION

Humans are visual beings. When we pick up the newspaper and turn to the editorial section, our attention seems to be drawn immediately to the political cartoon. In a matter of seconds a message (albeit largely symbolic) is communicated to us. Describing the same information in a word picture would require a time-consuming conversion from written symbols (words) to the mental picture that they would elicit. From infancy we learn to deal with spatial things, in the way that we perceive objects in three-dimensional space and interact with our world. Our sense of vision enables us to orient ourselves in space, control our posture, and provide other cues for locomotion and communication.

It can be conjectured that much of our analytical processing is also visual. As we read (largely a left brain function), we transform these word symbols of our language into mental images to which we can more easily relate. Speech also engenders the same mental imagery. Often as one speaker describes his own mental image, the listener is creating one of his own. (These two mental images may not match, which is often the root of communication problems.) Even mathematical concepts and numerical data often are more effectively communicated using visual means (e.g., graphs) than words. Visual thinking allows us to deal with magnitudes, dimensions, or properties associated with physical things.

The purposes of this paper are twofold: the first is to explore some factors associated with visual thinking in design tasks; and the second, to postulate a set of tools for enhancing the design of physical systems.

II. VISUAL THINKERS

Many of our civilization's most accomplished scientists have attributed their successes to observation and thinking. McKim (1980) observes that Sir Alexander Fleming turned a laboratory accident into the discovery of penicillin by his thinking about what he observed. In this case, Fleming noticed in a routine laboratory experiment that some plate cultures of staphylococci had apparently become contaminated and died. This observation had most likely been made by others who knew that some bacteria can interfere with the growth of others, but Fleming saw it in a way that eventually led to the discovery of penicillin.

Nobel laureate James D. Watson (1968) also attributes his discovery of the construction of the DNA molecule to the use of a three-dimensional physical model. He stated:

> Only a little encouragement was needed to get the final soldering accomplished in the next couple of hours. The brightly shining metal plates were then immediately used to make a model in which for the first time all the DNA components were present. In about an hour, I had arranged the atoms in positions which satisfied both the X-ray data and the laws of stereochemistry. The resulting helix was right-handed with the two chains running in opposite directions.

Benzon (1985) adds that even Albert Einstein's primary mode of thought was not words or mathematical symbols. He quoted the following passage from a letter by Albert Einstein:

> The psychical entities which seem to serve as elements in

Designing in Virtual Space 129

thought are certain signs and more or less clear images which can be "voluntarily" reproduced and combined....The above mentioned elements are, in my case, of visual and some of muscular type. Conventional words or other signs have to be sought for laboriously only in a secondary stage, when the mentioned associative play is sufficiently established and can be reproduced at will.

Benzon concludes that Einstein "thought in images and then translated those image-born insights into verbal or mathematical form." From these examples, we can better appreciate the role of visual thinking in the creative process.

III. WHAT IS VISUAL THINKING?

It has often been stated that we do not see objects, we see images. Indeed the rays of light which impinge on our retinae are converted into electro-chemical impulses which, traveling through various waypoints, finally arrive at our visual cortex. Our sensation of the light rays occurs at that point, and the spatial distribution of light constructs an image in our heads to which we may associate some meaning. But we can also construct an image in our heads with our eyes closed. This "imagined image" can have the spatial attributes of anything that we "see" in the world.

Still, there is a third kind of visual image which we can create through our own psychomotor functions. Imagine that you are standing in front of a chalkboard (formerly a blackboard) and holding a piece of chalk in your hand. Now, pretend to draw a triangle on the board by physically moving your hand. Now picture in your mind the triangle you have drawn. This third category of visual imagery can thus be connected to psychomotor functioning (i.e., moving your hand to conform to your mental image of a triangle). The process of drawing with a pencil on a pad is also a translation of a visual image into a physical one using this psychomotor connection. In this same regard, visual images can be formed from purely tactile stimuli as depicted by the following:

>Six wise men of India
>An elephant did find
>And carefully they felt its shape
>(For all of them were blind).
>
>The first he felt towards the tusk,
>'It does to appear,
>This marvel of an elephant
>Is very like a spear.'

The second sensed the creature's side
Extended flat and tall,
'Ahah!' he cried and did conclude,
'This animal's a wall.'

The third had reached towards a leg
And said, 'It's clear to me
What we should all have instead
This creature's like a tree.'

The fourth had come upon the trunk
Which he did seize and shake,
Quoth he, 'This so-called elephant
Is really just a snake.'

The fifth had felt the creature's ear
And fingers o'er it ran,
'I have the answer, never fear,
The creature's like a fan!'

The sixth had come upon the tail
As blindly he did grope,
'Let my conviction now prevail
This creature's like a rope.'

And so these men of missing sight
Each argued loud and long
Though each was partly in the right
They all were in the wrong.

(Quoted from Hampden-Turner, 1981)

This "illustration" of the blind men depicts three processes related to the science of design and vision. The first is that the observer always adds to his sensory stimuli (in this case, touch) a meaning which is built on his own experience. We seem to be always filling in the blanks, often based upon limited stimuli (such as viewing a house partially hidden by trees or piecing together a jigsaw puzzle). We still recognize the object as a house and even make assumptions about its size and shape by extrapolating from what we see. The second point is that even touch conjures spatial interpretation. We construct a visual object in our mind's eye to correlate with the object we have touched. Thirdly, most scientific description (and design!) can be approached from different points of view. Our ability to communicate is largely a function of understanding various perspectives.

Designing in Virtual Space

IV. THREE KINDS OF VISUAL THINKING

It can be concluded from the discussion above that there are three kinds of visual thinking: 1) the kind that we see from objects in the world; 2) the kind that we imagine, such as visual imaging while reading a book; and 3) the kind that we create ourselves when we write or draw. Humans (who are not blind) have various skills related to the interaction of these kinds of visual thinking. An illustrator for example may be effective at representing that seen in graphic form, whereas an artist may be more able to translate an abstract imagining to a canvas that conveys meaning in a symbolic way. Both examples require interactions between the kinds of visual thinking. Perhaps all three kinds of visual thinking interplay when one brings his own reorganization or interpretation of something that is seen, such as trying to draw the structural members which support a building by only observing the outside of the building.

Figure 1 is a symbolic representation of the three kinds of visual thinking discussed above. The creative process is most fully promoted when all three visual/psychomotor (kinesthetic) processes are active as shown by the overlapping region of the diagram. For some reason, the visual imagery can be strengthened when the efferents from the psychomotor activity are mixed with the afferents from the visual observation. (This may be what Einstein referred to as the "muscular type" of interaction.) Perhaps many of the problems which we have with design can be attributed to failure in the interplay of the three kinds of visual thinking. In our enlightened civilization, new design mediums should be developed to overcome these breakdowns.

Figure 1. Three Kinds of Visual Thinking. (Adapted from McKim, 1980, with permission).

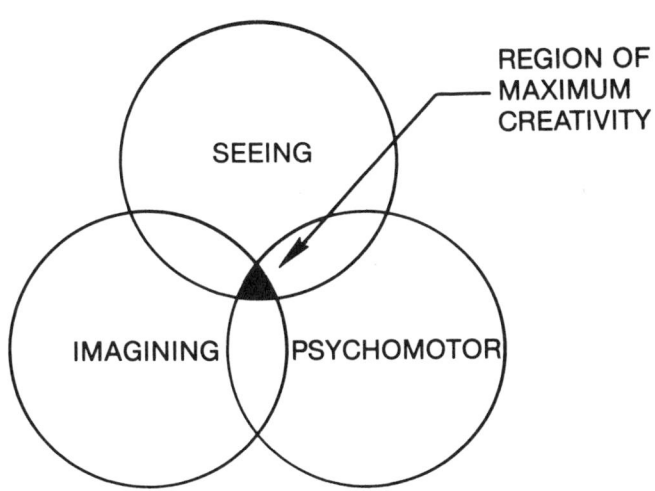

V. COMPUTER-AIDED DESIGN

The process of design can be thought of as forming a plan to organize or reorganize matter to carry out some purpose, usually that of changing some state in the designer's world. This changing of state may be to provide vehicles of conveyance for transporting individuals or other matter from one place in space to another, or to prevent others from conveying matter from one place to another (in the military realm), or just increasing our understanding about things. Design may also deal with the way that information is acquired, manipulated, and represented. Regardless, design is mostly visual.

Modern digital computing technology has provided us with marvelous tools for acquiring, processing, and storing information. More than ever before, the computer has the potential to extend the intellectual power of the human, but the degree to which computers can serve as tools depends largely on the quality of its interface with a human. Current methods of interfacing using cathode-ray tube (CRT) terminals and keyboards (and mice/light pens, etc.) restrict the flow of information and commands between the computer and the operator. Users are required to structure and channel communication to fit the programming languages of the machine. In turn, the computer portrays information in a highly coded form in two-dimensional space. Although natural language interfaces, once developed, will probably improve the flow of information from the human to the computer, these technologies do not increase remarkably the useful "bandwidth" of the interface and unlock the power of the combined intelligences and computing capabilities of the human/machine interface. The key to solving these problems is to make the machine more "human-like" rather than requiring the human to be more "machine-like."

In view of the visual thinking capabilities of the human, computing systems are especially limited. Although computer-aided design systems have facilitated design by allowing rapid "drafting" and perspective modeling of objects, the ultimate natural visual thinking capabilities of the human are not tapped because the existing interfaces do not enable the three kinds of visual thinking described above.

VI. A THOUGHT-CONTROLLED MATTER MANIPULATOR

Perhaps the best hypothetical tool a designer could have would be the means to translate his thought images instantly into physical realities with which he could interact using his other visual and psychomotor functions. Such a tool might be called a Thought-Controlled Matter Manipulator. Assume that by picturing an object in the mind's eye, this matter manipulator would organize matter in such a way as to cause a three-dimensional physical copy of the imagined object to be created in real time. The designer in this case could reach out and touch the object, hold it, turn it, and perhaps even operate it, given imagined functions for the object to perform. The idealized use of all three forms of visual thinking (i.e., seeing, thinking, and psychomotor) would be exercised. The designer could then easily modify the design by "rethinking" and watching those changes materialize before his eyes.

Our Thought-Controlled Matter Manipulator could also be connected to various data base "assistants" which would augment the knowledge of the designer. For example, if information regarding population physical anthropometry, materials science, and human information processing were accessible to the designer, the object of design could be manipulated with machine help to conform to the attributes of many users. Perhaps the designer would want to look at the object through the "eyes" of his specialists. In this regard, he would place in front of his eyes "specialist's spectacles" which would allow him to look at the object of design from the standpoint of his assistants. Human engineering spectacles would help him to consider the structure of the design so that it can be used by the 5th to 95th percentile of male and female populations or so that the dynamic tracking capability of the human provides sufficient precision for its operation. Structural engineering spectacles may point out the stresses, strengths, and types of materials which can be used to manufacture the object. Cost and reliability spectacles tell him he has to start over with the whole design.

Although hypothetical, the Thought-Controlled Matter Manipulator embodies the attributes of an ideal design tool. It truly makes the best coupling between the head, eyes, and hands of the designer in order to organize and manipulate matter in space. A more realistic alternative to this approach is to create a display medium in which synthesized images produce the same visual perceptions as physical objects.

VII. DEFINING VIRTUAL SPACE

The visual space which surrounds us can be thought of as a volume of infinitesimal luminous elements. These elements are the "atoms" of our visual world. The size of each element is equivalent to the limiting angular resolution of the eye. In a spherical coordinate system, the orientation of each element is a function of its angular coordinates relative to the head and eyes. The perceived depth or distance of an element from the eyes is generally a function of the differences in location of the light that each element makes on the retina of each eye (i.e., retinal disparity). Size, convergence, accommodation, and parallax cues also contribute to the perception of depth. With these angular and depth coordinates, the dimensional properties of visual space elements can be described. The luminous properties of each element can also be described using measures of radiance (e.g., hue, saturation, and luminance). The visual image of every physical object can then be described by an array of visual space elements (Hochberg, 1986).

Suppose that a device were inserted into the visual field that projected onto the retina a synthesized distribution of light rays equivalent to each element in our visual space. With the proper organization of the synthesized light rays, the sensations produced in the visual cortex would simulate those coming from a physical (three-dimensional) object. This synthesized image would then act as a "surrogate" for the original object (Hochberg, 1986). The image which is formed by our device is called a "virtual" image. Accordingly, the visual space occupied by virtual images is termed "virtual space." Although in virtual space a physical object does not exist, the perception of the virtual image is the same as if it does exist. Furthermore, if the instantaneous vantage point of the eyes relative to the synthesized object changes, the distribution of light in the virtual space can also be changed to reflect the new vantage point (thereby creating parallax cues). Such a virtual display could produce the same visual images as the Thought-Controlled Matter Manipulator (but they would represent non-physical objects).

VIII. A BASIC VIRTUAL WORLD GENERATOR

Figure 2 shows a system of functional components for producing three-dimensional "surrogate imagery" for the operator. A similar system was proposed by Sutherland (1968). The key to the system is special headgear which produces points of light on miniature television picture tubes (CRTs) and

Figure 2. Conceptual Virtual World Generator.

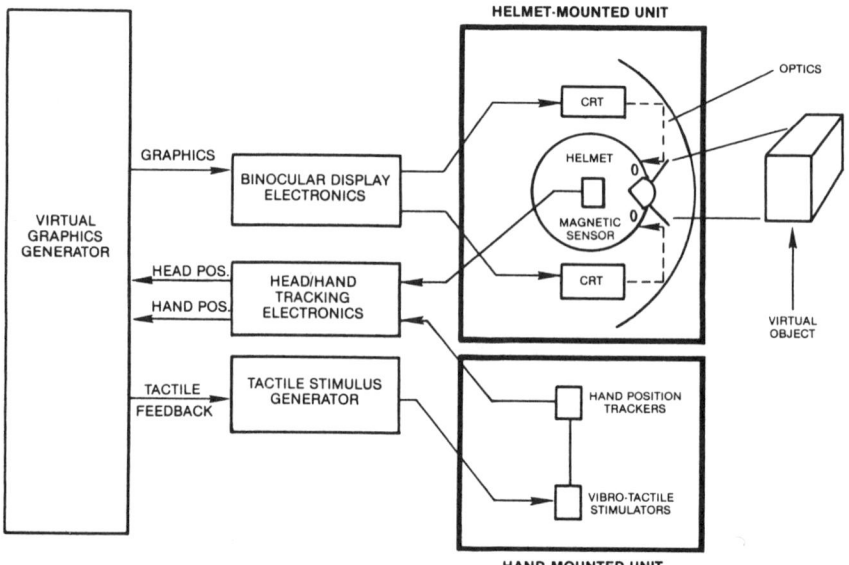

projects these light rays through optics into the eyes of the operator. The operator then perceives a virtual picture composed of the light points which appears in space in the direction of his gaze. Although the pictures are produced in miniature on the CRTs, they are magnified and projected into the eyes so that they appear as large scenes at optical infinity (i.e., they are collimated). This projected scene acts as a "window" into the larger virtual space. The window is positioned as the operator moves his head. The size of this window is a function of the instantaneous field of view of the optical elements which project the CRT images into the eyes. Since there are independent channels for each eye, retinal disparities can be produced to create a sensation of depth.

The original virtual space scenes are synthesized in a programmable graphics processor and input into the binocular display electronics where the signals are produced to create images on the CRTs. The instantaneous orientation of the head-aimed optical system, and hence the display window, is measured by a magnetic tracking system. These signals, in turn, are fed to the graphics generator so that new scenes are continuously recreated to compensate for head movement. In this way, the pictures are stabilized in virtual space or appear to have a fixed location relative to the observer. Since the head can also translate in virtual space, six degrees of freedom must be measured to properly position the synthesized images in virtual space.

In addition to the head-mounted unit shown in Figure 2, a hand-mounted unit is also used to interact with the virtual space objects. The hand-mounted position trackers measure the instantaneous position and orientation of the hands in three-dimensional space. These signals are also input into the virtual space generator causing images of the hands to be displayed in correspondence to their true position relative to the virtual objects being generated. As the operator's hand comes into "virtual" contact with the virtual images, vibro-tactile simulators provide a sensation of "touch" feedback. Although the virtual objects produced by the virtual display system above have no mass, the visual and some kinesthetic sensations are equivalent to the idealized Thought-Controlled Matter Manipulator discussed earlier.

The virtual hand tracker can also be used in reverse to "draw" three-dimensional virtual objects which appear as physical entities in front of the operator. This interaction with the virtual world is facilitated by adding to the basic virtual world generator (Figure 2) an eye line-of-sight measuring device and speech recognition controller as shown in Figure 3. Now the designer can move or alter virtual objects by giving verbal commands while simultaneously touching or looking at the objects within the world.

IX. CREATING IN A VIRTUAL WORLD

The essence of the machine in Figure 3 is to produce a circumambience of "virtual matter" which the designer can use interactively to create new designs.

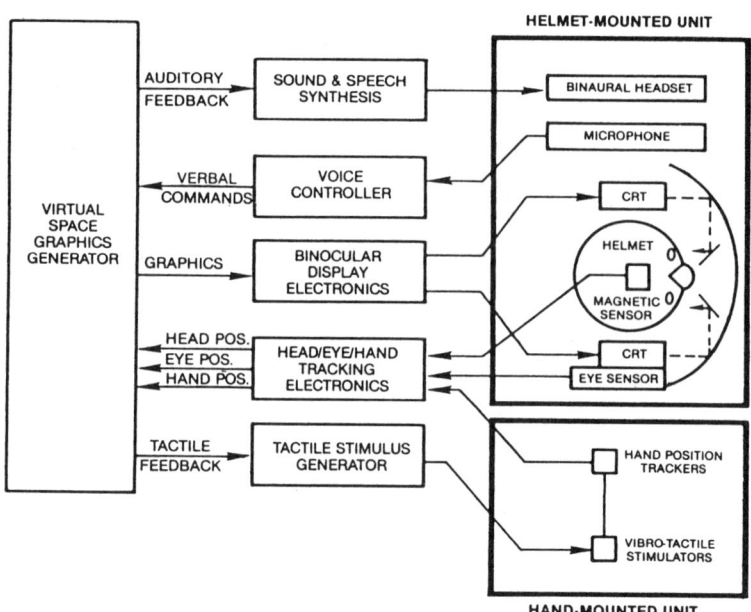

Figure 3. Enhanced Virtual Space Generator.

The virtual world produced by this system becomes a cognitive input/output port between the human and the machine to promote a more effective matching of the spatial skills of the human with the computing, graphics, and data base management power of the computer. It is in this virtual domain where the human and the computer "live" together. This space can be thought of as being like a visual "living room" in which computer-generated objects or symbols appear (as three-dimensional virtual images) to the designer as if they were physical realities. The designer can look around the room (e.g., the furniture, etc.) and reach out and "touch" these virtual objects. The operator can also translate within the computer-generated world by physically walking around in the "virtual room." If the optical system used to project the light rays from the miniature CRTs are made transparent to the outside, or real world, the designer's world can be superimposed over the real world. Thus, an empty real room can be filled with virtual furniture. Doorways can be moved, windows replaced, and carpet installed with the movement of the hands while drawing upon a virtual data base "catalog" of home furnishings.

Consider the application of the MacPaint or MacDraw programs for the Apple MacIntosh as shown in Figure 4. Used with a conventional two-dimensional terminal, it is necessary to compress a three-dimensional world into two dimensions. Perspective drawings convey a sense of depth, but true dimensional spatial interaction is limited because the objects of design cannot be

Designing in Virtual Space

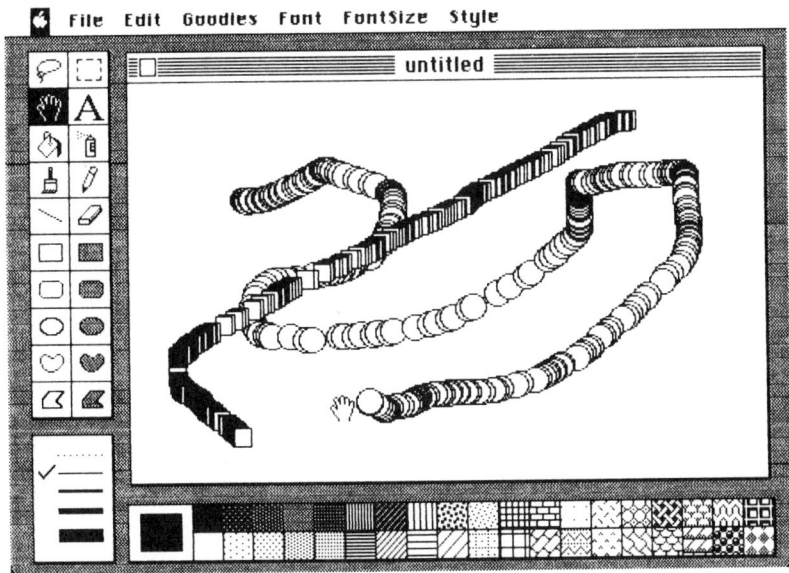

Figure 4. Simulated Three-Dimensional Drawing on an Apple MacIntosh Using MacPaint.

touched. Using the concept of MacPaint/MacDraw with a virtual terminal allows the designer to create a true three-dimensional virtual object, which can be realistically portrayed as it would exist as a real object. In order to create a virtual object, the operator would view an empty virtual space surrounded by icons and a virtual hand as depicted in Figure 4. He selects a mode by speaking a command (such as "plane"), or picking up an icon with his hand or eye (by looking at it). He then positions and orients his hand in space and speaks the word "stay." The virtual world generator then provides the signals to draw a representation of a three-dimensional plane in space. The operator then adds shading/color by picking up a virtual paint brush or spray can and virtually paints his plane. Many planes or surfaces of rotation are drawn the same way. Intersections and boundaries are shown. He can then change or smooth the corners of the object by saying "smooth" while moving his hand over the object, carefully shaving away the virtual matter to produce an object which has the desired contours. The operator can scale the design instantly by saying "zoom." He can walk around the object and see from different sides or views. By pushing the design with his hand and commanding "move," he can translate or rotate the object in space or group it with other objects so designed. Commanding "combine" causes the combination of designs to be treated as one object.

Functional operations can also be assigned to the objects of design. For

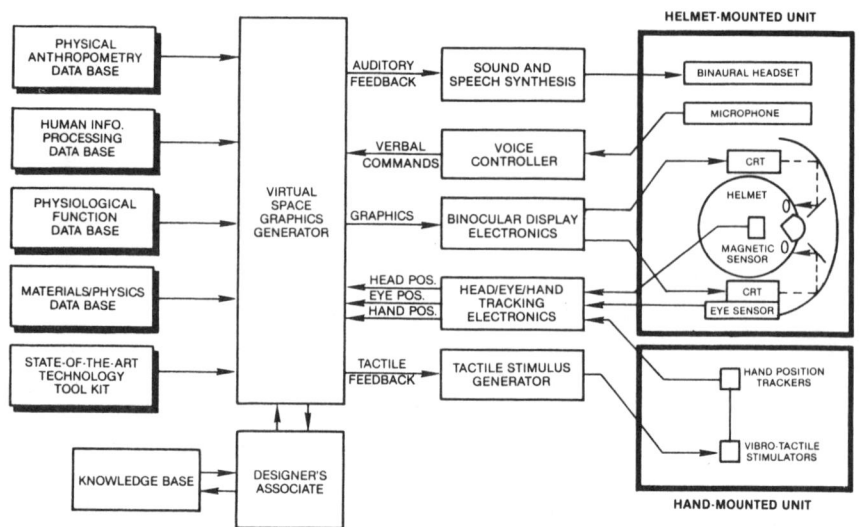

Figure 5. Virtual Space Design Terminal.

example, gear wheels with adjustable ratios will turn with each other. A virtual clock movement (the old kind without quartz and liquid crystals) can be designed and toured in virtual space.

Figure 5 is a further refinement of the virtual world generators in Figures 3 and 4 to provide a virtual space design system. In the specialized design application of the virtual world, data bases are added such as those described initially in the Thought-Controlled Matter Manipulator. Upon command by the operator, information from the data bases can be accessed and portrayed graphically or used to modify the designs in virtual space. The "Designer's Associate" is a machine aid to guide the design process and assist the designer in accessing the data bases. The characteristics of this associate will be discussed later.

X. MINDWARE

In his article on "The Visual Mind and the MacIntosh," Benzon (1985) discusses how a relatively simple computing machine has become a tool for the visual brain. Even though the MacIntosh has made inroads into user friendliness, the real breakthrough comes from the software, such as MacPaint, which provides the "visual power." The interface of the human with the virtual medium can be thought to exist in "direct" and "indirect" forms. The direct forms are those which are physical or involve the transfer of light, sound, or mechanical energy between the operator and the machine. The indirect pathways are the internal models that the operator develops regarding an understanding of the machine and its capabilities (and for that matter, the machine has about the operator). These models cause the data represented on the display to have meaning or "semantic content," that is, the level of meaning or information derived that can be associated with the data elements on the display. The direct pathways are concerned with the control and display of the virtual world. The indirect pathways are concerned with the composition of the machine intelligences that implement the cognitive port between the human and the machine. Both the direct and indirect pathways are important in communicating understanding between the human and the machine, and both require the development of new classes of software.

The software to implement the direct pathway is straightforward. In this case, virtual space is considered to be a three-dimensional volume to be bit-mapped. As objects are created in virtual space, the allocation and selection of these bits to portray the world must be computed. The solid geometric relationships based upon viewing perspective must be manipulated rapidly to provide a dynamic real-time portrayal of virtual objects. Computing algorithms will involve coordinate transformation matrices manipulated by array processors. The direct pathway software will also provide "drafting tools" for use in virtual space. This tool kit gives the designer a variety of graphical paint brushes, magnifying glasses, color pallets, sunlight, and functional associations (e.g., causing an object to obey a set of preprogrammed functions or movements) for use during the design process.

The indirect pathway software is more complex. As it pertains to the things which are going on in the designer's head, it can be appropriately called "mindware." Mindware assumes that there is a level of intelligence in the machine which understands to some degree the design problem and the intent of the designer. The mindware synthesizes the quality and personality of an "associate" for the designer. The associate first insures that both the machine and the designer agree on the perception of the problem, and then (gently) guides the design process, recommending design considerations as changes in the situation dictate, to achieve an efficient and successful solution to the problem. It is through this interface that the machine infers the intent of the designer, thereby allowing the machine to assist in accessing data and knowledge bases as shown in Figure 5.

The mindware also implements a form of "symbolic communication" between the designer and the associate through the medium of the virtual world. Symbolic communication provides a nonverbal two-way communication channel, allowing the system and the designer to signal understanding and intent to each other with simple integrated symbology that minimizes operator cognitive effort. The software functions by sensing and processing behavioral sequences to deduce current information needs of the operator. It also includes the delivery of the information via a communication architecture and display formats that capture relevant spatial relationships in an easily understandable form. This use of machine intelligence significantly extends the intellectual capacity of the designer by allowing the vastness of design data bases to be accessed in a friendly and expeditious way for refining the object of design.

XI. MAKING VIRTUAL DESIGNING A REALITY

Several concepts and machines have been developed over the last 20 years to explore the use of virtual space as a design medium. Sutherland (1963) describes perhaps one of the earliest using a binocular helmet-mounted display and head position trackers. Later, demonstrations of this work were also made (e.g., Sutherland, 1968). Krueger, Gionfriddo, and Hinrichsen (1985) also describe an "artificial reality" which serves as a visual medium through which the operator can interact with a graphics system. Termed "Videoplace," the authors have created a visual scene which includes a computer-generated silhouette of the operator and symbols projected simultaneously on a screen. The operator moves or interacts with the computer-generated symbols by the way he moves in the room and positions his limbs. The authors state that this artificial reality can be learned readily and used as a tool to interact in different ways with the computer.

In 1977, the Air Force's Armstrong Aerospace Medical Research Laboratory began a project to develop a virtual space simulator (Kocian, 1977). The impetus for this project was to provide an engineering research tool to investigate the portrayal of information in modern military aircraft. The basic design of the system was achieved by significantly extending the performance of previously developed helmet-mounted tracking and display systems (Furness, 1978). Designated the Visually-Coupled Airborne Systems Simulator, or VCASS, the system became operational in 1982 and since has sparked a revolution in the design of crew stations and simulators (Furness, 1985; Mills, 1985; "Virtual Cockpit's Panoramic Displays," 1985). Currently, different versions of the VCASS are being developed for airborne crew stations (i.e., virtual cockpit), spacesuit applications; visual performance test battery; command, control, communication; and computer-aided design (Furness & Kocian, 1986).

Furness and Kocian (1986) indicate that several psychophysical studies have been conducted with the VCASS, investigating the portrayal and interaction of information in virtual space, and that user response has been overwhelmingly positive. Head, eye, voice, and hand control interactions within virtual space have also been demonstrated.

XII. APPLICATIONS

Perhaps one of the most obvious applications of the virtual design terminal is for the design of crew stations. Such a design tool allows the cockpit designer to create a virtual work station which can (in real time) be manipulated and operated to perform intended functions. These prototypes may represent a conventional "physical type" cockpit using panel-mounted displays or the more advanced "virtual cockpit," wherein all cockpit information is portrayed in three-dimensional virtual space. Such a facility is rapidly reconfigurable to relate to any mission. When used with a terrain data base and computer-generated imagery system, the designer can become the pilot in a simulated operation of the crew station in a hypothetical vehicle (aircraft/tank/ automobile). Visibility outside the vehicle can be adjusted based upon the pretended size of the occupant (from the anthropometric data base). In essence, the human can be scaled relative to the system to see what it looks like if he were smaller or larger. In the case of an aircraft cockpit, the designer would select the parameters of information to be portrayed, indicate where the display is to be located, then fly the vehicle using that portrayal.

In the process of design, other data bases can be accessed to investigate human factors, materials, and other physiological design considerations. In all cases, the design session would be assisted by the Designer's Associate. In this application, the associate would guide the designer by asking questions: "Have you considered...? If not, here are some factors that should be taken into account." Other designers, consultants, or operators can be called into the design process. They too would be provided with a design terminal and able to react to the virtual crew station and add their recommendations. (It should be noted that the designers do not necessarily have to be in the same location; they could communicate designs through a modem.) At the conclusion of a design session, the design information would then be translated to punctilious dimensional drawings for manufacture.

The virtual space design system could also serve as tool to train designers and students of many visually oriented disciplines. Electronic engineering students can visually ride upon an electron while transversing the silicon space of an integrated circuit. A physicist could tour a virtual room of molecular structures synthesized from an X-ray diffraction machine. Indeed, he could reach out and touch the atoms within a crystal lattice, making adjustments and noting changes in the bonding energies. (James D. Watson would especially appreciate this application.) An entomologist, looking through the eyes of a scanning electron microscope, could take a tour of the eye of an insect, probing its structures while he walked around in a virtual room (while virtually connected to a scanning electron microscope). Electronic engineering students could study field theory by actually visualizing the three-dimensional nature of the electric and magnetic fields induced by the movement of charge in conductors or in space. The power of the visual image and psychomotor interaction can transform the teaching of advanced concepts to students wherein a virtual three-dimensional medium can be used. Even abstract mathematical concepts can come alive in visual space as the manipulation of equations changes the nature of n-dimensional space.

One of the most provocative applications of a virtual world system has been described as the "Knoesphere" (Lenat, Borning, McDonald, Taylor, & Weyer, 1984). Here the authors describe a scenario using equipment similar to that described in Figure 5, but with the additional feature of an expert system with encyclopedic knowledge. In this article, the reader is taken on a hypothetical tour through the data base subject of his choice, wherein the data is transformed into a "museum." The participant is assisted by an electronic guide whose personality and professional background are selectable. The participant can also adjust the perspective of his tour by using "spectacles" which filter out or stress various features of the tour to be taken. With these *a priori* adjustments, the knowledge base is unfolded as a museum tour, guided by the electronic personality. The authors speculate that the eventual impact of the Knoesphere will be on education, where not only facts can be presented, but principles taught.

XIII. CONCLUSION

There is something magic about being able to see and touch things. It brings a spatial reality to our consciousness. It is in our mindscape that we truly manipulate matter using our sight and touch as interfacers. Allowing the human and computer to relate spatially in a three-dimensional world opens vast new possibilities for improving crew stations, computer-aided design, and instructional systems. The challenge of the future is to develop the "soft things" which program the new virtual interface mediums described in this paper.

REFERENCES

Benzon, B. (1985, January). The visual mind and the MacIntosh. *BYTE*, pp. 113-130.

Furness, T.A. (1978). *Visually coupled information systems*. ARPA Conference on Biocybernetic Applications for Military Systems. Chicago, IL.

Furness, T.A. (1985, June 30). Virtual panoramic display for the LHX. *Army Aviation*, pp. 63-66.

Furness, T.A., & Kocian, D.F. (1986). Putting humans into virtual space. *Proceedings of the Society for Computer Simulation—Aerospace Conference*. San Diego, CA.

Hampden-Turner, C. (1981). *Maps of the mind*. Collier Books.

Hochberg, J. (1986). Representation of motion and space in video and cinematic displays. In K. Boff, L. Kaufman, & J. Thomas (Eds.), *Handbook of perception and human performance: Volume 1 sensory processes and perception* (chap. 22, pp. 1-5). New York: John Wiley and Sons.

Kocian, D.F. (1977). VCASS: An approach to visual simulation. *Proceedings of the 1977 IMAGE Conference*. Williams AFB, AZ.

Krueger, M., Gionfriddo, T., & Hinrichsen, K. (1985). *Proceedings of the Association for Computing Machinery CHI '85 Symposium*, (pp. 35-40).

Lenat, D., Borning, A., McDonald, D., Taylor, C., & Weyer, S. (1984). Knoesphere: Building expert systems with encyclopedic knowledge. *Proceedings of the International Congress for Artificial Intelligence*, (pp. 167-169). Karlsruhe, Germany.

McKim, R.H. (1980). *Thinking visually*. Wadsworth, Inc.

Mills, R.B. (1985, June 6). CRTs give new look to cockpit of the future. *Machine Design*, pp. 34-40.

Sutherland, I. (1963). *Sketchpad: A man-machine graphical communication system*. MIT Thesis.

Sutherland, I. (1968). *A head-mounted three-dimensional display* (FJCC AFIPS 33-1:757-764).

Virtual cockpit's panoramic displays afford advanced mission capabilities. (1985, January 14). *Aviation Week and Space Technology*, pp. 143-152.

Watson, J.D. (1968). *The double-helix*. Atheneum Publishers.

Chapter 11

THE DIFFICULTIES OF DESIGN PROBLEM FORMULATION

Ruston M. Hunt

Search Technology, Inc.
Norcross, Georgia

ABSTRACT

This paper discusses the need for automated support for design problem formulation. Several examples from the author's experience suggest that certain design problems are the result of poor problem formulation techniques. Designing with new and often unfamiliar technology places greater importance on the need for complete and thorough documentation from concept through development, and also requires new techniques for developing ideas such as rapid prototyping. Concepts are discussed for support of design problem formulation somewhat independent of the technical domain of the design. These concepts include data management, knowledge solicitation, and shallow domain expertise.

I. INTRODUCTION

Efforts on the part of those in the computer industry to make machines "user friendly" have resulted in a plethora of new human-machine interface devices. These devices range from the mouse and touch screens to voice recognizers and eye position trackers for command and control input. Typical computer output devices include high-resolution color, flat panels, and heads-up and helmet-mounted displays. As a person involved in the design of human-machine interfaces, I am particularly interested in understanding the appropriate applications and limitations of these new technologies. However, I have found that most of the problems encountered in designing with these new technologies are not intrinsic to the new technology employed. Rather, the problems are associated with understanding the design requirements, i.e., problem formulation.

Approximately five years ago, I participated in a project to develop guidelines for designing computer-generated display systems (Frey, Sides, Hunt, & Rouse, 1984). Initially, it was felt by those of us working on the project that these guidelines would focus exclusively on the technical details of using computer-driven display technology. As this information was assembled, it became clear that making proper use of this technology, just as for any other technology, required a well-conceived and thorough specification of what the designer was trying to accomplish. Hence, our guidelines eventually came to include a great deal of advice on problem formulation and front-end analysis.

As part of the effort to develop the aforementioned guidelines and on several occasions since, I have had the opportunity to participate in the design and/or evaluation of a number of different human-machine interfaces. This paper discusses briefly the approach my colleagues and I have developed for human-machine interface design, some observations I have made while participating in the design process, and some opportunities I see for automated support of design.

II. A DESIGN PHILOSOPHY

Throughout this paper when I say "design," I am referring to one part of design: new design. Much of what is often called design is actually redesign or modification of existing designs. Redesign is typified by explicit and purposeful acceptance of much or most of an existing specification with the intent of changing some portion to make it more effective, efficient, safe, or less expensive to use or manufacture. New design is typified by original or novel creation that requires, or should require, designers to establish objectives for the item to be designed and a concept for how the item will function. Any particular design project is likely to be a mixture of new design and redesign, but it is the process of developing new concepts which is the focus of this paper.

The interface design philosophy that we follow is based on three underlying principles (see Hunt and Maddox, 1986 for a detailed discussion of this process). First, in order to produce an effective and efficient interface, system users must be involved, in an appropriate manner, from the very first step in the design. In the context of system design, a "user" is anyone who will ultimately interact with the resultant system (e.g., operators, maintainers, managers, etc.). Appropriate use of these people requires that questions will be directed to those people most likely to know the answer based on experience or formal training. This seemingly simple rule is frequently violated by asking engineers to describe how operators do their job or by asking operators to explain how equipment is maintained. Only by involving the appropriate users from the beginning and throughout the design process is it likely that an effective, user-centered design will be produced.

The second underlying principle for this design process is that the entire process, from problem formulation through design implementation, should be documented. We suggest four documents which we refer to as System Objectives, System Requirements, Conceptual Design, and Detailed Design. Each of these documents is created early in the design process and then regularly updated and modified as necessary to reflect the current understanding of the design problem and its proposed solution.

The third underlying principle of this process is that constraints should be applied as late as possible. This is somewhat of a departure from the traditional top-down, systematic design process wherein constraints are identified "up front" with requirements. The purpose of this is to prevent the system from becoming "constraint-driven" as often seems to be the case in my experience. I discuss this further in the next section.

It has been suggested (Rouse & Boff, see Chapter 2) that any top-down, systematic approach to design might be unrealistic. This is based on the observance of real-world design as being typically heterarchical, opportunistic, iterative, and, in general, nonsystematic. Based on my observations, it is true that no design project ever flows as smoothly as the examples in a textbook on design. However, it is precisely for this reason that a highly structured, top-down process should be adopted. The more rigorous and well-defined the nominal design process, the easier it is to detect, identify, and understand deviations from the desired path and the impact on the design product. Furthermore, without a well-defined design process, automated design support will be difficult, if not impossible, to implement.

Parnas and Clements (1986) offer a unique perspective regarding the usefulness of systematic, top-down design and documentation. They suggest that even when the designer cannot follow the process as intended, the effort should be made to retroactively create the by-products and documentation that would have been generated in arriving at the resultant design. This is consistent with the notion that a design is not complete without the rationale for how it came to be. It also suggests that a design aid should require that all steps in the design process be completed eventually, but that any design aid should allow flexibility regarding the order in which the steps are completed.

The process described here does not represent a novel approach to design. It is, in fact, traditional, top-down systems engineering of the sort taught in most undergraduate engineering programs. It is my opinion that the top-down theory of design is a good one. However, putting the theory into practice is often regarded as constraining, impractical, or simply impossible because of the burden of documentation, change control, and communication among all of the parties involved in a large design effort.

I believe that automated design support will make it possible to put the theory of top-down design into practice. Furthermore, I believe that initial efforts should focus on the process of design rather than the subject of any particular design domain. Opportunities can be developed to support design in a fashion similar to the way word processing software supports writing an article. There are also, I believe, opportunities to greatly enhance design problem formulation with very shallow domain-specific knowledge bases. In the next section, I review several of my own design experiences to illustrate the need for domain-independent and shallow domain-specific automated support.

III. REAL-WORLD DESIGN EXPERIENCE

The purpose of performing design analysis is to establish what capabilities/features designers will need to incorporate in their design and to establish evaluation criteria to assess the effectiveness of the final design. However, anyone who has ever designed and built anything is likely to agree that some problems remain unanticipated until development is quite far along. Also, there are often many details of a design that seem not to warrant detailed analysis and specification before development begins. Many of these details are left to be handled by standard practices. Others will simply be worked out by developers when the time comes. One major problem faced by designers is when to stop analysis and begin development. Too much analysis may unnecessarily encumber and delay the development process. Too little analysis may lead to costly false starts during development or even the failure of the design product.

The product of design analysis is generally some sort of design analysis and/or specification document. It has been my experience that one of two things generally happen with these documents. In the first case, the analysis is completed but not adequately documented, or some document is written and placed on a shelf, never to be referred to or updated. There are some people that would argue the document has served its purpose as soon as it is finished. They reason that it is the process of analysis and not the product that will allow designers to proceed on the right path.

The alternative to shelving the document is to keep it alive via a system for making changes and updates to the original specification. By keeping the document alive, the designers always have a current specification of their requirements against which they can measure design progress and, ultimately, design success. One of the big drawbacks to keeping a design specification alive is that it will almost inevitably come to drive the design at some point. That is, there will come a time when the effort to make a change to the design documentation will outweigh the perceived advantage to the ultimate product, even though the actual change to the product may not require much effort at all.

A. The Effects of Shelving Analysis Documentation

I have had the opportunity to be involved in both of the situations described above. In the first case, a CRT-based system had been used to develop displays for augmenting existing, hard-wired displays for a process control plant. The displays, it was claimed, contained features that would support a radically improved approach to controlling the process. Over several years, this project had produced a number of technical reports, conference papers, and journal articles. At the time my colleagues and I became involved in this project, the displays were considered nearly complete. Our role in this project was to assess the adequacy of these displays from a human engineering standpoint. Toward this end, we requested copies of the design specifications from which the programmers were working. As it turned out, the documentation for this system consisted of the high-level reports, papers, and articles mentioned previously. These items were mostly conceptual and contained no formal function and task analysis or any other form of description of the operator's role regarding this system. Where details of the design were discussed in these documents, they were seldom the same as what had actually been programmed into the existing displays.

The authors of these documents were doing the programming themselves and, therefore, had never been forced to refine their intentions or communicate them to a third party. Consequently, the rationale for certain features had been lost over time, while other features discussed in the documentation were absent from the displays for unknown reasons. It quickly became apparent to everyone involved that these documents represented good conceptual work that had been put on the shelf and allowed to grow obsolete. It was also found they had very little backup material for the system they had actually produced.

As a result of our protestations, the design team suspended their development effort while they produced the documentation to support their current understanding of the problem. With this as a foundation, they then generated the documentation to support the details of their proposed solution. As a result of producing this documentation, the designers ultimately made substantial changes to the displays.

There are two important lessons to be learned from this experience. First, design documentation must be complete from concept to details. Any break in this documentation will make it difficult, if not impossible, to audit the quality of the design or the fulfillment of the requirements. Second, the documentation should be *used* by the developers of the system. This means that it must be updated, corrected, and modified so that it is always worth using.

B. The Effects of Living Documentation

On another occasion, I had the opportunity to assist in the design of the interface of a communication control system. From a technical viewpoint this project was very similar to the system in the preceding example: CRT-based system, single operator, wide variety of dimensions and variables to be monitored, highly automated system, and real-time response requirements. However, from the viewpoint of the design process, these two projects were very different. The developers of the communications control system were committed to following a rigorous design process that included a heavy requirement for documentation and configuration control.

At the beginning of the design effort, the developers produced a well-organized, reasonably complete, and fairly detailed design specification document. As development of hardware and software progressed, it was necessary to update the specification and make changes as problems arose. At first this was simple and straightforward, but, in a relatively short time, this became a large encumbrance to making changes. The developers used object-oriented programming techniques and tried to decouple software modules wherever possible. However, long before half the code had been written, we had reached the stage where the cost to document changes was far exceeding the time to make the change to the code. Furthermore, it became increasingly difficult to make sure that all implications of a proposed change had been considered before making the change. Also, it was difficult to know if all of the necessary changes had been made to accomplish an intended modification.

One might be tempted to suggest the way to avoid this problem is to "do it right" the first time and thereby avoid the need for changes. However, based on my experience, whenever new technology is involved, this is not feasible. It is inevitable that decisions will be made early in design for which the downstream implications are not understood until development progresses further. Unfortunately, some decisions such as equipment purchases can be extremely difficult to "undo" at a later date. By documenting how and why design decisions are made, designers maximize their chances of being able to support good decisions and recover from bad ones.

Based on the two experiences related above as well as others, I firmly believe that thorough and complete design documentation is of critical importance, and it must be dynamic. However, there must be a way to limit the encumbering effects of producing and maintaining documentation. Some specific ideas about automated support for this process are discussed in the next section.

C. Rapid Prototyping

In recent years, a practice known as rapid prototyping has become popular, especially for software design. In theory, the purpose of rapid prototyping is to help those specifying system requirements to visualize and perhaps manipulate a prototype to help them more fully understand their needs and communicate these needs to designers. Prototypes are particularly useful because as Alavi (1984) points out, "users are extremely capable of criticizing an existing system but not too good at articulating or anticipating their needs" (p. 557).

There are several potential problems with the rapid prototyping approach to design. First, rapid prototyping may be used as an excuse for not performing sufficient up-front analysis. In this situation, the designers may be allowing development to proceed or even substitute for analysis. This can lead to a situation I call "rabid prototyping," wherein the development of technology drives the design rather than the end user's needs. Schneiderman (1987) points out that "too often task analysis is done informally or implicitly" (p. 55). Consequently, system functionality may ultimately be determined by "design or implementation convenience" rather than by documented needs. There was a strong component of this in the process control design example discussed previously.

A second potential problem may arise from the fact that developers may resist throwing away the creative investment they have made in a prototype. In theory, the prototype will be discarded as soon as it has served its usefulness in helping to define the system requirements. However, the developers of the prototype have no doubt produced some useful work that they do not wish to throw away. Therefore, there may be a tendency to want to modify the prototype into the final product rather than throw it away.

A third problem with rapid prototyping is the tendency to focus on the proposed solution and forget about the problem. When a prototype is produced, there is often the inclination to fine tune details of the proposed system prematurely. This often has the result that features of a prototype may become de facto specifications for no good reason. After users and other potential customers of the eventual product have reviewed the prototype, they may have expectations regarding the product that the design team had intended to change.

In order for rapid prototyping to work, it must be pursued for well defined purposes and done within a controlled environment. The effort required to produce prototypes must be minimized to encourage multiple prototypes and make it less painful to throw away prototypes after they have served their usefulness. It would also be extremely useful to be able to link elements of the problem analysis design specifications to elements of the prototype in order to maintain focus on *the* problem and not *a* solution.

D. Requirement-Driven Design

It has been my experience that one of the most insidious and prevalent problems in problem formulation for design occurs when designers become constraint-driven rather than requirement-driven. One might argue that the difference between requirements and constraints is purely semantic. However, it is my opinion that this is not the case.

Requirements are those attributes of the final design that must be a part of any acceptable solution to the design problem. Constraints, on the other hand, apply only to particular solutions to the problem. For example, a common request early in design is to "use existing equipment." This is sometimes a requirement and sometimes a constraint. In order to know which it is, one must determine why it has been suggested. For example, one might suggest this for an application where space limitations simply prohibit the addition of more equipment. Alternatively, one might make this suggestion based on the assumption that obtaining more equipment will be necessarily more expensive than using existing equipment. In the first case, I might suggest that it is a requirement, whereas in the second case, it is a constraint. It is interesting to note in both cases that existing equipment is not the real issue at all. In the first case, it is space, and in the second case, it is money. This is important information to designers and it should not be hidden.

On one project, my colleagues and I were asked to design operator interfaces for a new security control center in parallel with the customer's own in-house communication specialists so that two designs could be comparatively evaluated. The in-house specialist pursued what I considered to be a constraint-driven approach. On the other hand, my colleagues and I chose to defer consideration of constraints and focus on requirements. In the end, the two resulting designs were so radically different as to make comparison nearly impossible.

For the in-house specialists, it was apparent that certain constraints, such as the need for compatibility with existing equipment, were guiding the search for solutions. The in-house design team began their search for solutions by considering only systems that were mechanically compatible with their existing equipment. By regarding the constraint as accurately stated and inviolate, the designers never considered the cost (functional or financial) of meeting the constraint. The requirement-driven approach revealed that the actual need was for functional compatibility among equipment and not mechanical compatibility. By viewing compatibility as a requirement rather than a constraint, everyone involved agreed that a wider variety of solutions could be considered.

In my experience, designers seem to spend an inordinate amount of time focusing on constraints. This may produce one or more of three undesirable results: 1) effort may be wasted considering a stated constraint that may, in fact, not be constraining the optimal design solution; 2) designers(s) may view satisfying the constraints as equivalent to meeting the system requirements; and 3) conversely, the value of violating a constraint (i.e., cost of meeting the constraint) may never be known.

A common occurrence in design efforts is to accept the first or most obvious question as an adequate statement of "the problem." With a question identified, the designer(s) may begin to focus on generating alternative solutions. This is typically more straightforward than trying to identify "the real problem." Focusing on "the problem" gives the appearance of getting down to business and moving forward. Questioning whether the real problem has been identified gives the appearance of backing up, out-scoping, or just simply not progressing. Unfortunately, the first statement of the problem is likely to be made up of constraints rather than requirements.

Many design efforts could be greatly improved by making all requirements and constraints known explicitly early in the problem formulation stage of design. Through careful consideration, it is likely that many perceived constraints can be identified with particular solutions and should not necessarily be included as part of the general statement of requirements. This will inevitably help designers to identify the real problem space and associated solution space in which to work.

E. Working with Familiar and Unfamiliar Technology

Designers often appear very constrained in their thoughts about a subject with which they are very familiar. Examples of functional fixedness abound whereby people naturally associate objects with their common functions and vice versa. This natural ability is, in fact, a necessity so that people do not need constantly to rediscover the world they live in. Unfortunately, it may stifle creativity and the ability to innovate, especially in very familiar domains. It is somewhat ironic that engineers and other subject matter experts who have invested a great deal of time and personal identity in the status quo are expected to innovate and produce "new" designs.

When it comes to working with unfamiliar technology, the potential problems are fairly obvious. For example, when selecting unfamiliar hardware, it is difficult to know what characteristics are important relative to your needs. This often leads to "over buying" in hopes that "too much" is better than "too little" when it comes to hardware. Unfortunately, in many cases, unneeded capabilities are more burden than benefit.

The problems I have observed and personally experienced regarding designing with unfamiliar technology seldom involve an inability on the designers' part to understand some technical aspect of the technology. Most often it is simply being unaware of the significance of certain details and when certain decisions should be made. A recent example from my own experience involved the design of our company's new office. Almost without exception, every problem that arose during design and build-out of our offices was a result of my not being aware of the effect and timing of certain decisions. Some things must be fixed very early and are nearly impossible to change. Other things can be changed at the last minute with little or no impact on cost or schedule. Some decisions are based on county codes, some on conventions within our building, some on the architect's perception of my desires, and others on the architect's own preferences. My unfamiliarity with designing in this domain meant that I considered some issues too early, some too late, and some not at all. However, in no case did I err because I could not comprehend some technical aspect of the domain.

It should be obvious that someone on the design team, at some point, must possess deep domain-specific knowledge on each aspect of design. However, in many cases, designers working on one aspect of a project need only be sensitized to important issues regarding technology unfamiliar to them.

Sometimes, simply knowing that an issue is important to some other aspect of the design and when to defer decisions to others is very important. I have observed that many conflicts among groups within a project are caused by one group unknowingly making decisions which arbitrarily constrain another group's capabilities. Some high-level, but fairly shallow, domain-specific knowledge regarding otherwise unfamiliar domains could greatly alleviate this problem.

IV. OPPORTUNITIES FOR AUTOMATED SUPPORT OF DESIGN PROBLEM FORMULATION

Based on my experience, I have concluded that design problem formulation is amenable to several forms of automated support. However, much effort will be required to completely define the detailed requirements of each form of support. In this section, I summarize three opportunities for support and discuss their conceptual requirements. These are presented in no significant order.

A. Opportunity Number 1: Data Management Tool

There is an old saying that creativity is one percent inspiration and ninety-nine percent perspiration. I believe these percentages also apply to design problem formulation. This process, which is often thought of as being highly creative in nature, requires a great deal of attention to detail. As a design evolves from concept to specification, there is a significant and potentially encumbering requirement to add, modify, and refine design data. Automated design support could provide a repository for this data and a mechanism for accessing and updating it.

Such a tool would need to support some structured design process, such as the one outlined in Section II, but would allow much flexibility within the process. Designers should be allowed to enter data at varying levels of abstraction/detail. They should also be allowed to attach flags to these data such as "incomplete," "unresolved," "firm," "soft," etc. to indicate how each item should be regarded and used by others on the design team. With the data entered, the designers should be able to establish relationships among data items such as "is constrained by," "is an element of," "depends on," "is required by," etc. This network of relationships would allow designers to ask "what if" questions when considering the implications of any proposed changes.

It has been my experience that few people who call themselves designers lack good ideas. However, many good ideas never make the transition to good products because of the difficulties of managing all of the details when making this transition. Many of the problems discussed previously in this paper can be traced to the absence of a good medium for design problem formulation. Knowing when to transition from analysis to development, modifying and

The Difficulties of Design Problem Formulation

updating a design specification, and discriminating requirements from constraints will all be made easier with a data management environment tailored to these purposes.

B. Opportunity Number 2: Knowledge Solicitor

When design problems are being "fleshed-out," it is often the case that some issues are defined in great detail while others are only superficially defined. As previously discussed, some issues are never fully specified until developers apply standard practices or personal preferences during development. On occasion, this may be acceptable and desirable, but on too many occasions, the failure of designers to be more specific is simply an oversight.

Take as an example this "specification" that was made by a systems engineer on one of the projects discussed previously:

> The operator will specify the command link frequency.

Answers to the following questions regarding display design were left up to the programmers to infer or ferret out themselves:

- How many character places does this require?
- Is the frequency alphabetic, numeric, or both?
- Is this actually the frequency or an arbitrary label?
- What is the acceptable range of frequencies?
- Is there a list to choose from? If so, how many are on the list?
- How does the operator choose one to specify?
- Is there a default?
- What if the operator makes a mistake or changes his mind?
- How often/rapidly will this task be performed?
- When and why will this task be performed?

As it turned out, the system designers had not intended to give this much latitude to the programmers/developers. They simply had not realized the degrees of freedom left in their specification.

In order to avoid this problem, automated support, such as the questions above, could be keyed to certain types of specifications in the evolving design data base. The designer would not be forced to answer these questions but would be given the opportunity to consider them where appropriate. There might be some required effort on the part of designers to use a somewhat constrained and thereby recognizable vocabulary so that the system would know when to ask questions. However, it would be up to the designer to decide whether to conform to the system so that it can provide support or choose not to conform and receive less support.

This form of support would be similar to the role filled by knowledge engineers trying to extract information for developing an expert system. The purpose is not to add more information to the design but rather to extract more information from the designers where appropriate.

C. Opportunity Number 3: Shallow Domain Expertise

One of the most obvious and often discussed forms of automated support is in the form of domain-specific expertise (i.e., an expert system). I am not suggesting that the system be capable of automated design. Instead, the system would provide expertise, upon request, regarding technical issues, pitfalls, and problem areas that would help designers understand unfamiliar technologies.

It has been my experience that many large systems fail as a result of poor integration of parts that are each individually well-designed. Large design teams often evolve into collections of small groups focused on narrow design issues. These groups may easily loose sight or simply never be aware of issues which are important to other groups working on the same project.

The expert system needed to address this problem would be very different from typical expert systems. In general, the expertise would be broad and shallow rather than narrow and deep as with typical expert systems. In fact, this need might be handled with a simple relational data base with a user interface suited to probing by designers.

V. CONCLUSION

In this paper, I have considered some of the issues associated with design problem formulation. In particular, I have focused on problems that are often not thought of as central to the purpose of design. I believe there are at least two good reasons to begin with automated support in these areas. First, there will be broad applicability for general problem formulation support. This is in contrast to most automated support which is typically narrow or domain-specific. The second reason for introducing automated support in seemingly peripheral areas is to enhance the likelihood of user (i.e., designer) acceptance by not usurping tasks which the user likes and values doing personally.

REFERENCES

Alavi, M. (1984). An assessment of the prototyping approach to information systems development. *Communications of the ACM, 27*(6), 556-563.

Frey, P.R., Sides, W.H., Hunt, R.M., & Rouse, W.B. (1984). *Computer-generated display system guidelines, volume 1: Display design* (Report No. EPRI NP-3701). Palo Alto, CA: Electric Power Research Institute.

Hunt, R.M., & Maddox, M.E. (1986). A practical method for designing human-machine system interfaces. *Proceedings of the 1986 IEEE Conference on Systems, Man, and Cybernetics.* Atlanta, GA.

Parnas, D.L., & Clements, P.C. (1986). A rational design process: How and why to fake it. *IEEE Transactions Software Engineering, SE-12*(2), 251-257.

Schneiderman, B. (1987). *Designing the user interface: Strategies for effective human-computer interaction.* Reading, MA: Addison-Wesley.

Chapter 12

THE ROLE OF MAN IN THE SYSTEM DESIGN PROCESS: THE UNRESOLVED DILEMMA[1]

Edgar M. Johnson

U.S. Army Research Institute
Alexandria, Virginia

> "Nearly all men die of their remedies not of their illnesses."
>
> Moliere

ABSTRACT

In the design and development of military systems, man is becoming recognized as a critical element that contributes significantly to system performance. A number of trends in the systems development and acquisition process have contributed to this view. The results of case studies are used to illustrate deficiencies in considering man as a design element and also to illustrate the complexity of the issue of integrating human requirements into systems design. The broad strategy to consider man as a design element is simple; the tactics are complex. Observations are made concerning some of the key elements influencing the shift to this emerging view of man's role in systems. Recommendations concerning the system development process and the field of human factors are made for translating what we know about human factors into action.

I. INTRODUCTION

The role of man as an element of military systems occupies an increasing degree of attention and resources in the research and development community and an increasing amount of the time of military and civilian program managers and

---

[1] The findings in this report are not to be construed as an official Department of the Army position unless so designated by other authorized documents.

executives in the defense community. The past, incorrect view of man in the system design and development process is shifting from one which considers man to be a homogeneous, noninteracting element of a system; an owner-operator-maintainer, standing apart, to be added simply to manipulate the system to meet its mission requirements. Man is now being recognized as one element of the set of elements forming the system. In system design, man is not only an interactive element to be integrated into the system, but also an element that varies on many dimensions and contributes significantly to system performance. The underlying basis for this belated recognition of the role of man has been the recent advancements in methodology and technology that can incorporate human factors effects into the design of systems leading to superior system performance in the field.

One might believe none of this is new, and the need to integrate human requirements into system design has been required for a long time. This is true. What is new, therefore, is that it has finally been realized that lack of serious attention to human requirements during system design can no longer be tolerated, that such an oversight is an inadequate design process in terms of overall effectiveness and overall costs.

This shift in view has been accompanied by a seemingly endless series of study reports which review weapon system design deficiencies ("horror stories"), provide recommended patent solutions, and which are usually followed by management initiatives, large or small, announced with appropriate fanfare.[2] The acquisition process is then shifted, albeit slightly, to incorporate the changing view of man as a system element, and the systems acquisition community continues on course until the next round of study reports. Given the growing consensus over the criticality of integrating man as an essential, interactive element in the design of systems, why does change occur so slowly and what are the prospects for the future? This paper provides 12 observations on some of the key elements influencing the consideration of man as a design element by the acquisition community. These observations are from the perspective of an applied research psychologist working in a defense laboratory supporting the development process.

II. THE ROLE OF MAN IN SYSTEMS: WHAT IS THE PROBLEM?

The goal of the weapon system acquisition process is to field new military units; that is, to produce the required improved military force effectiveness to meet potential threats. Viewing man as an element of the system, the basic issues are

---

[2]Many of these reports are unpublished staff studies that are not generally available to the public. Recent publicly available reports which illustrate the genre include General Accounting Office (1981a, 1981b); Naval Research Advisory Committee (1980); and Office of the Director of Defense Research and Engineering (1967).

The Role of Man in the System Design Process   161

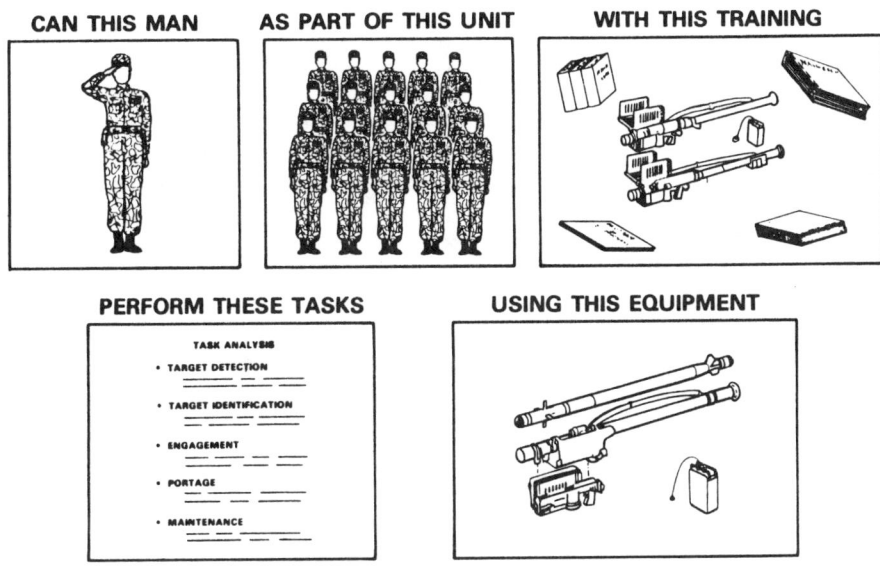

Figure 1. The Basic Question of Man as a Design Element.

variants of the question pictured in Figure 1; viz., "Can *this* individual as part of *this* unit with *this* training effectively operate and maintain *this* equipment to accomplish the mission?"

During the last few years, we have conducted a number of case studies of the development of specific existing weapon systems (Kane, 1981; Arabian, Hartel, Kaplan, Marcus, & Promisel, 1984; Hartel & Kaplan, 1984; Marcus & Kaplan, 1984; Kane, Bean, & Kirchner-Dean, 1986) to determine how man was considered as a design element in the acquisition process and the relationship of people factors to the field performance of the weapon system. A brief review of the most recent series of four case studies—STINGER, Multiple Launch Rocket System (MLRS), BLACK HAWK (UH-60A), and the Fault Detection and Isolation System of the M-1 tank (summarized in Promisel, Hartel, Kaplan, Marcus, & Whittenburg, 1985)—provides a rich, and necessary, context for a discussion of man as a design element.

The basic case study methodology, labeled "reverse engineering," involves a detailed, systematic audit of the development and acquisition of a specific weapon system using the following approach:

o   The system is defined and described.

o   Requirements documents are reviewed to determine how system performance was specified.

o   Test and evaluation data are analyzed and compared to performance criteria.

o   Problem areas in system performance are identified.

o   People factors are examined for their impact on the problematic aspects of system performance.

o   The acquisition process for each system is reviewed to identify features that contribute to people issues.

The results of these case studies confirmed the existence of a series of explicit deficiencies in considering man as a design element linked to operational system performance. For example, in the case of STINGER, failure to allocate tasks effectively between the hardware and the gunner resulted in an engagement task sequence more complex than in the predecessor system, the REDEYE. The engagement sequence is so complex that lower mental category soldiers, constituting a large portion of the current population of gunners, cannot operate STINGER to meet the designed effectiveness. The case studies confirmed the obvious: people problems are readily found in weapon systems. Our results were briefed in July 1984 to General Maxwell Thurman, the Vice Chief of Staff of the U.S. Army. General Thurman subsequently initiated an ongoing, major Army-wide initiative, *man*power and *per*sonnel *in*tegration (MANPRINT), to "fix the problem." MANPRINT is an umbrella concept to identify, address, and impose human factors, manpower, personnel, training, system safety, and health hazard considerations across the entire acquisition process.

We then took the next step of synthesizing this data base of explicit deficiencies to identify causal factors contributing to these deficiencies (Figure 2). Although the types of people problems found in these case studies are not new and the causal factors are not surprising, their impact on weapon system performance was greater than anticipated. Man has become more tightly coupled to the hardware, and the total system is more complex than in prior years or in earlier weapon systems. (For a general analysis of this issue, see Perrow, 1984). System characteristics which lead to this conclusion include the following:

o   The existence of unintended or unfamiliar feedback loops in system operation.

o   Invariant task sequences with an increased number of discrete steps required for effective performance.

o   Increased specialization of duty positions, thereby limiting

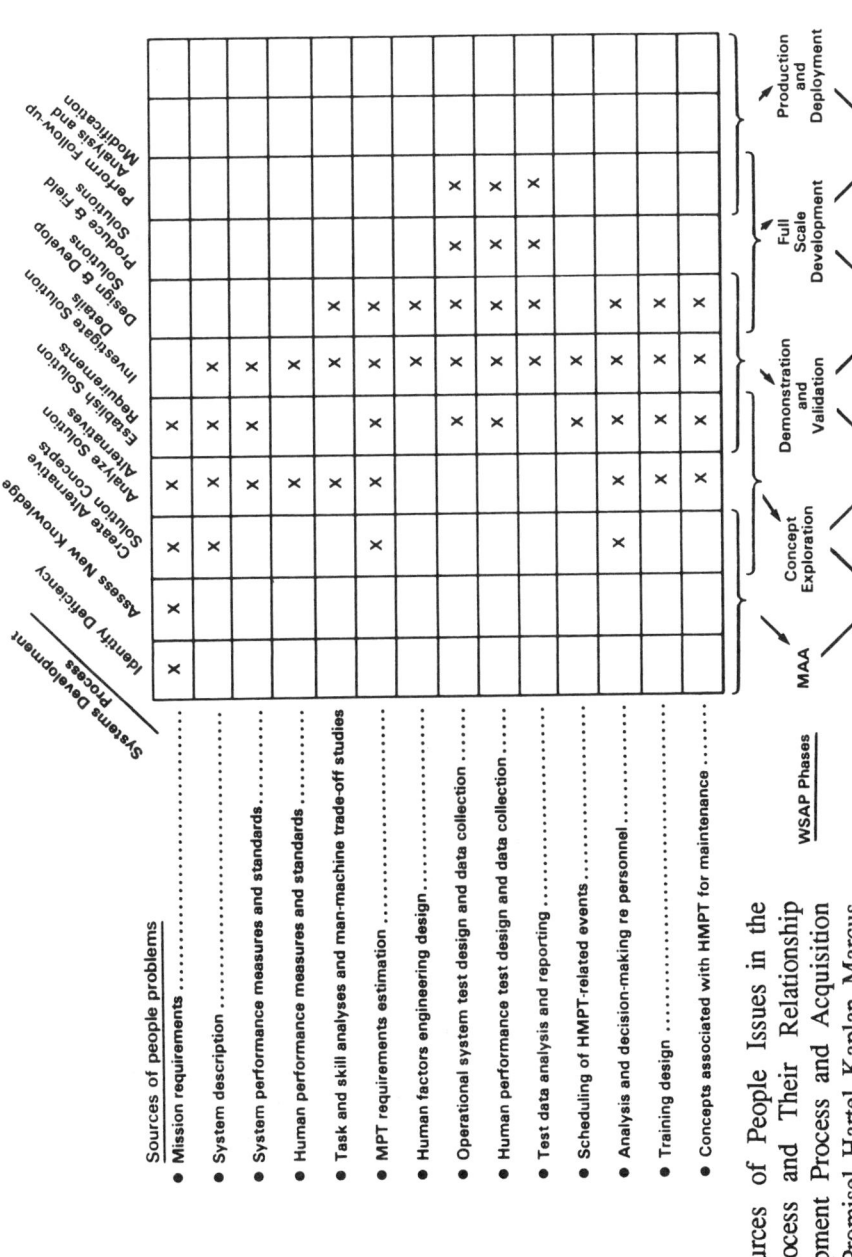

Figure 2. Sources of People Issues in the Acquisition Process and Their Relationship to the Development Process and Acquisition Phases. (From Promisel, Hartel, Kaplan, Marcus, & Whittenburg, 1985.)

awareness of task interdependencies by operators or maintainers and reducing personnel substitutability.

In addition to shifts in system characteristics, changes in the military personnel system made deficiencies in considering man in the design process more visible in the fielded system's operational performance. For example, slack in the number of military personnel was reduced as the number of military personnel or end strength was constrained, and actions were taken to increase the ratio of combat soldiers to support personnel—the "tooth-to-tail ratio." This reduced slack made the consequences of unanticipated, increased manpower demands or increased skill level requirements more disruptive. As the tooth-to-tail ratio was increased, there was a shift in training from the schoolhouse to the unit, reducing initial skill levels achieved in school training, and increasing the need for additional training in units.

The result of these and other trends was that system performance and unit capability became visibility limited by people factors. Man could not be considered just another of the large number of elements comprising a complete weapon system. As he had been all along, man was now seen as a critical element. Man could no longer compensate for hardware design deficiencies. If not considered as a design element, he became a constraint on system effectiveness. People problems of the type found in these case studies are symptoms that something is broken.

Observation 1. There is a consensus that the current system acquisition process fails to adequately consider man as a design element, but there is no agreement on what specific aspects of the system acquisition process are broken—there is not an agreed upon "fix" or "set of fixes."

A number of solutions have been proposed to ensure the inclusion of man as a design element. These solutions can be characterized as fragmented, piecemeal, and short-term oriented. They have engendered high emotion from all sides. The following examples briefly provide the flavor of these solutions:

o   Technology Fix. A missing element of human factors technology is required to blend with engineering technology and will ensure total systems development.

o   Rules and Regulations. The Army regulations and directives governing the acquisition process are revised to require greater coverage of all people issues anyone can think of to ensure total systems development.

o   Checklists and Guidelines. Long lists of people issues and discrete items are created on the assumption that if all items are checked off or completed, total system development has occurred.

Although change has occurred, each of these purported solutions has fatal flaws. Available human factors technology is frequently not applied or used in the system acquisition process. As Kerwin, Blanchard, Atzinger, and Topper (1980) accurately pointed out, there is no traffic cop to ensure that rules or regulations concerning man as a system element are followed. Further, such checklists and guidelines are often devoid of substance and can frequently be met independently of system design.

Observation 2. The problem of man as a design element cannot be solved by a simple one-shot fix—a successful solution will be evolutionary, long-term, and multifaceted.

III. THE SYSTEM ACQUISITION PROCESS

The weapon system in the field is the end product of a complex acquisition process involving diverse players. The weapon systems acquisition process is a formal, explicit management strategy which was created to prescribe a sequence of events and phases of program activities and decisions leading to efficient and effective, fully supportable systems responsive to validated Army requirements (Department of the Army, 1975, 1984).

The final, fielded system is the cumulative product of decisions by many individual participants, agencies, and activities representing different elements of the total system. A network of regulations and directives governing the acquisition process details the role and contribution of each participant. These regulations and directives have evolved with the objective of weaving together the loosely coordinated web of multiple participants with their numerous disjointed, and sometimes conflicting, subobjectives into an effective total system.

It is a complex process created to ensure that all aspects of the total system are considered in the development process. A review of regulations and directives governing this process indicates that the human element is, in fact, covered in large measure (Rhode, Skinner, Mullin, Friedman, & Franco, 1980). If all of the stated requirements concerning man as an element of a weapon system were or could be met in the acquisition process, the current problem would be the limitations of human factors technology, not the implementation of existing human factors technology.

Observation 3. Man is recognized as a design element in the formal process governing weapon systems acquisition.

Weapon systems are designed by industry. In the acquisition process, the government states requirements. Using the requirements, industry designs and manufactures the hardware components of the system and much of the supporting equipment and documentation. The government then uses the hardware (and associated elements from industry, and system elements from a variety of individual government activities), and packages and fields the total system—new units. Although there is a rich informal dialogue both within the government and between government and industry, explicit requirements are stated by contract, and government-industry interactions are constrained by statutes and directives governing the contract process. Development of a specific weapon system is an interactive process in which the principals are the government program manager on one side and the prime contractor on the other side, with the contracting officer filtering and monitoring the interaction.

Observation 4. For man to be considered as a design element by industry, the requirements must be part of the contractual process: stated in the procurement, a stated factor in the source selection, and funded in the contract.

If the government does not require and fund it, there is no incentive for industry to consider man as a design element. To state the obvious, industry is in business and can stay in business only as long as it makes a profit. A corollary is that the bulk of the profit is from the manufacture and sale of weapon systems, not their design. Thus, industry will consider man as a design element only to the extent that it is paid for by the government and such consideration contributes to the eventual sale of the hardware.

Observation 5. If the government buys hardware where man was not considered as a design element, there is an implicit incentive for industry to reduce costs by not considering man.

The rules that guide weapon systems design are essentially political. Their legitimacy comes from congressional mandates and actions, defense policy and directives, military needs, and precedent that tie military systems explicitly to a program manager whose charter is based upon an operational requirement and whose success is based upon satisfying both an operational user (or a user's representative) and the requirements of the contracting process. In contrast, the rules that shape systems engineering are methodological. Their legitimacy springs from a set of conventions that experience has demonstrated to be successful in the development of complex hardware systems. Industry designs, and the military acquires weapon systems whose design and development represent the best, not of the field of systems engineering, but of "public engineering." Public engineering is that part of systems engineering endorsed by the key participants in the system acquisition process. It represents a consensus of what is considered generally acceptable systems design practice or technology.

Observation 6. Public engineering has been used successfully to design, manufacture, and field systems where man was not considered as a design element. Human factors has not been a part of public engineering.

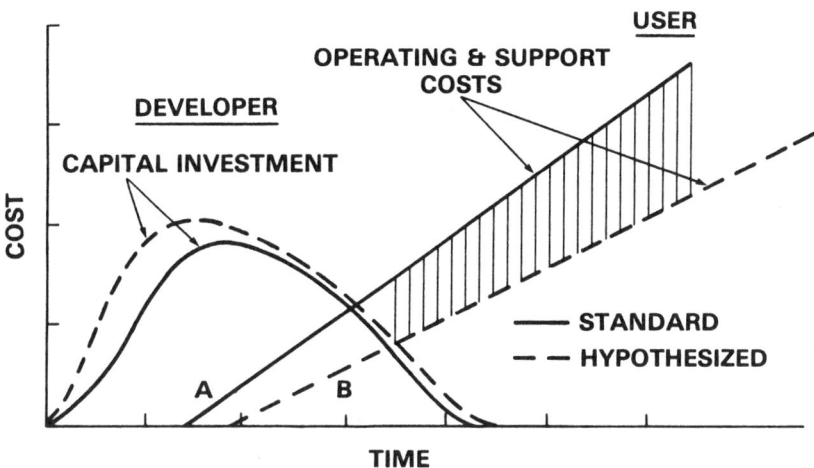

Figure 3. Alternate Investment Strategies in Man as a Design Element.

Inclusion of a technology in public engineering is not measured by either the number of people working on the element or by resources expended on the element. Rather, the metric is one of influence on systems design and acquisition decisions. Thus, public engineering appears to have had the effect of defeating the best efforts of the logistics community. However, there are signs that human factors may become part of public engineering. An example is the 1985 appointment of a human factors psychologist, Dr. John L. Lauber, as a member of the National Transportation Safety Board.

Consideration of man as a design element is done most effectively early in the acquisition process. People problems (Figure 2) have their origin in inadequate or incomplete analysis of system requirements or the use of inappropriate assumptions during the concept and exploration phase. This is the phase where the greatest pressure is to demonstrate hardware technical performance. The acquisition process is success-oriented, focused on individual system cost and technical performance. Investment in man as a design element has its greatest payoff at a point in time where it has the least impact on measures of component or prototype system performance. Thus, increasing the investment in human factors is, in effect, shifting resources to overcome people problems from the field (the user's account) to the front end of the design and acquisition process (the program manager's account). This concept is illustrated in Figure 3 by the difference between line A and line B; the expectation is

---

[3]Resolution of this issue is complex because procedures for estimation of system life cycle costs, especially manpower costs, are primitive. The results are often more dependent on scenarios and analysts than on the details of the specific system.

reduced total life cycle costs. (For some data supporting this hypothesis, see West, 1984). The issue is one of who pays the bills: the user through operation and support costs or the developer through development and acquisition costs; and whether larger "today" costs offset smaller "future costs."[3]

Observation 7. Consideration of man as a design element increases cost of weapon system acquisition (the front-end cost) and decreases the operating and support costs—and the total life cycle cost.

As indicated above, the focus of the acquisition process is cost and technical performance. Technical performance is measured through engineering tests of hardware performance, and the data are used in a variety of analytic procedures and simulation models to defend the requirement for the system and the capability of the system to meet the requirement. Common in all of these analytic procedures and simulations is the discounting (or absence) of man as a performance element in the system. Although front-end assessment of expected system performance does not now include man as an element, methods are available which would allow the inclusion of man.

Observation 8. Failure to consider man as a design element of weapon systems is not apparent in early assessments of weapon systems (and frequently is not apparent in later assessments and operational tests).

IV. HUMAN FACTORS

Human factors is a multidisciplinary field of technology concerned with man as an element of systems. (Human factors in this context is used as a convenient shorthand for the set of human-oriented technologies including ergonomics, training, organizational design, safety, and other specialties.) It is useful to distinguish between micro human factors (design of the interface between the hardware and the man or the design of specific tasks) and macro human factors (design or adaption of all manpower, personnel, training, and organizational components of the system).

Observation 9. Where applied, human factors has focused on micro issues to the virtual exclusion of macro issues.

The human factors community has been reactive, not proactive, in dealing with man as a design element. Growing out of the need to solve man-machine interface problems in World War II, human factors has continued to focus upon design of the man-machine interface. (For example, there are over 500 experimental studies of keyboard design.) In fact, it often appears that design of the man-machine interface is an end in itself. While interface design appears deceptively simple, there are a sufficient number of examples of poorly designed man-machine interfaces to demonstrate that it is not an easy part of the system to design. Although relatively immature (knowledge is patchy and sometimes contradictory), design of a good man-machine interface is well within the state-of-the-art of human factors technology, as demonstrated in the aviation industry with the Boeing 757/767 or Lockheed 1011.

Improvements in micro human factors technology (e.g., rapid prototyping, dynamic anthropometric models, etc.) are focused upon increasing the ease of interface design and reducing interface design to usable standards and design practice. These types of development lead to incremental improvements worth seeking but not dramatic change in systems design. Focusing on micro issues, the program may often be deep into design and development without knowing the impact on manpower, personnel, or training issues. Conversely, improvements in the capability to relate personnel and manpower issues along with other system elements to system performance, and to perform tradeoffs of human factors issues (answer "what if" questions), may lead to dramatic change. The manpower, personnel, and training tools to deal effectively with these issues in system development generally are laborious or do not exist. As human factors begins to deal with macro issues of prediction of performance and manning requirements instead of comparative evaluation, there is the opportunity to consider man as a design element with certain parameters instead of a design constraint.[4]

Observation 10. Improvements in micro human factors technology will not change the degree to which man is considered as a design element.

Human factors is an applied technology; it is inevitably driven by hardware technology. Technology is concerned with improved systems and processes which must be salable if the technology behind them is to survive. Although people research is often a slower process than hardware research and less money is spent on it, the issue of human factors application is not only one of time and dollars, but of salability. While there are numerous "horror stories" of failure to consider human factors, there are few success stories where man was effectively considered as a design element. Real successes in human factors design are not readily observable because they do not lead to any problems in human performance that might draw attention. One of the deficiencies of human factors methodology has been its inability to easily measure the magnitude of benefits that various trivial human factors improvements might bring over the life cycle of the system being developed. Human factors tends not to make the central case (which I believe is feasible) that some proposed human factors modifications in the design of a developing system have significant impact, in

---

[4]For an outline of some potential research efforts oriented toward macro human factors issues, see DePuy and Bonder (1982).

objective terms of performance and cost.[5] Human factors as a technology to effectively consider man as a design element requires success stories to be salable as a part of public engineering and as a key player in the systems acquisition process.

Any technology to be included in the acquisition process must have a record of success, and it must meet the general constraints of cost and schedule required of other technologies included in the process. Although human factors is an applied technology, it is frequently not subject to the same costing and scheduling discipline of other technologies. Researchers, practitioners, action officers, and managers often emphasize the differences between people and hardware, and human factors and engineering. They use these differences to defend the inability to accurately estimate or meet cost and schedule constraints. Man is, in fact, more complex and variable than hardware. There are clear differences which require different approaches and methodologies. However, in the absence of a specific schedule and cost discipline, managers and evaluators concerned with the human element of systems resort to arm-waving and vague promises of things due tomorrow, practitioners become uncertain of design milestones, and research is completed too late to be implemented.

Observation 11. Human factors must develop success stories and be managed using an identified set of explicit, verifiable items which are scheduled, costed, and integrated into the acquisition process if man is to be effectively considered as a design element.

A related issue is the research orientation of the human factors community. Drivers in the field of human factors, as in other fields of technology, are related to improved systems and processes, not to the solution of human factors issues using the existing knowledge base and tools. There is little tradition of moving technology into a user product as in engineering. Rather, the tradition is one of moving technology. As a consequence, there is a small base of human factors practitioners as opposed to researchers. Consideration of man as a design element requires the transition of what we know into what we do—development of an industrial base of human factors practices that employ the results of the government/industry base of human factors research and development.

Observation 12. Effective consideration of man as a design element requires development of human factors practitioners, not just researchers.

---

[5]For examples of a proposed methodology—impact assessment—to measure the contribution of human factors to military system development, see Price, Fiorello, Lowry, Smith, and Kidd (1980); and Sawyer, Fiorello, Kidd, and Price (1981).

V. CONCLUSION

The strategy for including man as a design element in the systems acquisition process is simple:

- o We must know what we want.
- o We must describe what we want, then schedule and cost it.
- o We must know what we are getting, then show how it impacts the overall system performance and life cycle cost.

Although the broad strategy is simple, the tactics are complex. The issue of man as a design element is an unresolved dilemma. The acquisition process is executed by well-intentioned, highly skilled people who (despite "horror stories") have a long record of success. The system acquisition process, in fact, provides weapon systems with the level of consideration of man as a design element which we are willing to buy. What is not clear is the extent to which the key players in the system acquisition process want and are willing to pay for increased consideration of man as a design element. Human factors technology is more often seen as the icing, not as part of the cake.

Yet, change is occurring—the desire to consider man as a design element has increased. More change is possible and efforts are increasing. However, these efforts are often fragmented, based on reaction rather than vision, and to a large extent, individually initiated. There is no comprehensive strategy or derived master plan—a vision of the future—for man as a design element. The results of increased consideration of man as a design element will be reduced total life cycle costs, more effective weapon systems, and improved force effectiveness. The observations presented in this paper lead to several conclusions or assertions:

- o Man is not effectively considered as a design element because of historically stated organizational goals, objectives, and issues; not because of limitations of human factors technology.

- o Improved consideration of man as a design element requires an evolutionary, multifaceted approach supported by recognized key actors/players in the acquisition process. The human element in the weapon system needs a "daddy"—there is no simple, patent solution.

- o Changing the acquisition process will not change consideration of man as a design element unless:

    - Mindsets are changed.
    - Program managers are rewarded (given incentives).

- Industry is required and funded for it in the contractual process.
- Human factors issues influence acquisition decisions.

o Increased investment in man as a design element requires that, at a minimum:

- Human factors technology becomes part of the political process.
- Human factors be included in public engineering.
- The consequences of human factors issues be stated in terms of system effectiveness.

Weapon systems are designed to extend man's capabilities. Thus, man should be the basic element in the design of systems. In many systems, man optimizes system performance by compensating for hardware design deficiencies. As systems have become more complex and man more tightly coupled to the hardware, man can no longer compensate for design deficiencies. System effectiveness has become limited by the failure to adequately consider man as a design element. We have begun to move from hardware-dominant to man-dominant systems—from a focus on manning equipment to equipping the man. The challenge is to develop a systems acquisition process which will translate what we know about human factors into what we do; that will translate knowledge into action.

REFERENCES

Arabian, J.M., Hartel, C.R., Kaplan, J.D., Marcus, A., & Promisel, D.M. (1984). *Reverse engineering of the multiple launch rocket system: Human factors, manpower, personnel, and training in the weapons system acquisition process* (Research Note 84-102). Alexandria, VA: U.S. Army Research Institute.

Department of the Army. (1975, May). *Life cycle system management model for army systems* (Pamphlet No. 11-25). Washington, D.C.: Headquarters Department of the Army.

Department of the Army. (1984, February). *Systems acquisition policy and procedure* (Army Regulation 70-1). Washington, D.C.: Headquarters, Department of the Army.

DePuy, W.E., & Bonder, S. (1981, March). *Integration of MPT supply and demand and the system acquisition process* (Research Note 82-16). Alexandria, VA: U.S. Army Research Institute.

General Accounting Office. (1981a, January). *Effectiveness of U.S. forces can be increased through improved weapon system design* (Report No. PSAD-81-17). Washington, D.C.

General Accounting Office. (1981b, December). *Guidelines for assessing whether human factors were considered in the weapon systems acquisition process* (Report No. FPCD-82-5). Washington, D.C.

Hartel, C.R., & Kaplan, J.D. (1984). *Reverse engineering of the Black Hawk (UH-60A) helicopter: Human factors, manpower, personnel, and training in the weapons system acquisition process* (Research Note 84-100). Alexandria, VA: U.S. Army Research Institute.

Kane, J.J. (1981, January). *Personnel and training subsystem integration in an armor system* (Research Report 1303). Alexandria, VA: U.S. Army Research Institute.

Kane, J.J., Bean, T.T., & Kirchner-Dean, E. (1986, April). *Utilization of human resources data in battlefield automated systems* (Research Note 81-30). Alexandria, VA: U.S. Army Research Institute.

Kerwin, W., Blanchard, G.S., Atzinger, E.M., and Topper, P.E. (1980, August). *Man/machine interface - A growing crisis* (Army Top Problem Areas Discussion Paper No. 2). Aberdeen Proving Ground, MD: U.S. Army Material Systems Analysis Activity.

Marcus, A., & Kaplan, J.D. (1984). *Reverse engineering of the M1 fault detection and isolation subsystem: Human factors, manpower, personnel and training in the weapons system acquisition process* (Research Note 84-101). Alexandria, VA: U.S. Army Research Institute.

Naval Research Advisory Committee. (1980, December). *Man-machine technology in the navy* (Report NRAC 80-9). Washington, D.C. Office of the Assistant Secretary of the Navy.

Office of the Director of Defense Research and Engineering. (1967, October). *Study of manpower considerations in development: Vol 1. findings and recommendations of the study group* (AD No. 326438). Washington, D.C.: Department of Defense. (This report is sometimes referred to as the "Nucci Report" after the Study Group Chairman).

Perrow, C. (1984). *Normal accidents.* New York: Basic Books, 1984.

Price, H.E., Fiorello, M., Lowry, J.C., Smith, M.G., & Kidd, J.S. (1980, July). *The contribution of human factors in military system development: Methodological considerations* (Technical Report 476). Alexandria, VA: U.S. Army Research Institute.

Promisel, D.M., Hartel, C.R., Kaplan, J.D., Marcus, A., & Whittenburg, J.A. (1985, January). *Reverse engineering: Human factors, manpower, personnel, and training in the weapon system acquisition process* (Technical Report 659). Alexandria, VA: U.S. Army Research Institute.

Rhode, A.S., Skinner, B.B., Mullin, J.L., Friedman, F.L., & Franco, M.M. (1980, October). *Manpower, personnel, and training requirements for materiel system acquisition* (Research Product 80-27). Alexandria, VA: U.S. Army Research Institute.

Sawyer, C.R., Fiorello, M., Kidd, J.S., & Price, H.E. (1981, July).

Measuring and enhancing the contribution of human factors in military system development: Case studies of the application of impact assessment methodologies (Technical Report 519). Alexandria, VA: U.S. Army Research Institute.

West, H.M. III. (1984). Capitol investment can reduce weapon system manpower costs. *Defense Management Journal, 20*, 17-23.

Chapter 13

ANALYTICAL VERSUS RECOGNITIONAL APPROACHES TO DESIGN DECISION MAKING

Gary A. Klein

Klein Associates
Yellow Springs, Ohio

ABSTRACT

How do designers solve problems and make decisions? There is a tendency to assume that the equipment design process is primarily analytical, and that where designers depart from analytical strategies, they will need tools to bring them back into line. Systems approaches are inherently attractive because they appear to transform chaotic problems into orderly ones. However, it is argued that this appearance may be misleading. Many difficult types of design problems are ill-defined and will, therefore, resist formal analytical methods. Moreover, skilled designers appear to rely on their experience with analogous problems. Their recognitional capabilities are of key importance, including their ability to recognize which analytical methods to use and how to use them. This paper describes a recognitional approach to design decision making and argues that design support tools should help users identify and extrapolate from comparable items of equipment.

I. INTRODUCTION

The goal of this position paper is to describe how experienced designers solve difficult problems. I am skeptical of some of the standard accounts of the design process. Such approaches are often misleading because they portray the

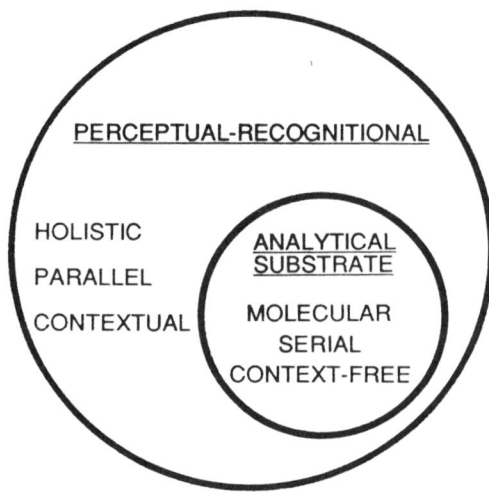

Figure 1. Reliance of Analytical Strategies on Perception/Recognition Processes: Integrated Framework for Human Performance.

designer as overly analytical and continually performing tradeoffs and calculations.

In contrast, it seems to me that much of the strength of an experienced designer comes from his ability to recognize the types of problems encountered, to recognize the typical ways of handling such issues, and to recognize the implications of contextual nuances. Figure 1 represents this view. As can be seen, perceptual-recognitional processes are dominant for providing situational awareness. The overriding strength of a skilled designer is in knowing what problems are typical and what reactions they call for, along with what problems are unique and what approaches are most promising.

Figure 1 presents the designer's reliance on analytical methods (e.g., costs versus benefits, and calculations of sizes, durations, and intensities) as an analytical substrate that depends on the way the designer perceives and recognizes the problem. This is because the proficient designer must *recognize* when analysis is called for, what type of analysis is appropriate, and when the analysis has been sufficient. These things cannot be calculated. As deGroot (1978) found for chess mastery, calculations are essential for evaluating the quality of moves, but the factor that separates mediocre players from good players is the ability to recognize which moves are worth evaluating.

I am not claiming that analytical techniques are dispensable. Our current level of technological sophistication could not have been achieved without powerful analytical methods. I am only claiming that analysis by itself is insufficient. The analogy is to a camera. The power of the camera to achieve sharp focus (like the power of the analytical substrate) is essential for clear pictures. But even the best automatic focusing system will be useless without a photographer who can sense where to aim the lens, which lens to use, and when to press the shutter control.

In this paper, I will 1) present a recognitional account of problem solving and decision making within the context of design engineering, and 2) present implications for design support tools.

II. RECOGNITIONAL ACCOUNT OF PROBLEM SOLVING AND DECISION MAKING

A. Well-defined and Ill-defined Problems

Consider the way Rouse and Boff (see Chapter 2) have described the typical analytical approach to design: 1) formulation of problems, 2) generation (or synthesis) of alternative solutions, 3) evaluation (or analysis) of alternatives, and 4) selection (or optimization) among alternatives. They point out that "artistic" qualities are needed in the first two steps, and "analytical" qualities are needed during the last two steps.

This approach is the standard recipe for solving *well-defined* problems. However, it does not capture the ongoing goal clarification needed for *ill-defined* problems. For these, the "artistic" qualities of designers may be required throughout. It is an easy way out to assert that the artistic processes will mysteriously generate alternatives, and that the powerful analytical engine of the mind can take over from there. This amounts to an admission that we do not know how options are generated and, therefore, treat them as random and of varying quality. Because we do know the types of formal analyses that can be conducted, we put our faith in these. All we want from the option generation phase is that a large enough number of options be produced so there is a good chance a high quality one will be included. Then the "actual" analyses can begin.

Even for well-defined problems, this approach has difficulties. One is that it creates the substantial burden of presenting exhaustive sets of options. When does one stop? How many are enough? Is this a good use of time since it does not contribute to goal clarification and may be inappropriate if the goal is misunderstood? Moreover, there is the additional burden of assessing all of the options. It is for these reasons that this strategy is not always the preferred one: it is time- and energy-consuming.

For the present purposes, a more important limitation is that many difficult equipment design problems are ill-defined: there is no general agreement about what would constitute a specific solution. As Klein and Weitzenfeld (1979) have noted, ill-defined problems are those where the goal is unclear, so there is no way to follow the standard prescription to move from step 1 (defining the goal) to step 2 (generating options). The task of defining the goal can never be finished if the goal is, by definition, ill-defined. The only way to make progress with ill-defined problems is to press toward a solution with the expectation of increasing the goal clarification along the way. Therefore, for ill-defined problems, goal clarification must occur in parallel with the search for solutions, and these two processes must support each other. As potential solutions are evaluated, new goal properties can be recognized and added to the evolving goal specifications.

Part of the designer's skill is in understanding what the design is really supposed to accomplish. A mistake novices often make is to fixate on the original statement of design criteria. In doing so, novices miss opportunities for breakthroughs and fail to notice factors that render the original design criteria inappropriate. This is why the distinction between well-defined and ill-defined problems is so important. If designers treat ill-defined problems as if they were well-defined, their problem solving can become ineffective.

We found many examples of ill-defined problems in a recent study we performed (Klein & Brezovic, 1986). The purpose of this study was to identify the human performance data that training device designers needed. We conducted interviews with 42 experienced simulator designers from the DoD and industry. The interviews focused on identifying cases where human performance data were needed but were not immediately available, and tracing the designers' efforts to obtain such data or work around the lack of them. A total of 76 design decisions were identified and probed.

Most of the problems we studied were ill-defined. Even for those that appeared to be well-defined, there were ill-defined stages. That is, the original goal was well-defined but the solution involved rejecting this goal and replacing it with another well-defined goal. For example, in one case, the requirement was to follow MIL-STD-1472 to determine the size of CRT letters. The CRT was to present information in a classroom, and students in the back of the room were to be able to read everything. This was a reasonably well-defined problem. However, some of the graphs to be shown on the CRT would have required legends so large that the entire graph might not have fit on the CRT. The designer reconceptualized the problem from "how large to make the letters so that they could be read in the back of the room" to "how to present text to increase the effectiveness of instruction"—an ill-defined problem. The designer requested and received a waiver from MIL-STD-1472, determined which aspects of text were really important for the students to read clearly, and then clarified the problem as "how to present the relevant text so that students in the back of the classroom could read it." This was again a well-defined problem.

Another example involved a training device that had some difficulties with the control dynamics. Subsequently, it was found that pilots had trouble flying formation in the device. This was interpreted as evidence that the control problem was not yet solved. However, after much effort, one person suggested that maybe the problem was that the model of the lead aircraft was painted white, against a light blue sky, providing too few orientation cues during formation flying. Identification markings were painted on the model, and the problem disappeared. Thus, the original problem was well-defined but mistaken, and the breakthrough came in clarifying the goal as providing perceptual cues rather than cleaning up the control dynamics.

Ill-defined problems require two simultaneous processes: 1) goal clarification, and 2) option development. Designers cannot always wait for goals to become perfectly well-specified before starting to work. They must expect to learn more about their goals during the design process itself. This helps to make the design process challenging and maddening. The designer is trying to give the sponsor/user a product, and at the same time, is trying to learn more about what the user really wants and to help the user go through this learning cycle.

B. Goal Clarification

The search for a solution is a source of trial-and-error learning about options and goals. It provides failures, lessons, and occasional successes. As the goal becomes better understood, the chances of success increase. Analogues play a prominent role in goal clarification by presenting previous concrete cases with explicit sets of goals, causal relationships, and effective strategies.

How do designers clarify goals? In the study by Klein and Brezovic (1986), we found that the primary sources of goal clarification were the use of analogous pieces of equipment for the devices being designed, as well as informal experiments (use of mockups and breadboard, brassboard, and prototype models) that provided trail-and-error learning.

Analogue use was quite frequent—it was found in about one-third of the decision points we studied. In its simplest form, a designer needed a specification for an instructor/operator station, and used a previous specification for an earlier training device, making only the obvious changes in nomenclature.

In another example, a panel training device was portraying hydraulic lines using an artificial color code. One designer remembered an analogous device that had left users confused when they saw the actual equipment. He therefore recommended including a photograph of the actual system on the training device itself.

Informal experiments were even more frequent than analogues. In approximately half of the design decisions, we found that designers relied, to a large extent, on the use of mockups and informal experiments that allowed them to observe users interacting with new systems. In a typical example, a designer was concerned with the computer refresh rates and studied different rates to see how they affected cue perception. The study enabled the designer to understand the types of benefits obtained from using faster refresh rates and to find alternative ways of achieving these benefits without the expense of a faster computer system.

C. Recognitional Processes

Recognitional processes play a key role in decision making as well as problem solving. This is illustrated by the results of a recent study of decision making (Klein, Calderwood, & Clinton-Cirocco, 1986). Our subjects were experienced fire ground commanders with an average of over 20 years of experience. They were responsible for directing operations during urban fires, and they made decisions about the allocation of resources. Obviously, lives and property depended on these decisions. We specifically asked them about their most difficult decisions, about the options they considered, and the bases for their selections. To our surprise, they denied ever making decisions in the usual sense. They denied contrasting different options and trying to select the best. In our probing, we found that they were in fact moving through "decision points" where several options were available. The way they represented these situations made it obvious to them what actions to take.

They were not calculating the strengths and weaknesses of one option versus another. In the interest of efficiency, they matched a situation to a standard or typical case, which then had a standard form of reaction. If the fire ground commanders had enough time, they often evaluated the action to see if it would work, if it needed to be modified, or if it had to be rejected in favor of the next most typical reaction. Usually, the evaluation relied on imagery to run the action through its imagined paces and to examine it for flaws. The strategy also generated expectancies so that the fire ground commanders could anticipate what was going to happen next and prepare for that eventuality.

The power of this Recognition-Primed Decision strategy is that it allows people to perform rapid decision making. It views experience as creating a bank of analogues that lets a skilled performer rely on familiarity matches to know immediately the best way to proceed. Under these circumstances, it would be counterproductive to generate a large number of options, which would then have to be evaluated. This takes time and energy. I am suggesting that when dealing with experts, we can respect their ability to immediately access an effective option.

A full account of problem solving and decision making has to portray the interaction between recognitional and analytical approaches. This paper does not attempt to offer a full account. The goal here is to describe the importance of perceptual-recognitional processes because these have not been adequately appreciated and because it is so easy to develop inadequate design tools that are predominantly analytical.

In describing the decision making of designers, Rouse and Boff (see Chapter 2) point out that designers "satisfice" and do not try to generate an exhaustive set of options. My reaction is that to do so would require an exhaustive amount of evaluation for uncertain return. In this sense, satisficing is an intelligent strategy rather than a human limitation to be transcended. Designers' efforts go into developing at least one effective option. If they can find a typical way to proceed, they are satisfied. If they must modify a typical approach or synthesize a new approach, they do so cautiously. Their creativity goes into making a dominant approach work rather than in thinking up new and imaginative untried/risky options.

In the study by Klein and Brezovic (1986), we found little evidence for the systematic use of decision analysis or multi-attribute utility methods, whereby decisions are factored into options and evaluation dimensions and are evaluated according to relative advantages. Indeed, it is only in rare cases that the effort of a formal analysis is required. Such cases may be recognized by a breakdown in the recognition of typicality, they may be mandated by orders from supervisors, or they may be necessitated by the requirements of working together in a design team and communicating the rationale for given design decisions. When the strategy shifts from a recognitional to a calculational one, the Recognition-Primed Decision mode no longer applies.

One of the rare examples of decision analysis in our study involved the selection of data input and output mechanisms for a computer-driven instructor station. There were seven alternative output devices and these were evaluated along 14 dimensions. However, this evaluation was not seen as conclusive. It resulted in a final set of four output devices, and these were then evaluated subjectively using real equipment in an informal study.

The same design group also conducted a second analysis of input devices. There were five alternative input devices which were evaluated along more than 50 dimensions. The analysis took six to eight weeks and resulted in nine different combinations. These final candidates were then brought in for hands-on study.

III. RECOMMENDATIONS FOR DESIGN AIDS

What types of information do equipment designers seek and find valuable in making design decisions? In this section, I will examine the importance of analogues, imagery, research data, and rapid prototyping.

A. Analogues

From the perspective of a recognitional model, we would expect that designers would want information about analogous pieces of equipment. They would want to learn about the characteristics of this equipment to see if it really is a good analogue. They would want data on the strengths and weaknesses of this equipment to learn how to improve it. They would want information about the construction of this equipment to avoid a duplication of effort.

If the designers were going to be asked to factor in human performance constraints through a program such as MANPRINT, they would want human performance data from analogous systems.

The power of analogical reasoning for generating new ideas and hypotheses has been appreciated for fields ranging from art to science (Klein & Weitzenfeld, 1979). Analogical reasoning is a prime source of new ideas in a given domain, and it is used heavily by design engineers. Where they may need help is in identifying and applying analogues most effectively.

First, we could provide the designers with data banks consisting of analogous cases. Many information systems combine many cases to yield overall summary statistics, but these are not helpful for analogical reasoning. What is needed is the detail contained in the cases themselves. In fact, it would be advantageous if the information systems were organized around case study data to allow designers to identify the best analogues, to obtain the fine-grained details for these cases, and to identify people to contact for more background. In this way, the data bank would be literally a collection of analogues available for multiple purposes; it would be a repository of field experiences. The coding of information would have to be carried out so as to maintain the necessary levels of detail, and the accessing system would have to be designed to let designers find analogues from entirely different fields to use as a rich new source of ideas.

Second, we could provide the designers with a strategy for using analogues. Weitzenfeld (1984) has described the logical basis of analogical prediction. Many of the principles he presented have been used to formulate a structured approach called Comparison-Based Prediction, which applies analogical reasoning to front-end analysis (Klein, John, Perez, & Mirabella, 1986). Basically, the logic of reasoning by analogy involves identifying an appropriate analogue, collecting the relevant data for the analogue, determining the ways that the analogue and target case differ, and using these to adjust the analogue data so that they fit the target case. During the stage of idea generation, there are no logical criteria for analogue selection; anything can be helpful and a good analogue is one that suggests new goal properties.

The power of analogical predictions is that they can capture the influence of factors that cannot be guessed, let alone be included, in a formal prediction model. For example, if I want to predict the reliability of subsystems on an advanced tactical fighter, I could use the operational data for the F-16 and adjust them. In doing so, my prediction is capturing many qualities about the maintenance and logistics system in TAC. If I had tried to formally identify all the variables that might have affected maintenance, I never would have succeeded, and I would have been unable to quantify most of those I identified.

We use the same comparison logic when we set up a control group for an experiment. That is, we seek to ensure that all the causal factors are matched between the experimental and control groups, except for the independent variable which can then be linked to the outcome measures. We are able to draw conclusions about some causal factors even though we do not know all the variables affecting the phenomenon we are studying or how they operate and interact. An analogue is serving as a control group, but it cannot ensure a match on all the causal factors. Instead, we try to identify areas of significant causal mismatch and artificially adjust for them.

Designers who want to use analogical reasoning more effectively might benefit from the Comparison-Based Prediction structure. They might be helped by guidelines for selecting analogues. For example, if you consider a set of causal factors that affect a prediction target, some of them will allow easy adjustment (e.g., the analogue has twice the surface area as the target case; therefore, the mean time between failures should be roughly half the rate for the analogue), and others will be hard to adjust (e.g., the analogue is a cargo aircraft, flying a very different type of mission than a bomber as far as the auxiliary power unit is concerned). The optimal strategy is to select analogues that match on causal factors that are hard to adjust, and then to work with those factors that are easiest to adjust.

There are other guidelines for using multiple analogues to obtain convergence and other types of strategies. The major point is that analogical reasoning appears to be a natural form of inference for designers, and there are ways to improve its power and applicability.

There is another way to view a data bank of analogues, and that is as a form of second-hand expertise. The development of expertise is time consuming. If there were ways of compiling case studies from a large number of designers, this could be useful. A model might be the magazine, *Consumer Reports*, which tests various models of different products. Often, these models are no longer available by the time a reader goes shopping for that product. Nonetheless, the case study analyses presented in *Consumer Reports* alerts readers to critical dimensions for evaluation, possible drawbacks, and so on. Simply reading the article may improve expertise by sensitizing readers to causal implications of different features and by helping them learn to make finer discriminations.

A case study format may be more useful for this type of learning than an abstract presentation of information. This is illustrated by typical lists of design guidelines. Often the advice contained in such lists seems too general to be helpful. Yet, when that same advice is embedded in a set of illustrative case studies, these act as concrete comparison cases and make it easier for designers to recognize when they are in a situation calling for guidance.

B. Imagery

We have seen in chess, firefighting, and equipment design that when a person recognizes a likely course of action, the next step is to evaluate it through mental imagery by playing out what will happen, by anticipating the dangers, and by considering what new opportunities might exist. Good design also depends on effective imagery; experience, whether direct or second-hand, must support imagery. Designers working out the implications of new approaches may benefit from support here. Graphics illustrations have certainly been valuable and could be augmented by a pictorial means of representing contingencies. Computer-assisted design as a means of supporting design imagery is an important tool that is becoming increasingly available.

C. Research Findings

We would expect that designers would want basic research information in order to conduct evaluations of proposed options. However, in our study of training device designers, research reports and military standards resolved very few of the design questions, and some of these cases were questionable. For example, one designer was trying to determine the necessary letter size for a CRT display. He had two choices and reported a preference for the larger letter size. He claimed that he found one article that resolved the question in favor of the larger letters. However, further questioning revealed that his article actually reported no performance difference between letter sizes. What he did was search through the fine details of data until he found some trend suggesting that the larger letter size was better. In other words, the data went against his bias, but he kept digging until he found something to support his bias. In a second example, a designer reported locating a study that supported his preferences, and he used it as a justification. These instances suggest a pattern of using preference as a justification. Otherwise, findings are selectively interpreted. If an interpretation does not work, the data are often rejected. An example involved a designer who needed to know how much gravity a pilot could pull before entering visual blackout. A recent technical report had answered this question, but he disagreed with the answer and rejected it. He also determined that the problem with the technical report was in using naive subjects rather than experienced pilots. He worked to have the study redone, and it subsequently yielded the "correct" answer. We found four cases where data were rejected because they did not agree with experience.

For the rest of the cases, the subjects felt that the design of the studies was so different from their own situation that they could not extrapolate the research findings. In our research, this constituted the largest barrier to using research studies—the inability to generalize beyond the parameters selected for a study. Here again, it appears that the problem is in knowing how to extrapolate from a data base, and an approach such as Comparison-Based Prediction might be valuable. Designers would need ways of accessing the appropriate research findings, and then the analogical reasoning structure might enable them to systematically adjust the research data to take into account causal differences between their situation and the research paradigm. In this way, they would be better able to apply research findings to their immediate requirements.

D. Rapid Prototyping

The hands-on, experiential approach taken by designers should make it very attractive for them to use rapid prototyping methods. These methods should allow the designers to quickly get a sense of what they are proposing, and to help them imagine the overall design and how it is affected by the changes they are continually making. In addition, other capabilities that let them develop demonstrations, at whatever level of realism, will fit in with their strategy of concrete and holistic evaluations of their planning.

IV. CONCLUSION

It is likely that there are different types of support for analytical and recognitional skills. By becoming sensitive to both of these skills, we should be in a better position to provide meaningful help during the design process.

The danger I am responding to in this paper is the use of design aids that require system designers to apply formal analytical methods to problems that are best handled through recognitional skills. These design aids can really become design barriers. One such barrier is to force designers to use forms of decision analyses (e.g., Bayesian or multi-attribute utility theory) that are unnatural, time-consuming, and never-ending. Rarely will anyone seek to make a decision based on a decision analysis; therefore, the process may just eat up time and postpone the real decision.

A second barrier is to expect to always use a systems approach to move from goal specification to option generation, and so on. This produces neat progress charts but can miss the real dynamics of ill-defined problem solving. Systems analysis can be an adjunct of the design process, and it can help with the planning, but it does not always present meaningful guidance for actually performing hard design tasks.

A third barrier is to present so much data requiring analysis that the designer's workload becomes too large and the ability to stand back to review the recognitional perspective is diminished.

The optimal design support would support both analytical and recognitional processes so that the designer is in the appropriate mode as often as possible. For example, if I am a novice photographer, I will need help in focusing and composing pictures. An ideal camera might have automatic focusing to take that analytical burden from me. It might also have a built-in light meter to help me recognize when the overall light level is too low or high to make an effective picture. It would also have an automatic lock to prevent me from taking the picture if the shutter speed is too fast, the aperture setting is too small, or the ambient light level is too low. Similarly, an effective design aid will help designers maintain and possibly increase their recognitional sensitivity while supporting the selective and powerful use of analytical methods.

ACKNOWLEDGMENTS

The preparation of this paper was supported by Contract No. F33615-82-C-0513 1054-10B with MacAulay-Brown, Inc. I wish to acknowledge the helpful comments and suggestions provided by Kenneth Boff, Roberta Calderwood, Helen Klein, and Beth Crandall for earlier drafts. Responsibility for any remaining flaws in logic remain with the author. I also want to thank Christopher Brezovic for his help in preparing and reviewing sections of this paper.

REFERENCES

deGroot, A.D. (1978). *Thought and choice in chess* (2nd ed.). New York: Mouton.

Klein, G.A., & Brezovic, C.P. (1986). Design engineers and the design process: Decision strategies and human factors literature. *Proceedings of the 30th Annual Human Factors Society Conference.* Dayton, OH.

Klein, G.A., Calderwood, R., & Clinton-Cirocco, A. (1986). Rapid decision making on the fire ground. *Proceedings of the 30th Annual Human Factors Society Conference.* Dayton, OH.

Klein, G.A., John, P.G., Perez, R.S., & Mirabella, A. (1986, August). *Comparison-based prediction of cost and effectiveness of training devices: A guide.* Prepared for the Army Research Institute.

Klein, G.A., & Weitzenfeld, J. (1979). Improvement of skills for solving ill-defined problems. *Educational Psychologist, 13,* 31-41.

Weitzenfeld, J. (1984). Valid reasoning by analogy: Technological reasoning. *Philosophy of Science, 51,* 137-149.

Chapter 14

UNIFIED LIFE CYCLE ENGINEERING

Bernard Kulp and Anthony Coppola

Air Force Systems Command
Andrews Air Force Base, DC

ABSTRACT

Unified Life Cycle Engineering (ULCE) is a new computer-based technology application to impact design engineering. ULCE unifies, at the computer-assisted engineering-design workstation, not only the relevant data bases and models for product performance and manufacturing, but also those for product reliability and maintainability (R&M). Thus, with ULCE available, the design engineer will be able to evaluate the tradeoffs between the complex and often conflicting demands of performance, manufacturing, and support for the products he is designing. Eventually, other data bases and models will be added for additional "ilities" like those for personnel availability and trainability. The ULCE development program is designed, for the next eight years, to develop and integrate the reliability and maintainability data bases and models, to evaluate the integration, and to provide two independent technology demonstrations.

I. DESCRIPTION OF THE TECHNOLOGY

A. Introduction

Unified Life Cycle Engineering (ULCE) is a new computer-based technology, the development and application of which will provide revolutionary changes in the design engineering process. ULCE unifies, at the computer-assisted engineering-design workstation, not only the relevant data bases and models for product performance and manufacturing, but also those for product reliability and maintainability, as well as, eventually, data bases and models for personnel availability and training. This type of support will enable the design engineer to evaluate the tradeoffs between the complex and often conflicting demands of performance, manufacturing, and support for the products he is designing. ULCE will thus provide the capability of meeting the 1984 challenge from the Secretary of the Air Force and the Air Force Chief of Staff: "For too long the reliability and maintainability of our weapon systems have been secondary considerations in the acquisition process. It is time to change this practice and make reliability and maintainability primary considerations."

Current engineering practice for the design of a component such as a jet engine disk almost always involves a serial approach. The disk is first designed for maximum performance (e.g., minimum weight). The design is then sent to the manufacturing department for definition of the appropriate production approaches; the manufacturing department may request some redesign to facilitate manufacturing. Finally, some of the product support (or "ilities") considerations such as reliability and maintainability and repair receive attention, but by then it is often too late to make any substantial change or impact on the design.

A significant development of the past decade has been in Computer-Aided Design (CAD): the ability to use computers not only for drafting, but also for rapidly and effectively analyzing the performance of many design options. Progress in this "performance" domain has been rapid and is continuing. Similarly, significant progress, although at an earlier stage of development, is being made in the "manufacture" domain through Computer-Aided Manufacturing (CAM). Other models are being developed for the combination of unit operations into manufacturing centers and, as in Integrated Computer-Aided Manufacturing (ICAM), for an entire factory.

Modeling in the "supportability" domain (which includes "ilities" such as reliability, maintainability, inspectability, accessibility, repairability, and even personnel availability and trainability) is in its infancy, but its potential is indicated by earlier Computer-Aided Support (CAS) work that permitted prediction of the probability of detection of flaws. CAS modeling permitted the evaluation of inspectability as a function of design details such as the specifics of the disk geometry and size of the flaw.

Given tools such as these, the designer can consider performance, manufacture, and supportability essentially concurrently rather than serially. A large number of design options can be rapidly and accurately evaluated from many points of view. For example, the geometry of the disk might be changed, sacrificing a lower weight to increase inspectability or to reduce the cost of manufacturing. Thus, the goal of effectively considering the "ilities" during initial design can be approached.

The above simplified example has served to illustrate the concept of ULCE. However, development of ULCE is not without its problems and challenges: 1) analytical models may not be available or appropriate; 2) the input information may be in the form of qualitative models such as a computer-generated mockup, graphical or tabular data, or various forms of symbolic data such as lists of "do's and don'ts"; and 3) the growing ability to develop fast numeric and symbolic computation techniques is taken for granted, but many of these techniques will require special efforts and vectoring to wed them to compatible methods for representing the diverse types of data needed to simulate the phenomena representing the performance, manufacturing, and supportability domains. Moreover, the problems of integrating the three domains have received very little attention.

On the other hand, computer-aided engineering workstation utilization is growing at a rate of 30% to 50% per year, and one industry estimate is that by 1995, 80% of all designs will be performed on workstations. Indeed, very large-scale integrated circuitry is presently designed exclusively on workstations. The principal opportunity provided by such automated design is the creation of analysis modules that will automatically evaluate the impact of design options while the design is still fluid. Once such modules are in place, it will be no great effort to account for the "ilities." Further development of the ULCE concept can provide a new paradigm for engineering in which the nature of the design process will change and expand to provide a capability to simulate and examine many more options and to evaluate all the impacts of these options. This will both permit and guarantee, through a semiautomated integration discipline, effective attention to the "ilities" and enhanced management visibility of design impacts on supportability.

B. Needed Technologies

The development of models and other data bases for the performance and manufacturing domains is receiving considerable attention. Work in the performance arena is well advanced and making rapid progress; work in the manufacturing arena has made a good start. They will not be discussed further, except regarding the heterogeneous nature of the data base characteristics of the different domains and the difficulties this heterogeneity poses for the integration problem discussed below. Thus, for manufacturing and supportability, it will be desirable to find a unifying structure and representation approach that can represent broad ranges of manufacturing information in a tractable, efficient manner, and that can permit them to be fed back to design. The development of quantitative and qualitative models and data bases appropriate to supportability is in its infancy and must receive accelerated and intensive attention. A partial list of research topics is given in Table 1.

The testability allocation module would apportion testability requirements among the system components. The test point location module would determine where test points must be placed and the Built-In Test (BIT) selection module would determine the best BIT designs for the points. The Automatic Test

Table 1. Supportability.

Life Prediction
Durability
Repair
Maintenance Manpower
Testability Evaluation
Design Rules Checking
BIT Selection
Test Point Selection

Pattern Generator (ATPG) module provides the test vectors for automatic test equipment. The analysis module evaluates the testability inherent in the design throughout the process. Each of these modules is useful in itself, and the aggregate assures a testable system. To ensure a reliable, maintainable, and testable system, it would be necessary to include all the modules listed in Table 2. The logistics community could probably add several of their concerns to this table.

Deterministic life-prediction functions for specific hardware components or subsystems, if they can be developed, can be extremely valuable tools. They will be of the following form:

Table 2. CAD Reliability, Maintainability, and Testability Modules.

Reliability Prediction	FMEA
Reliability Allocation	Maintainability Prediction
Thermal Analysis	Maintainability Allocation
Electrical Analysis	Accessibility Evaluation
Mechanical Analysis	Testability Evaluation
Sneak Circuit Analysis	Testability Allocation
Design Rules Checker	BIT Selection
o Derating	Test Point Selection
o Parts Selection	ATPG
o Testability	Test Sequencing

Unified Life Cycle Engineering 191

> Life equals a function of definable characteristics of the service conditions and environment, and of measurable characteristics of the state of the hardware.

Definition of such functions for structural components has provided the basis of damage-tolerant design and for rational maintenance strategies such as Retirement for Cause. In the latter, non-destructive inspection of hardware, together with service stresses, provides the input to the life-prediction function from which the decision can be made to retire or to continue the use of a *specific* piece of equipment. The Air Force Wright Aeronautical Laboratories (AFWAL) are attempting to develop such functions for electronic equipment and to determine if they are generalizable over enough cases for practical use.

Models of inspectability can be derived from analysis of the interaction of the inspection probe with different flaws. For example, adequately accurate analysis of the manner in which a flaw scatters a defined ultrasonic energy wave, and the manner in which component geometry also deflects and scatters it, provides the basis for the predictions. Such quantitative modeling of nondestructive inspection has been shown to be feasible and should be accelerated.

The development of expert systems to capture, organize, and represent the experience and lore of field maintenance and repair people, followed by the first steps of infusing it into design, offers much potential. Corrosion prevention, where the problem often involves adequate care in the design and use of appropriate protection (rather than a need for improved materials or protective processes), would be a typical example.

A universal manpower, personnel, and training requirements model, automatically derivable from contractor-generated task-analytic Logistics Support Analysis Requirements (LSAR) data, or perhaps eventually directly from design concepts or details, can be envisaged. It would be a powerful tool to evaluate the manpower implications of different design options, as well as facilitating the acquisition and training of the necessary operators and support personnel.

Since modeling or other information from the supportability domain will have its greatest impact when integrated into design or manufacturing as early as possible in the development cycle, it is considered essential to develop it as part of an overall ULCE program rather than independently. The integration of inputs from the three domains of performance, manufacture, and supportability will require pioneering fundamental research. A partial list of areas in which research on ULCE-oriented or ULCE-unique problems are important (as distinct from general advances in computer or information sciences) is given in Table 3.

Executive expert systems will be required to provide an intelligent design environment to support the creative capabilities of the human designer. They must prompt access to appropriate data bases and guide optimization approaches among the multitude of diverse inputs from the three domains. They should also

Table 3. Integration.

Executive Expert Systems and Meta-Models
Optimization Algorithms
Expert Computing
Heterogeneous Computer and Data Base Interfacing
Data Base Structures and Representational Approaches
Expert Systems Including Heuristic Capabilities
Sensor Fusion
Solid Geometric Modeling
Integration of Numeric Computation with Symbolic Processing

provide the framework for a disciplined approach to ensure that no significant "ilities" have been overlooked. The development of the human-computer interfaces for such systems will require an understanding of the creative design process and may even involve heuristics, in the sense of providing guidance in terms of an individual's perceived experience base.

Extensive research will be needed to define the data characteristics necessary to integrate the unique aspects of the design and manufacturing domains, to say nothing of the supportability domain, within a distributed data base environment composed of heterogeneous computers. New approaches, other than current brute force methods, for the extraction of knowledge and the development of expert systems for the manufacturing and supportability domains will also be required. Geometric modeling (beyond wireframe representations) that will provide a complete and unambiguous representation of the geometry in three dimensions still requires attention. If this can be done in an object-oriented manner such that the geometric features used imply appropriate manufacturing processes, it will provide a very useful tool for design-manufacturing integration. Such feature modeling could provide a library of three dimensional objects that could be manipulated and combined to form a complete three-dimensional model of a part. Each object making up the part model would carry with it, invisible to the designer, all associated knowledge regarding the feature. By using available knowledge in considering various tradeoffs regarding performance, quality, cost, etc., the best design possible will result. An additional benefit to such an approach is if the design representation is complete, it can be provided to manufacturing systems for complete automated manufacture because all the required information/knowledge is, by definition, associated with the design as it was originally used.

Unified Life Cycle Engineering

The integration or fusing of knowledge-based/symbolic processing with numeric computation will require extensive effort. For example, when solving a problem using both symbolic and numeric processing and inputs in either shared or redundant strategies, how do we integrate results, how do we operate on combined results? The demands of ULCE imply the need for more powerful and robust optimization methods encompassing the much broader requirements involved in optimal design for performance as well as for manufacturing, reliability, maintainability, etc. Although a human designer can effectively optimize in one design parameter relative to an initial design point, he quickly saturates as the number of design parameters increase. This applies to most fields of modern engineering, where powerful tools for analysis have been developed, but the optimization aspects of the analysis-design loop remain primitive. Only recently have techniques been developed that can realistically handle such a problem. An important issue for the success of the ULCE program is the coupling of these optimization strategies to Air Force design application areas.

The integration capability may generate results that are difficult to compare. For example, one design may have extremely high reliability and another extremely low weight. If each one is satisfactory, which is better? Hence, system level meta-rules will have to be developed; effective interaction with the designer is essential.

Large cooperative government laboratory, university, and industry efforts will doubtless be required for the full development of the ULCE technologies. Otherwise, the fragmentary efforts will continue, each providing a valuable service to those who can use it, but with increasing difficulty in integrating the programs created by separate disciplines. Consortia that provide multidisciplinary centers of activity can be used to provide the extensive science and technology base that will be required. The first three to five years should be devoted to expanding the fundamental science and technology base, but the efforts should be directed toward carefully selected "windows" that will allow generic advances while ensuring relevance. Each "window" must be a specific, realistic product that someone will actually use when the ULCE process completes its design. Each must provide a focus for interdisciplinary cooperation and an "integrating emotional issue" to overcome institutional barriers. Each must require "adequately eloquent" theoretical analysis and generate early user interest and support. Potential windows include an electronic subsystem, an advanced composite structure, or a metallic structural assembly for corrosive environments.

C. Current Research

Although activities in the individual domains of performance, manufacturing, and supportability are growing, the integration of these activities beyond the transfer of CAD to CAM is in its infancy. Extensive contact with major universities such as MIT indicates little effort, probably much less than $1 million per year throughout the country.

A microcosm of the ULCE concept exists in the reliability and maintainability community. Under the coordination of the working panel of Reliability and Maintainability in Computer-Aided Design (RAMCAD was formed by the Joint Policy Coordinating Group on Logistics Research, Development, Test, and Evaluation), efforts in all three services were identified, and a plan for a unified approach was developed at a workshop with the aid of the American Defense Preparedness Association and the National Security Industrial Association. The following are the recommendations resulting from the workshop:

1. Ongoing RAMCAD efforts should continue with emphasis on integrating the existing elements.
2. A strong focal point should be established.
3. A major research effort is required.
4. The formal requirements for RAMCAD systems of the future, including interactions with systems engineering and the support community, must be defined.

Furthermore, the panel proposed establishing a national RAMCAD consortium of government, industry, and academia to provide consolidation and transition of RAMCAD products.

Industry automation efforts have largely been confined to their own proprietary systems. However, the need for communication among different vendors' products has spurred some standardization. Electronic Design Interchange Format (EDIF) is a standard format for the exchange of design data for electronic circuitry. It was created by a committee sponsored by Daisy System Corporation, Mentor Graphics Corporation, Motorola, Inc., National Semiconductor Corporation, Tektronix, Inc., and Texas Instruments, Inc. Using EDIF, a designer can transfer information to other manufacturing facilities despite differences in the CAD/CAE tools.

Under the leadership of General Motors, significant activity has been undertaken by industry to provide communication links between different computers and networks. These efforts, triggered by industry economic considerations, will be of immense benefit to ULCE, but they currently address only the CAD to CAM flow, with little attempt to provide feedback to CAD from CAM and CAS.

In addition, in September 1985, the Deputy Secretary of Defense, William H. Taft IV, organized a steering group headed by Russell R. Shorey to oversee implementations of a program called Computer-Aided Logistics Support. Its objectives are to accelerate both the integration of reliability and maintainability design tools into contractor workstations and the automation of contractor processes for creating logistic information products, and to rapidly increase the

Unified Life Cycle Engineering 195

military capability of using logistical data in digital form. The intent is to accomplish these ends for systems entering production in FY 1990 and beyond.

D. Feasibility Demonstrations

The value of individual analysis modules has been widely demonstrated. For example, Ground-Launched Cruise Missile (GLCM) maintainability of the power generator was improved by a graphics program used by General Dynamics under sponsorship of the Air Force Human Resources Laboratory. Sneak circuit analysis, a Boeing proprietary program, is widely used to find hidden errors in both hardware and software. Testability analysis making use of new models has cut maintenance time of one electronic countermeasure system by 75%. The Retirement for Cause Program for engine components is expected to yield over $500 million in savings for the F100 engine by the year 2000. Thus, individual analysis modules have demonstrated their ability to provide valuable results, and the number of available modules is growing rapidly.

Industry efforts have evidenced the ability to provide communication between different machines to tie together design and manufacturing and, by implication, different design modules. The computer revolution continues unabated, providing a tenfold improvement in computing power every eight years. The integration problem is large and complex, but no fundamental limits are evident. All that remains is to get on with the job.

II. APPLICATION OF THE TECHNOLOGY

A. Range of Application

The ULCE technology can be applied to any hardware development. Examples have been given of analysis modules for mechanical and electronic applications, and for manufacturability, producibility, and testability. Any design, production, or support consideration that can be reduced to a model can be handled. Qualitative models (e.g., CADAM) for such considerations as space for maintenance activities can be used as well as quantitative tradeoff models. The range of application is limited only by the effort put in to develop and integrate models.

B. Examples

A designer of an engine disk would have the programs illustrated under the CAD, CAM, and CAS labels resident on or available to his workstation. Guided by an expert controller and fusing all the information, this workstation would interact with the designer to generate and evaluate design options until one satisfactory in all respects is produced. On acceptance by the operator, the design data would be read into all relevant data bases, thus permitting the automated generation of drawings and parts lists, the programming of automated machinery for milling the part, the computer-aided generation of technical orders and logistics data, and the automated creation of the inspection test requirements. Should a design change be made later, all data bases would be updated to keep all those involved working in unison.

As an electronic example of a ULCE application, consider a designer creating a printed circuit board. His workstation would generate design options and evaluate their impacts on the incorporated models. The computer would find wire routing options and determine the impact of these on manufacturability. The thermal profile of each design option would be computed, and its impact on reliability predicted. Test points would be located, permitting rapid and unambiguous fault location, and test requirements would be generated. Stresses on the board would be computed for different component layouts and different environments such as vibration. Sneak circuit and failure modes and effects analyses would be performed for each option. Design rules, such as part selection and derating, would be enforced and deviations flagged. Upon completion, the designer would have the design of a board that meets his performance requirements; is also easy to manufacture; and is reliable, testable, and durable. The design data would be forwarded to data bases for maintenance and logistics, and technical orders would be printed out automatically. The data would also go to manufacturing where automated production machinery would use it to create the boards.

C. Resulting Capability

Using the ULCE process, the designer will have all the tools needed to create a fully satisfactory design in one operation. At present, defective designs are produced because considerations outside the designers experience are ignored. An iterative process of review of specialists in various disciplines is used to identify these defects, which must then be corrected by redesign actions. The process is cumbersome, at best. The time required to redesign is alone a problem sufficient to discourage corrective action after a flaw is discovered. The overhead required for the design reviews and the myriad of considerations affected by design changes guarantees that some aspects will be overlooked or ignored. As a result, a fully satisfactory design for any reasonably complex system would have to be considered a random success.

Though front-end consideration of manufacturability and supportability is mandated by high-level direction throughout the Air Force, the complexity of the development process has generally precluded its successful accomplishment. To be successful, it must be embodied in technology, not merely mandated by managerial direction. The enabling technology is provided by ULCE. Because of its automated analysis capability, ULCE will cut design time by at least 20%, even with all considerations actually included in the process. The elimination of manual analysis, feedback, and reiteration of design will reduce costs even more, perhaps by 30% or more over current practice, and without the current omissions of supportability considerations. Further, as a result of doing it right the first time, there will be substantial savings in logistics costs and significantly improved mission-capable rates. Systems will be down less often and for shorter durations. The spin-off to the private sector will be revolutionary.

III. PROGRAM PLAN

The ULCE research and exploratory-development program will include eight years of technology-base development and exploratory model integration evaluation. The development and evaluation will be performed by two multidisciplinary centers of activity, each associated with a university-industry consortium to speed transition of both individual and integrated products. The centers will develop analytical modules, integrating programs, fusing algorithms, data base management schemes, and integration methods. The selection of windows to guide this work will be an important near-term activity. By the third year, the level of effort will be 100 man-years per year for each center.

The ULCE advanced development program will include two parallel demonstrations, one by each center, applying ULCE to an electronic subsystem and a mechanical assembly of a weapons system at 200 man-years each and performed in the sixth through eighth years of the exploratory-development effort. At the end of the demonstration, ULCE should be proven and because of the joint industry-academic involvement, it should already be transitioning into broad use.

Chapter 15

INFORMATION TECHNOLOGY AND OTHER FACTORS IN SYSTEM DESIGN

L. W. Lassiter

Lockheed-Georgia Company
Marietta, Georgia

ABSTRACT

The rapidly developing field of information technology provides society with many opportunities to understand better the nature of interrelated processes. A cornerstone of this technology is automation which is, itself, composed of many constituent technologies. This chapter deals with how automation is impacting system design as well as some of the aspects of the organizational and management functions. The characteristics and importance of integrating the islands of automation, data base considerations, expert systems, data base management, future development, and other system design factors are discussed with supporting examples.

I. INTRODUCTION

Today's computer resources are directed toward solving specific problems. Thus, technologies have been built around the capabilities of computers to provide information to individuals as rapidly as they can use it. In the technical world, these technologies have primarily taken the form of CAD, CAM, CAE, and CIM, among others. Each acronym means different things to different people. However, other than for the convenience of discussions, the definition of these terms means very little, per se. The important point to note is that they represent vast potentials for rethinking and redoing the design process.

Traditionally, each designer's technique is, to a large extent, used in isolation from others. The approach to laying out the loft lines for an airplane, for example, may differ markedly from designer to designer. Therefore, the challenge to advanced concepts in automation is to provide a set of specified tools to each and every user which allow a variety of means for their utilization.

Engineers collect and utilize data as a source of information in different forms, in different quantities, and by different techniques. The data is typically input to a computer system, stored, and retrieved in a logical and consistent form. It is quite possible for designers to generate data more rapidly than it can be effectively used. That is the paradoxical nature of automation. That is, automation should be utilized as a controller of data and its associated information, yet it can be the source of the data/information explosion which exacerbates the control problem. In effect, it is a feedback problem.

Thus, the tools of automation have moved beyond our ability to use them efficiently. The 32-bit CPU, basic drawing software, high resolution displays, color, and a degree of computer-independent standards are here today and offer sufficient interactive power to serve the "computational" needs of 95% of our engineering design problems. What, then, are the problems that remain unsolved? What information should be readily accessible to the designer? What kind of support should be provided? These questions and others arise and can be addressed in light of both the existing computational tools and the emerging concepts of integration.

II. INTEGRATION

Figure 1 shows a serial flow of data from engineering through manufacturing to product support and then to operation and utilization. There is one additional feature shown in this figure that is not standard in today's systems, although it is very important in the total product life cycle. This is the feedback of requirements from the downstream functions to the upstream engineering design functions. With this kind of information, the designer can extend the basic engineering functional requirements to incorporate both producibility and supportability, and life cycle costs can be more readily determined as a function of the many life cycle functions.

While much automation in design has occurred during the past decade, it has occurred primarily as isolated activities within functions. That is, to a large extent we have what is called "islands of automation." To achieve significant productivity improvements, integration of activities within and across functional areas must now occur.

The need for more complete integration is based on such considerations as:

1. Improved product quality through error reduction.

Information Technology and Other Factors

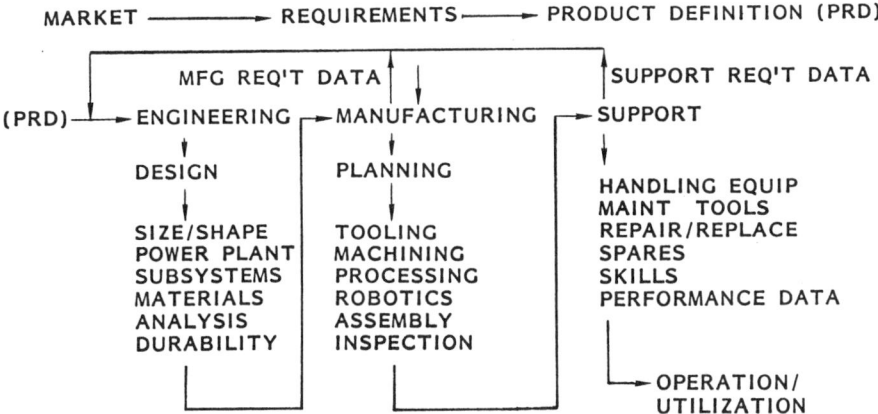

Figure 1. Today's Advanced Integration Processes Provide the Means for Making Vital Downstream Data More Accessible to Designers.

2. Prototype simulations and evaluations prior to manufacturing and support.
3. Shorter design time to meet customer requests.
4. Evaluation of more alternatives.
5. Better control of materials and other resource requirements.
6. Improved basis for tracking manpower and project activity.
7. Better comprehension of the nature of design in the context of product options and their impact on downstream functions.

These and other desirable features require a major commitment of personnel, equipment, and financial resources. Thus, the many benefits that should be achieved must be viewed with respect to the state-of-the-art in CAD/CAM/CAE/CIM, to the corresponding commitment of resources, and to the task priorities. From the designer's point of view, the desired objective is to support the designer in a fashion such that he can use the hardware/software facilities with a minimum of special training requirements. That is, the system should be easy to use through the user operating functions, and the necessity to become trained in the support functions of programming, geometric modeling, networking, data base management, etc. should be minimized. The goal is to provide any information the designer might need without having to know the source or the means of information retrieval.

Figure 2 shows the generic connectivity from the basic workstation level to the complete company activity. An engineering workstation can be configured to operate independently of other stations and/or computers through one or more networking schemes provided by vendors. The objective is to provide the workstation operator with information quickly and accurately without his being concerned with its source. Similarly, he can input information at the

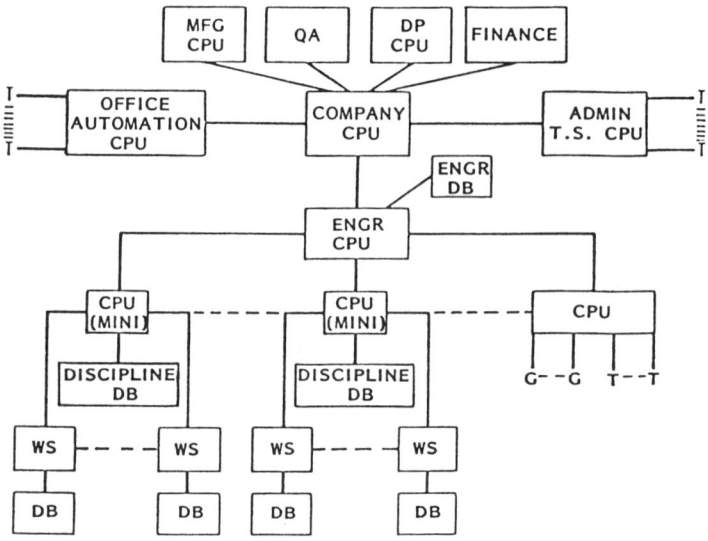

Figure 2. A Comprehensive Network that Provides an Effective Means of Communicating Information Throughout the Operation, Regardless of Diverse Specialty Hardware and Software, is an Ideal Target.

workstation which becomes available to other parties. The system should provide the means to accomplish the objectives through data base management, networking, and communication facilities. The total integration, as represented by Figure 2, is an idealized target which is still in the earliest stages of development at the more progressive companies.

The common integration element throughout the entire operation is data. The output of one process becomes the input for other processes. Thus, the key activity for achieving true integration is the efficient management of data and the flow of data. There are many issues then surrounding data base design and data base management systems which must be addressed.

At each organizational level a data base may be developed, as opposed to a single company data base that might contain everything. This proposed networked approach for data bases as well as for workstations is viewed as a mechanism for providing an efficient means of data storage, data processing, and data retrieval.

III. DATA BASE CONSIDERATIONS

The development of effective data bases and the efficient use of data base management systems are still in their infancy. Although they are needed to support designers' needs, their specific characteristics defy definition in a rigorous sense. Much research and development is required to understand how the varied needs of designers can best be satisfied and what characteristics are needed by the data base management system. The designer needs to be able to define those parameters that should be maintained within his personal data base, as well as what should be maintained at higher levels (for example, at the engineering discipline level). In this way, we are asking the designer to be much more introspective than he has ever been in the past. Thus, it will probably take several iterations and a considerable design effort before a complete understanding of what should be contained within the various data bases is developed. There are certainly a number of functional capabilities and data that are characteristic of design and should be made easily accessible to the designer. They include computational aids, report formatting, plot formatting, historical data, translators, tutorial messages, help instructions, applications programs, and basic geometric modeling functions, as well as a massive amount of analytical and geometric data. The form and nature of these characteristics have only in recent years taken on major importance, and that is due to the great increase in both computational power and computer memory at substantially lower costs. There is a good deal to be learned about what should be specifically provided and to whom, with due respect to the complexity of the human in the automation loop.

Particularly important, but unresolved at this time, is what the general geometric tools for design should be. Much has been written on the subject, and both industry and vendor investments in geometric modeling R&D have been huge. The reason for this is the enormous impact on the product life cycle that is projected. Visionaries are well aware that control of product design and the simulated "movement" of the computerized definition throughout the life cycle represent a component of a new industrial revolution. Concepts of wire frame models, solid modeling, and surface modeling each have their niche, but no technique is efficient for all classes of problems. However, basic geometric modeling can be useful today. Such basics can be built around the design input of points and tangencies (such as on a manual drawing) to produce certain classic shapes—i.e., conics, polynomials, splines, lines, arcs, etc.

IV. EXPERT SYSTEMS

As experience is gained, the nature of design and its relationship to other factors such as geometric modeling will become better understood. The best modeling "tools" and their efficient, easy-to-use capabilities will become increasingly apparent. When this knowledge is captured and put into methods for interactive design, a major step will have been taken toward the objective of freeing the designer to concentrate on higher-level cognitive functions while relegating computerized design to handle the lower-level activities. Industry must seek computer vendor products that approximate their needs. However, it must be recognized that vendor software may have to be modified and/or in-house software may be required to meet projected automation objectives.

A good example of knowledge-based design is the Graphics for Advanced Design Engineering (GRADE) function at Lockheed, which is represented in Figure 3. In Figure 3, a wire frame representation of a fuselage section is shown along with construction lines and tangent lines. The designer works in two-dimensional projection planes at the workstation display, just as he would do on a drafting board. When points and tangencies are specified at designer-defined critical locations, the program automatically draws the appropriate curves and automatically develops the corresponding mathematical equations. Based on knowledge of the curve shapes that traditionally represent the transport aircraft class of shapes, such computations can be preprogrammed into the design program. The designer then needs only to inspect the resulting curve (or curves) to verify that they meet the desired result based on designer experience and expertise. If they do not, easy-to-control input modifications can be used for shape adjustments. When the desired frame is created and viewed from any desired vantage point, the program automatically "fills" in the surface between the frame boundaries. The surface-defining process is stored within the computer to provide quick and easy section views, areas, volumes, and other

Figure 3. The Automatic Development of Descriptive Equations Requires the Specification of Algorithms Based on Expert Knowledge of Product Shape Characteristics.

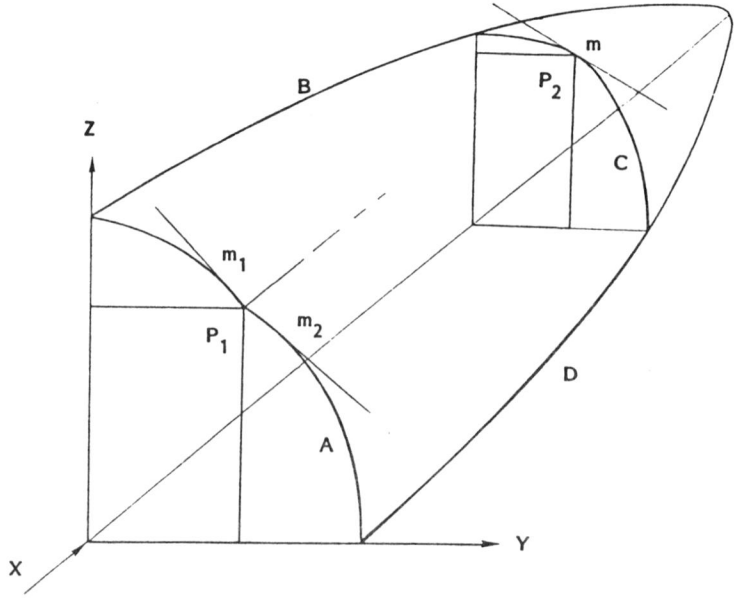

Information Technology and Other Factors   205

output which might be used for further downstream design, analysis, and manufacture.

How is the surface automatically created? It is a knowledge-based expert system which builds upon the knowledge of shape characteristics, shape ratios, point and tangency transitions, etc. and upon the analyst's knowledge of how such characteristics translate into mathematical models. This knowledge is programmed into the system and provides electronic design by the designer input of data, much as it would be done on a conventional drafting board. The captured "intelligence" and character of the design, however, provides so much more information content and availability than does the drawing, per se. Thus, a good designer can layout an entire airplane of the Lockheed transport type in two hours instead of two weeks, and has the great added value of the computerized definition—much as if the plane itself resides within the data base.

The typical use of this type of preliminary design is for aircraft sizing and for structural analysis programs. Output from GRADE and input to such programs must be carefully planned by experienced designers. Such application interfacing is one function that must be painstakingly planned by the experts. Much detail regarding parameters must be provided. Whether parameters should be treated as input constants, as variables, or as preprogrammed constants are typical of the input decisions that must be made. No system can automatically make such decisions unless the decision-making process is, itself, specified in the appropriate context of the design process.

V. DATA BASE MANAGEMENT

Like so many supporting technologies, the efficient management of data within the computer is a process that is not well understood. In economics, even the experts seem to understand only a small portion of all the considerations. There is an analogous problem in engineering. In fact, it is even difficult just to list all design-affecting parameters that should be considered. In spite of this, progress is being made.

To use existing DBMS systems requires a certain commitment and a certain human discipline. Thus, the person who generates data is obliged to conform to the DBMS input requirements initially requiring extra designer time. The benefits of this will derive from improved interrogation capabilities of the data base, making it easier for programmers to work with their applications. They may have to know what data and how much data should be stored in the data base, but the internal formatting would be a function of the DBMS, thus relieving applications programmers of many time-consuming, error-prone tasks. Of primary importance is the fact that programmed interfaces will not have to be created between each pair of applications for which data will be passed.

Thus, in the earlier discussion of GRADE, it would not be quite so important for the basic design system to have specific programmed interfaces to structures, to aircraft sizing, to the weights group, to lofting, etc. In fact, it is not always possible to anticipate all future needs of a particular application such as GRADE. Therefore, it may take extra time and effort to ensure that the airplane definition, provided by GRADE, will be stored with sufficient detail and documentation such that any using discipline may create its own translators (interfaces) from GRADE to the discipline. This philosophy should be followed for each application from which data will be extracted for other functions. It is fundamentally easier for each user of GRADE data, for example, to specify the discipline interface requirements than it is for GRADE to provide interfaces to all users—especially since all users may not be identified at the outset.

However, there are some negative aspects as well. Most DBMS systems require a lot of memory and computer processing time for storage, search, and retrieval. There is, of course, some cost for DBMS software, but that cost is minimal with respect to the functional pros and cons of DBMS systems. With expanding memories, faster processors, and more efficient DBMS software, the tradeoff balance from today's capabilities will shift and lessen the negative factors in the future.

As in geometric modeling, the payoff seems well worth the cost. There does not seem to be any other practical approach than to seek improved DBMS standards and procedures. Direct interfacing, as an alternative between all users, is virtually impossible.

VI. FUTURE DEVELOPMENT

There are many other technically complex areas that bear on the integration process, such as the varied modes of communication, artificial intelligence, etc. It should be recognized, however, that the direction of such developments depends somewhat upon the understanding of just what we want to do. A lot is known about the design function, but much of design is an art. Therefore, when developing design support systems, considerable attention must be given to the complex combination of art and science. Such developments are further complicated by the fact that design and designer needs are continuously shifting and many functions will change from an art form to a science as our use and understanding of design support systems matures. The human element is a major mystery in the evolution of automation systems. Nonetheless, the human interface is the primary consideration in defining synergistic systems for today and for the future.

VII. OTHER SYSTEM DESIGN FACTORS

In addition to the usual physical equipment and electronic support systems, the designer's work environment consists of numerous practices, procedures, and

Information Technology and Other Factors   207

policies that directly affect the design process. Of the many factors involved, it is instructive to examine six which, in my opinion, have a direct influence on the creative output of the designer. These are:

o   Requirements
o   Specifications
o   Specialties
o   Controls
o   Organizational Structure
o   Management

These factors take on special impact and significance when the primary customer is the Federal Government, and the Department of Defense in particular. Since most of my experience has been in this type of monopolistic procurement environment, my comments are primarily confined to the designer in this context.

A. **Requirements**

This is the customer's statement of desired product characteristics, principally performance related. Requirements can be overstated to limit the solution to one that is already known or one that is favored by the particular group generating the requirements. As an example of the latter, we recently received a Request for Proposal (RFP) from one of the Services for an aircraft to perform a mission which one of our Lockheed airplanes currently does very cost-effectively. Although there had been no substantial change in the mission requirements, the RFP called for speed and attitude performance that is beyond the limits of our airplane. We concluded that certain parties in the Service were asking for high performance in order to bring a "new" airplane into the inventory and not because it was justified by the mission.

More recently there has been a strong movement, particularly within Government circles, to expand requirements to include contractor guarantees in such areas as maintainability, reliability, and availability. The result is that the designer is less likely to seek innovative approaches and is more prone to apply conventional ones for which there is a forecasting data base. For example, electric primary flight control systems have not gained acceptance even though there are large potential benefits in the areas of reliability, maintainability, power consumption, and life-cycle costs. The primary reason for this is that there are insufficient data for assessing the financial risk that management must assume to meet a contractually guaranteed requirement.

B. **Specifications**

Specifications are not only a further statement of desired product characteristics, but include program procedures imposed in an attempt to control product quality. Over the years, the DoD has embedded its historical experience and lessons learned into its body of regulations, specifications, and standards. These data were very useful in the days of mutual trust since justified deviations and waivers did not invoke the current notion of fraud, waste, and abuse. Because of the recent era of criticism for the DoD and the aerospace industry, these regulations, specifications, and standards have become in themselves inflexible documents which demand compliance under the penalty of being nonresponsive to RFPs or nonconforming to contract requirements. Because of embedded design solutions, they serve to dictate answers and discourage innovation. There is, I might add, a developing belief in the DoD that this situation needs relief.

C. Specialties

In recent years, the galloping complexities of technology have required a perceived, substantial amount of specialization leading to the question of who is the designer. The answer all too often is the Project Designer plus the Preliminary Designer, Reliability, Maintainability, Availability, Safety, Human Factors, Design Assurance, Preservation and Packaging, Materials and Standards, Structural, Aerodynamics, *ad infinitum*. A dozen personnel out of the engineering organization will leap forward to claim the title or at least share in the title of designer, and to their chagrin, find themselves surrounded by manufacturing engineers, industrial engineers, quality assurance engineers, *et cetera*.

In many instances, these specialists have preconceived opinions and/or contributions based on analytical methodology and prior experience that tend to limit creativeness. When very complex computer programs are used to assess design alternatives, the analyst may lack a level of understanding of the program that will permit recognition of its limitations when applied to innovative solutions. When this occurs, his assessment is likely to be in error. Even when these limitations are recognized, he will often still resist an innovative solution. This may be because he lacks the applicable tools for assessment or because of pressures from his functional organization which disdains risk associated with uncertainty in analytical methodology.

D. Controls

Most government procurement programs are structured around a cost, schedule and performance value system. Indeed, the inflexibility in these programs is so great that it tends to rule out the more daring concepts which typically entail uncertainty and risk. Schedule, in particular, has a significant influence on innovative solutions since it limits the time available to derive and evaluate alternatives. The result is that it is both easier and safer for the designer to

Information Technology and Other Factors 209

propose an established solution rather than risk pursuing a new one that might require a longer time span to achieve.

E. Organizational Structure

A company's organizational structure is created primarily to establish management authority and to provide for coordination and communication among the various activities that are relevant to the product.

Taking an historical approach to engineering organizations, one observes that most of them began with a small cadre of dedicated engineers who performed their work within an informal structure that permitted maximum coordination and a large degree of flexibility. If the group's output met with "success" in the marketplace, there followed the inevitable requirement for additional personnel and growth of the enterprise. Characteristically, this growth takes place in two ways. First, as the number of specialized support functions increases, there is a horizontal expansion of the organization. Second, this horizontal growth brings about the need for additional managerial and administrative groups and this, in turn, leads to a vertical expansion.

Both of these growth patterns have had a significant impact on the role of the designer. Specialization has meant that many engineers limit their professional development to their area of specialization. The specialized individual therefore comes to have a concern for only his portion of the task, often without a full understanding for the impact of his actions on the total project. The vertical growth patterns serve to reduce the influence of the engineering designer by placing additional layers of middle management between him and the decision making levels of the organization. Overall, I think it is fair to say that the way engineering organizations have grown has resulted in a reduction of the impact of the individual designer within the organization and, to a great extent, the loss of control over the final configuration of the product.

In general, some of the best design work that I have seen has been developed in the "project" style environment. There are the ever-present hazards associated with dividing functions between project and functional organizations, but these can generally be avoided by a clear delineation of the manager's authority over the project and in his dealings with functional organizations.

Many current organizational structures do not necessarily accommodate the basic needs, talents, and authority of the designer. To achieve the quality of design we desire, we must create an organizational environment that recognizes and supports these important aspects of the designer's personality.

Perhaps the best example of a good project-type structure is the "skunk works" organization that has been used so successfully at Lockheed. The advantages of this type of environment are many, but for the designer the most important is the climate of intellectual freedom, individual responsibility, and flexibility that it generates. The design engineer is able to follow his design from concept to hardware without the usual encumbrance of complex procedures, functional fences, and other bureaucratic barriers. In a sense, he becomes an "entrepreneur" and is able to satisfy a personal drive to implement his visions and be in control of the situation.

F. Management

Perhaps the most pervasive factor in the design environment is the influence of management, for it is here that the policies, procedures, and practices of the organization are established. The direction and productivity of the design effort are largely determined by those in charge. Many times the extent to which a design is conventional, innovative, cost-effective, producible, and supportable is a reflection of management policy and practices. One of the most effective things that management can do to enhance creative activity is to mitigate the constraints imposed by the bureaucratic processes present in the designer's work environment.

VIII. CONCLUSION

Information technology is rapidly becoming a science of extraordinary proportions. As progress is made in networking, communications, data base management, and software/hardware standards, the mechanism to reach high levels of integration becomes more of a reality. Thus, the engineering designer is being provided with the power to create sophisticated designs, evaluate alternatives, and assess the life cycle cost of a hypothetical product before material, tools, and other resources are committed. Integration and automatic data handling errors are reduced, product quality is more consistent, and time is saved. Also, a wide range of new products can be conceived and assessed while the designer sits at his workstation.

There is no doubt that this capability provides impetus for us to reassess the system design process to learn what procedures can be improved, what processes should be merged, and how to better organize data toward effective integration. Accompanying the increasing application of computer technology is an expanded emphasis on customer requirements and specifications, designer specialization, and contractor organization and controls. It remains a management challenge to structure an environment in which these factors support rather than inhibit the innovative design process.

Chapter 16

ON NATURE OF DESIGN AND AN ENVIRONMENT FOR DESIGN

Larry Leifer

Center for Design Research
Department of Mechanical Engineering
Stanford University
Stanford, California

ABSTRACT

In the domain of programmable electromechanical systems design, there is an acute need to enhance the productivity of designers schooled in both the art and science of design. However, productivity is increasingly dependent on the rate at which new knowledge (information) can be incorporated and the facility with which complex trade-offs can be quantitatively assessed. Design processes are particularly dependent on access to cross-disciplinary expertise. In an attempt to grasp the complexity of this situation and to define a strategy for creating better design environments, I would like to introduce an "aviation metaphor" to guide the introduction of computational "expertise" in computer-aided design, manufacturing, and service.

The success of our endeavor to enhance designer productivity will depend on how well we understand design processes. With no generally accepted "theory of design," and considerable reticence among design professionals to even consider the notion of "design science," it is asserted that the study of design should, for now, focus on the development of quantitative observational methods that effectively instrument designer behavior and design tool performance in computer-aided design environments.

I. INTRODUCTION

It is often heard in design education circles that playfulness is an important attribute of good designers at work. In essence, design processes operating within a single mind are free to explore both the breadth and depth of a problem domain. And, while there may be considerable structure to the process, it is most notably different from institutional design in the freedom and facility with which attention can be focused on different aspects of the problem; perspective may be taken from different view points; information can be dealt with at various levels of abstraction; and judgment may be invoked, withheld, or filtered (see Figure 1). The "apparent" lack of structure in these design processes is evidence of our lack of understanding rather than inherent randomness.

Although design is a venerable human occupation, there is no generally agreed-upon body of design theory. The rather large amount of literature on design may be roughly divided into three categories. The earliest material comes from the Arts. It is typically narrative and concerned with the psychology and behavior of individual designers (Amabile, 1983; McKim, 1980; Osborn, 1963). In its most rigorous form, it is concerned with the taxonomy of design methods and strategies (Jones, 1980; VanGundy, 1981). A second body of literature comes from Engineering. It typically deals with creativity (Adams, 1984), optimization (Wilde, 1978), decision making (Siddall, 1972), and axiomatic design (Rinderle, 1982; Suh, 1978). The third and most recent source of thinking on design comes from Computer Science, especially the field of Artificial Intelligence. These researchers are typically interested in the development of formal models of design (Mostow, 1985) and representation of knowledge in the domain of design (Bobrow, 1985). They are concerned with language, search, management of goals and constraints, and human-computer interactions (Card, Moran, & Newell, 1983).

At least one simple lesson can be taken from this richly heterogeneous literature. Design is a set of many processes whose orchestration is personally, socially, and technically context sensitive. It is convenient to use information as

Figure 1. The design is a flexible, but seemingly chaotic activity when observed in the behavior of an individual designer.

On Nature of Design and an Environment for Design

the underlying medium of these processes and decision making as the fundamental operator. However, having said this, the task remains to give explicit form to ideas that could shape design tools and environments of the future.

II. MANAGING DESIGN

It is widely perceived in industry that there is no greater leverage for managing product cost, quality, acceptance, and service than that afforded by better design at the conceptual stage. At the same time, as devices become more complex and sophisticated, their design becomes increasingly difficult and time consuming. At some point, limited understanding of design and design technology constrains the quality and cost-effectiveness of our design and manufacturing enterprises.

Product and manufacturing complexity have tended to force engineers, scientists, and management into narrow disciplines linked together as simple "linear" organizations (see Figure 2). While this approach to complexity and

Figure 2. The institutionalization of design activity is typically structured and highly compartmentalized. While rendered observable, perhaps for management purposes, it also loses a great deal in the process.

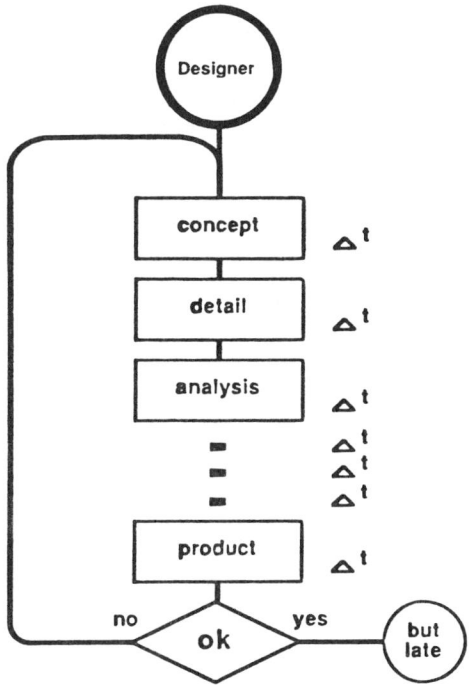

information management does work, it can also be subject to excessive information propagation delays. As the natural processes of design iteration and sequential dependency interact with institutional delays, timeliness becomes the dominant factor affecting product survival and correctness. Better understanding of design will require a larger allocation of computational resources in support of design activity, especially conceptual design. In the ideal design environment, it will be especially important to manage the relationship between design information and design experience.

III. EXPERIENCE AS KNOWLEDGE

Of the various dogmas associated with design, few are as widely held as the notion that good design comes with experience. This is curiously contradictory to another dogma that suggests that all children are creative in the beginning and become less creative with experience. Excellent designers often show evidence of benefiting from experience while maintaining an ability to be "childlike" in their design activity. Without proof, it is asserted that a good percentage of the experience that counts for good design is motor-sensory, not "symbolic" information. There is a danger that in relying exclusively on computer data bases for information, we may emphasize one form of representation to the exclusion of other modalities, especially direct physical experience.

Even if there is truth in the above speculation, how can one incorporate direct sensory experience within a computer-aided design environment? The first, and preferred choice, must be to build and test real devices. When full implementation is not feasible, build "working" prototypes. Most importantly, one should build with facility, iteratively prototyping and refining one's ideas until features and trade-offs come into balance. When people are expected to interact directly with the device, e.g., cockpit displays and operator consoles, then real-time simulation (animation) becomes the preferred choice. High resolution flight simulation provides a reasonable analog to the kind of product simulation that should be done routinely in computer-aided design environments. This and other lines of reasoning have led me to see aviation, especially flight simulation, as a metaphor for the design environment of the future.

IV. DESIGNER AS PILOT

Imagine the designer as a pilot who is charged with delivering his craft safely on time, within cost constraints, and on target. Key avionic subsystems are dedicated to this task (see Figure 3): standard flight protocols, a co-pilot (navigator), standardized flight controls and displays, flight instrumentation, a mechanic, and vehicle performance standards. The designer is similarly charged

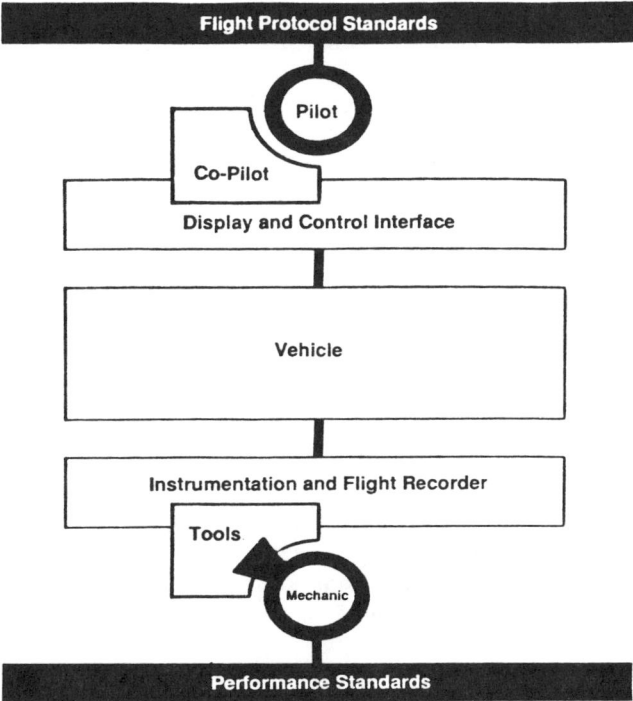

Figure 3. The Designer/Pilot Aviation Metaphor.

with delivering his safe product on time, within cost constraints, and on target (market). Flight simulation is now so good that a pilot can be fully certified to command loaded passenger craft without previous experience in that particular vehicle. Can you imagine launching a major new product development effort with people who had no previous design experience in that product domain?

V. DESIGN ENVIRONMENT AS AN AVIATION SYSTEM

In looking upon the designer's environment as an aviation system, the search for an analogous function leads one to consider information management as the analog to flight motion. The importance of decision making is somewhat equivalent in both domains. While temporal issues are clearly different, the quickening pace of product development lends an air of "real-time design" to many situations. To a large extent, the computational tools of computer-aided design are analogous to the flight vehicle, controls, instrumentation, and flight recorder. In keeping with the experience that digital flight simulation has contributed profoundly to our understanding of pilots and ability to train them effectively, it is imperative that both research and production computer-aided design environments be thoroughly instrumented. Data derived from these instruments can drive adaptive features of the proposed "designer's associate," provide insight into designer methodology, and facilitate design education through detailed design mission scenarios that reveal the intent as well as the final form of systems and devices.

If aviation and flight simulation bear some useful resemblance to the designer and the design enterprise, how might this analog be implemented in practice? Preliminary results with the application of expert systems technology in design (Curran & Leifer, 1985) and examination of the role of information in design processes (Rouse, 1986) strongly suggest that key features of the aviation metaphor can be implemented with machine intelligence (knowledge-based expert systems).

VI. "MACHINE INTELLIGENCE" IN THE DESIGN ENVIRONMENT

Even as it became clear that knowledge-based expert systems technology is relevant to the architecture of design environments, it remained unclear when and where to apply these programming techniques. A comparison of Figure 3 and Figure 4 will convey the sense in which it is anticipated that machine

Figure 4. Knowledge Engineering for the Aviation Metaphor.

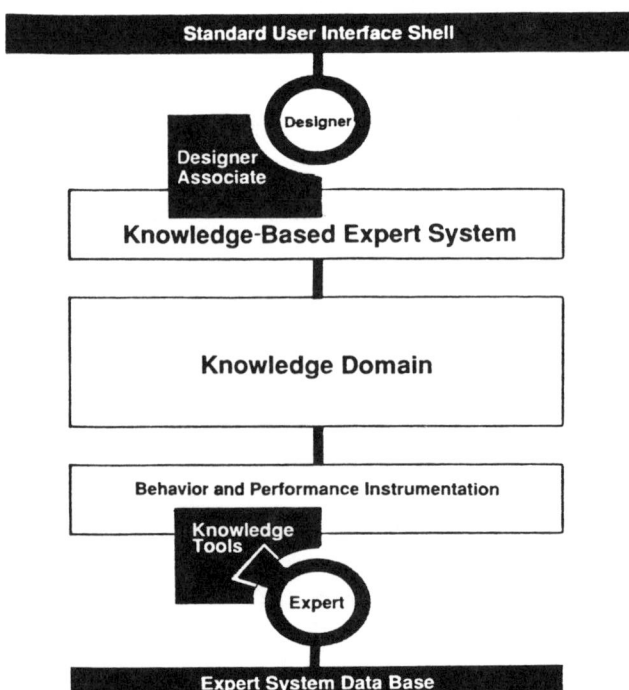

On Nature of Design and an Environment for Design 217

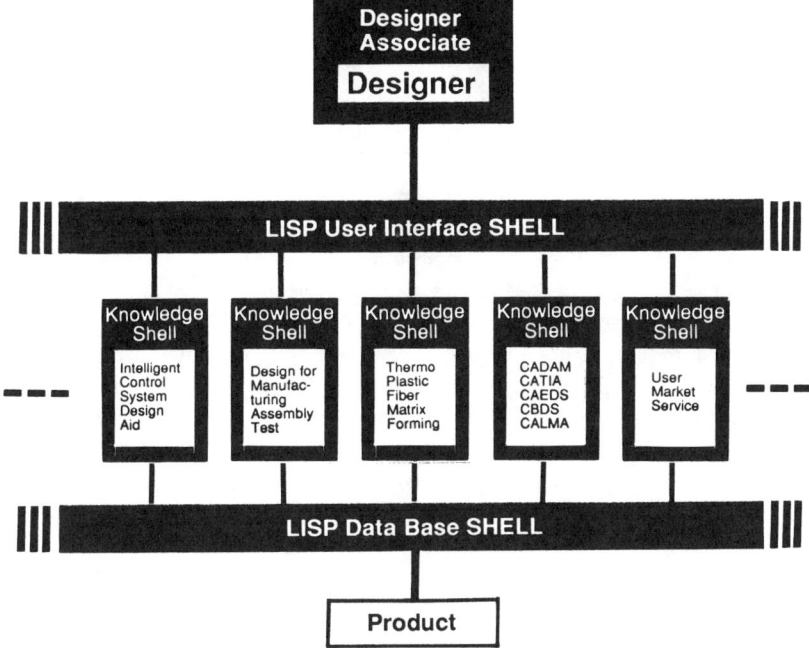

Figure 5. An environment for concurrent design engineering is under development at the Stanford Center for Design Research. The architecture is based, in part, on the aviation metaphor and the information management potential of knowledge-based systems engineering. One of the key objectives of this endeavor is to create, in an institutional setting, the kind of designer fluidity and power of association found in individual endeavor.

intelligence will fulfill key "roles" in the design environment (shaded areas). The co-pilot role, typically handled now by occasional contact between the designer and a project advisor (supervisor) will be built into the "designer's associate," an expert system dedicated to augmenting the performance of individual designers. The designer's associate must know about individual designer expertise, style, and objectives. The designer's associate is responsible for navigating through the complexity of design environment resources. It "travels with the designer" and is available at all times to provide strategic advice and manage information access.

A key task for the designer's associate is to interface expert systems associated with specific knowledge domains (i.e., specific computer-aided design packages). As represented in Figure 5, these domain-specific expert systems may deal with new areas such as design for thermoplastic composite fiber matrix materials, or they may encapsulate large bodies of existing code (e.g., CADAM, a two-degree-of-freedom drafting package written in Fortran).

Designer behavior and design environment performance must be instrumented to provide an objective record for subsequent information utilization and decision making analysis. The motivation for this specification is largely based on successful use of digital flight simulators in pilot training. It is also based on the use of interaction history lists, multitasking, and interactive debugging features of modern artificial intelligence programming environments.

The building and maintenance of knowledge engineering systems (designer's associate, knowledge domain expert systems, and expert system instrumentation) require development tools for the human experts who will create and maintain the designer environment. In a sense, knowledge engineers and human experts are expected to create the environment for other people to design (fly) new systems and devices.

The aviation metaphor has helped shape a strategic design environment plan at the Stanford Center for Design Research (see Figure 5). The common LISP programming environment is used to develop a consistent user interface shell and to manage information exchange between an open-ended set of "concurrently accessible" knowledge domains. Each area of expertise will have its own domain-specific shell. Navigation through this complex environment will be managed by the designer's associate. Specific projects deal with each of the areas referenced in Figure 5.

Based on experience to date, we have chosen to implement the designer's associate strategy implementation presented in Figure 6. IBM and DEC medium-scale computer systems dedicated to computer-aided design and manufacturing have proven to be unsuitable for concurrent knowledge engineering development. We now use "artificial intelligence workstations" for designer's associate and expert system research and development. While we originally expected to port mature programs to the main computers, it is increasingly clear that we will need to keep artificial intelligence workstations in the configuration shown in Figure 6.

VII. CONCLUSION

Ideas presented in this position paper require considerable refinement and validation. To this end, several projects within the Stanford Center for Design Research are underway to gain an objective understanding of the issues. The current strategy calls for a balanced program (see Figure 6) of 1) industry-sponsored design/development projects, 2) development of "intelligent" computer-aided design and manufacturing programs, and 3) quantitative study of designer behavior and computational tool usage patterns. There is a heavy (hopefully not dogmatic) emphasis on artificial intelligence methodology in the area of knowledge-based expert systems, intelligent data bases, and the development of a designer's associate. The designer's associate is also responsible for navigating through an increasingly complex array of design tools, expert systems, and data bases. The "aviation metaphor" has been of

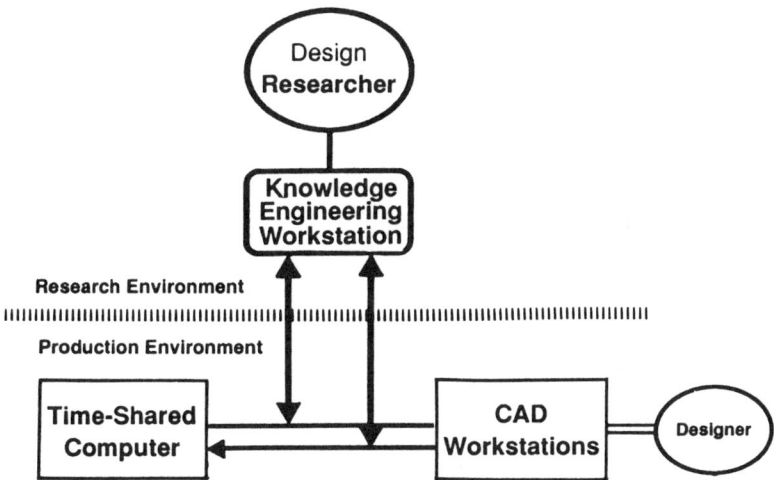

Figure 6. Design researchers need powerful artificial intelligence workstations to "observe" while others "design."

value in identifying key design environment issues amenable to knowledge engineering methodology.

ACKNOWLEDGMENTS

Work referenced in this paper has been supported in part by the following agencies and institutions: Stanford Center for Design Research (a member of the Stanford Institute for Manufacturing and Automation); Veterans Administration Rehabilitation Research and Development Center; and International Business Machines Incorporated. Individuals who have contributed to the thinking that underlies this paper include the following: John Tang, Allen Curran, Mark Cutkosky, Elliott Levinthal, Marty Tenenbaum, David Dungan, Ren Curry, Bill Rouse, Fred Lakin, Walter Wang, and Don Brown.

REFERENCES

Adams, J.L. (1984). *Conceptual blockbusting.* Norton and Co.
Amabile, T.M. (1983). *The social psychology of creativity.* Springer-Verlag.
Bobrow, D.G. (Ed.). (1985). *Qualitative reasoning about physical systems.* The MIT Press.
Card, S.K., Moran, T.P., & Newell, A. (1983). *The psychology of human computer interaction.* Lawrence Erlbaum Associates.

Curran, A.R. & Leifer, L.J. (1985). An intelligent control system design aid. *Proceedings of the ASME International Computers in Engineering Conference* (pp. 100-104). Boston, MA.

Jones, J.C. (1980). *Design methods: Seeds of human futures.* New York: John Wiley & Sons.

McKim, R.H. (1980). *Visual thinking.* Lifetime Learning Publications.

Mostow, J. (1985, Spring). Towards better models of the design process. *The AI Magazine,* pp. 44-57.

Osborn, A. (1963). *Applied imagination.* New York: Scribner and Sons.

Rinderle, J.R. (1982). *Measures of functional coupling in design.* Ph.D. Thesis, MIT Laboratory for Manufacturing and Productivity.

Rouse, W.B. (1986). On the value of information in system design: A framework for understanding and aiding designers. *Information Processing and Management, 22,* 217-228.

Siddall, J.N. (1972). *Analytical decision-making in engineering design.* Prentice-Hall.

Suh. N.P. (1978). On an axiomatic approach to manufacturing systems. *Journal of Engineering for Industry: Transactions, ASME, 100,* 27-30.

VanGundy, A.B. (1981). *Techniques of structured problem solving.* Van Nostrand Reinhold Co.

Wilde, D.J. (1978). *Globally optimal design.* New York: John Wiley & Sons.

Chapter 17

TOWARD A MORE SYSTEMATIC, EFFICIENT DESIGN PROCESS: THE POTENTIAL IMPACT OF INTELLIGENT DESIGN AIDS

Edward A. Martin

Training Systems Division
Aeronautical Systems Division
Wright-Patterson Air Force Base, Ohio

ABSTRACT

It is postulated that design aids which would 1) appropriately support rapid access to design-critical information, and 2) provide for efficient transfer of information across government, industry, and/or academic agencies involved in a project would mitigate some of the severe deficiencies now existing in the design process. It is argued that current information retrieval technology does not adequately support the extraction of information from domains in which the designer lacks expertise. Consequently, it is concluded that a friendly, cross-disciplinary interface is essential to optimizing the design process.

 A friendly, cross-disciplinary interface would enhance the ability of designers to consider much more relevant data which are not currently factored into the design process. The availability of this interface would also open the door to new prospects for enhancing communication in the design process. Potential design process enhancements, realizable by establishing intelligent communication linkages between designers and a central design team "corporate memory," are considered. It is proposed that these enhancements would not only facilitate the communication of information among design team members, but would support the formation of a readily accessible, auditable history of the design process. One anticipated result is a better focus on the needs of the ultimate user of the product throughout the design process.

I. INTRODUCTION

Design is conceptually viewed as a structured, multiphased, iterative process which transforms a need into a product (Pahl & Beitz, 1984; Ray, 1985; Rouse, 1986). The inability of designers to identify and retrieve all existing design-critical information in a timely fashion, and inadequate communication among design team members[1] are major impediments to the orderly and effective progression of the design process. These two broader deficit categories include some specific problem domains which repeatedly surface in the design process and which are amenable to support by appropriate design aids. Based on my experience as a designer, I believe in order to effectively enhance these areas of the design process, design tools or aids should 1) provide each design team member with a means to identify and access germane information rapidly, 2) provide access to guidelines/regulations regarding corporate and customer policies and procedures, 3) permit the sharing of a corporate knowledge base, 4) genuinely foster communication among design team members/agencies, 5) enhance the involvement/consideration of the ultimate user or customer in all phases of the design process, and 6) improve documentation. These are not in any particular order of importance, but they are roughly grouped in accordance with their relationships to the need for improved access to relevant information and the need to effectively communicate that information.

II. IMPROVED ACCESS TO RELEVANT INFORMATION

A systematic approach to design requires the identification and retrieval of information about the system under consideration. Such "information seeking" is an integral part of problem solving (Pahl & Beitz, 1984; Rouse, 1986). To the extent that this aspect of design may be enhanced, the efficiency of the overall process will benefit. Clearly, efficient access to valuable information[2] is required. This, of course, requires communication links to the sources of information supporting the numerous disciplines involved in the design process *including relevant standards, regulations, and policy guidelines*. Perhaps less obvious—but no less essential—the user interface must be sufficiently friendly so that the time and effort required to access and retrieve valuable information does not exceed what the designer is willing or able to expend. In this context, a friendly interface is one in which the cognitive complexity of the interface does not require extensive training or experience for appropriate human interaction—i.e., it must be easy to use. It must support the rapid identification

---

[1]The design team for any given project is assumed to consist of all relevant research, procurement, support, and vendor activities that cut across government, industry, and academia.

[2]For information to be "valuable," it must have the potential to reduce uncertainty, be relevant, and be in appropriate form (Rouse, 1986).

of information critical to the design and the delivery of this information in a form optimal for the designer's assessment of its value. Furthermore, it must be sufficiently flexible to account for differences in designer experience, cognitive style, and information needs at the various stages of design.

A. Cross-disciplinary Information Access

1. Identification of Design-critical Information

A critical constituent of information seeking is the identification of valuable information. The designer, faced with limitations on time, resources, and resident expertise, is typically unable to identify all information germane to the design problem. The usual consequence of this situation is a tendency on the part of the designer to draw heavily on past approaches and what the current technology will support, tempered primarily by cost/performance tradeoffs and externally imposed constraints (Boff & Martin, 1980; Caro, 1977; Rouse, 1985). As a result, much information which could contribute to a particular design problem is never considered.

One manifestation of this problem in the Aircrew Training Device (ATD) design process is the continued focus on designing simulators to mimic the aircraft and its environment rather than to train. Caro (1977) suggests that this apparent lack of attention to behavioral and training data exists because a suitable way to communicate these data to design engineers is lacking. There is more to the problem than simply establishing a link between engineers and appropriate human factors data, however. There is a *general* lack of adequate design tools to support the efficient cross-disciplinary access of information.

The design of complex systems normally requires the integrated efforts of a multi-disciplinary design team encompassing a number of highly specialized areas of expertise. The ATD design process, for example, requires expertise across a wide range of domains including aircraft operations, training, contract administration/law, cost/schedule projection, logistics, engineering, ergonomics, and human perception and performance. The cross-disciplinary nature of the problem exists along multiple dimensions, of which linking the engineer to human factors data is but one. Along any dimension, current information retrieval technology does not adequately support designer attempts to extract information from a domain in which he or she has little or no prior training (and perhaps no knowledge of the appropriate key words). Tools are needed which will aid in structuring the search for information (perhaps even helping formulate the objective of this search) through domains which are foreign to the designer's prior experience.

2. Information Transformation

Another problem confronting the designer is that the information which is accessed in such a foreign domain is likely to be in an inappropriate form and couched in unfamiliar jargon. For example, the research literature is generally written from a perspective, and with an emphasis, which restricts its usefulness to the applied community (Rouse, 1985). Potentially valuable research results may be buried in verbiage and expressed in a technical vocabulary not well understood by design engineers. There may, in fact, be terms with entirely different meanings in the research and applied communities which could lead to misinterpretations on the part of the designer. This can severely constrain the value of that information to the designer. Appropriate transformations must be applied to accessed data to assure that the information retrieved is at the level of knowledge and understanding of the receiver, in order for it to be properly interpreted and appreciated (Boff, Calhoun, & Lincoln, 1984; Ray, 1985).

B. Adaptive Interface

The interface between the designer and the design aid needs to be adaptive. Information needs vary with designer skill, set, and information processing characteristics as well as with specialization. Information needs also differ with the nature of the design problem in the various stages of the process, across the various domains of expertise (Boff & Martin, 1980; Rouse, 1986). (Clearly the information needs for detailed engineering design are quite different from those required for defining system specifications or for managing system configuration.) An effective design aid will have to be sufficiently adaptable to account for this variance in information needs both within a discipline and across disciplines. It must simultaneously be sufficiently flexible to permit efficient access for designers possessing varied degrees of experience and style, while not unduly burdening or discouraging the more experienced designer with the additional prompts/aids required by the less experienced.

III. IMPROVED COMMUNICATION OF INFORMATION

Design aids would make a major contribution to the efficient progression of the design process if they would aid in the development of a design team corporate memory (or design process history), and in the communication of this to appropriate design team members. The ability to trace through the total design process efficiently would be of considerable benefit, especially for purposes of communicating a history of 1) the requirements definition rationale, 2) alternatives investigated and design decisions made at critical junctures in the design process, and 3) user feedback regarding the utility of the product.

A. Visibility of the Genesis of Requirements

Currently, not all design team members have a means for readily obtaining sufficient knowledge of, or insight into, the rationale surrounding the identification of requirements. As a result of this lack of visibility, the requirements established at one phase of the design process may be improperly understood by the designers responsible for the synthesis, analysis, and selection among alternatives at later stages of the process.

Why does this condition persist? The diversity of specialized expertise required for the design of complex systems frequently results in a distribution of responsibilities among agencies staffed by specialists who can efficiently perform the functions of their respective agency. Most often these various agencies are rated in terms of their adherence to budget and schedule constraints rather than in terms of the utility of the final product. Taken together, there results a tendency for design decisions to be made in isolation from the user or other agency responsible for initial identification of the requirements (Caro, 1977). When design decisions must be made by specialists who lack an understanding of the users' needs, and in relative isolation from the agency formulating the requirements, it is not surprising that the product sometimes differs from expectations.

If a tool or aid were available which would permit designer insight into the requirement formulation decision process—at a level of detail and knowledge consistent with the immediate need—one might expect a product more in line with the *intent* of the specified requirements. Ideally, such a tool would carry this process one step further, and (with appropriate management controls) assure/encourage a greater level of design team interaction during the formulation of requirements and *throughout the design process*. This would benefit all agencies involved, for effective communication is essential throughout the design process; to the extent that communication breaks down, the process will not proceed efficiently and it is likely that improper decisions will be made (Pahl & Beitz, 1984; Ray, 1985).

B. Visibility of Design Decisions

A design tool should support enhancements to the process of *recording* design decisions/agreements made by the design team, along with supporting rationale. Current procedures make it difficult for all agencies to efficiently trace through this particular aspect of design history. Often solid records cannot be located when needed, and designers are forced to resort to fragmentary notes or to memory, a situation which sometimes leads to misunderstandings and even animosities among team members. Certainly, such a tool would have strong potential for improving government-contractor relationships, while simultaneously affording enhanced opportunity for using agency involvement throughout the project.

There is potential for yet another benefit. It may happen that the trajectories through the design space, necessary to the establishment of an auditable design record, will essentially capture the collective knowledge of the more experienced design team members. Such a knowledge base, in conjunction with friendly information access aids, could prove an invaluable aid for training and providing guidance to the less experienced designer.

C. Visibility of Product Utility

A tool is needed to enhance feedback from the operational/user environment. This information should become a part of the design team corporate memory knowledge base. An effective, formal vehicle permitting design team visibility into the utility of the product is frequently lacking. The design team requires feedback regarding both successful and unsuccessful aspects of the product. To the extent that new product design is a derivative of what was done in the past, this feedback is critical to the designer attempting to capitalize on past successes and avoid past mistakes. Too often these data are not now available to the designer in a format or at the appropriate time to properly impact new design.

D. Improved Design Team Communication

The corporate memory architecture should be designed to enhance communications across the design team. In the process of generating the corporate memory, "pointers" or references to those data critical to the design should be recorded as the various members of the team identify that information which they deem valuable. These references (along with cross-disciplinary information access aids and appropriate visibility into design decisions and requirements) would serve to greatly enhance the effective communication of information so essential to efficient design (Pahl & Beitz, 1984; Ray, 1985).

IV. CONCLUSION

It is argued that there are a number of ways in which appropriate design aids could enhance information seeking, documentation, and communication throughout the design process. The essential support need, threading through all areas addressed, is a requirement for effective, cross-disciplinary information access aids. To the extent that design aids can facilitate access to valuable information in multiple technical domains, they will enhance the probability that potentially design-critical information is factored into the design process. In the absence of tools permitting the designer to find and understand the best information objectively available, inappropriate design decisions are invariably going to be made.

The availability of an effective cross-disciplinary information access tool introduces new prospects for enhancing communication and documentation throughout the design process. By affording all design team members better visibility into the genesis of requirements, the rationale behind major design decisions, and the utility of design team products, a better understanding of the system requirements, design goals, and available alternatives will be fostered. It is expected that this will result in products which better satisfy the *intent* of customer requirements, more favorable customer-vendor relationships, and a more efficient design process.

REFERENCES

Boff, K.R., Calhoun, G.L., & Lincoln, J. (1984). Making perceptual and human performance data an effective resource for designers. *Proceedings of the NATO DRG Workshop (Panel IV)*. Royal College of Science. Shrivenham, England.

Boff, K.R., & Martin, E.A. (1980). Aircrew information requirements in simulator display design: The integrated cuing requirements study. *Proceedings of the Second Interservice/Industry Training Equipment Conference*, 355-362.

Caro, P.W. (1977). *Some current problems in simulator design, testing, and use* [HumRRO-PP-2-77 (AD A043240)].

Pahl, G., & Beitz, W. (1984). *Engineering design*. New York: Springer-Verlag.

Ray, M.S. (1985). *Elements of engineering design: An integrated approach*. Englewood Cliffs, NJ: Prentice/Hall.

Rouse, W.B. (1985). On better mousetraps and basic research: Getting the applied world to the laboratory door. *IEEE Transactions on Systems, Man, and Cybernetics, SMC-15*(1), 2-8.

Rouse, W.B. (1986). On the value of information in system design: A framework for understanding and aiding designers. *Information Processing and Management, 22*(2), 217-228.

Chapter 18

A COGNITIVE THEORY OF DESIGN AND REQUIREMENTS FOR A BEHAVIORAL DESIGN AID

David Meister

U.S. Navy Personnel Research and Development Center
San Diego, California

ABSTRACT

The development of a support system to aid engineers in the solution of design problems (those with behavioral implications) has two prerequisites: The first is the construction of a meaningful theory of the engineer's cognitive processes in design; the second is the availability of empirical human performance data associated with design variables. The following paper outlines the beginnings of such a theory and describes the characteristics a design support system should have.

I. THE NATURE OF DESIGN

A. Introduction

It is impossible to assist the designer effectively without understanding how he designs because any aid provided must match his design processes. The repeated failures of human engineering guides to elicit designer interest and use (Meister & Farr, 1967) may well have resulted from ignorance of those processes.

In this section of the paper, the author presents a conceptual model of the design process which, although apparently reasonable, is highly speculative because empirical data describing how the engineer designs are sparse. The little data collected by the author, his associates, and others (Lintz, Askren, & Lott, 1971; Meister, 1971; Meister & Farr, 1967; Meister, Sullivan, & Askren, 1968; Meister, Sullivan, Finley, & Askren, 1969a, 1969b) have been supplemented by personal experience in system development and logical deduction. Still, it is quite risky to pretend certitude about the design process.

It is also necessary to distinguish between design viewed as an abstract phenomenon (the "ideal"; the way in which design *should* proceed) and design as actually practiced in an engineering facility (reality). Reality deviates from the ideal in many ways.

Along with Rouse (1986), the author conceives of design as a problem solving process. In this formulation, there are four major stages: 1) formulation of the design problem; 2) generation of alternative design solutions; 3) analysis/evaluation of these alternatives; and 4) selection of a preferred alternative.

This is fine as far as it goes, but to schematize the process in this way is not to say a great deal. The four stages are merely the scaffolding of the design process and must be filled in with a great deal of lath and plaster. In practice, the four stages are not sharply delimited and tend to overlap a good deal. Design in real life is more complex than the single thread categories listed above. Moreover, they describe *what* happens but not *how*, and it is the *how* that is critical.

B. Cognitive Design Elements

The major elements in the design process consist of the following (not in order of importance):

1. The initial statement of the problem is represented by the design requirement and is expressed either in verbal or written form. The system requirement describes what the system is supposed to accomplish and may include information about development criteria and constraints. Often the requirement is vague and may change over time.

2. Criteria (either explicit in the system requirement or implicit in the designer himself) describe the factors to which the designer will give highest priority in solving his design problem. These criteria may include performance capability, reliability, feasibility, cost, development time, producibility, etc. Unless the criteria are explicitly called out, the designer will utilize his own internalized criteria which he has developed on the basis of his development experience.

3. Constraints (primarily time and cost) on the designer's freedom to

select optimal solutions. Time constraints are discovered by examining the development schedule and cost by "not-to-exceed" figures.

4. The engineer's design style. For example, some designers are more logical or deliberative or intuitive than others. Some designers are more realistic than others. Nadler (1985) categorizes designers as inactivists, reactivists, preactivists, and interactivists.

5. The engineer has specific items of information that relate to the design problem or knows where he can get information.

6. The designer's experience is represented by the design problems he has successfully solved (he will probably have repressed the less successful ones). As a result of his experience he has certain design skills. It is possible that because of his experience, he has certain solutions to classes of design problems or at least a preferred design strategy or mode of attacking types of design problems.

7. Certain elements of the system may have already been decided upon by others, and the designer can modify these only slightly, if at all. For example, a sensor may have already been specified for a surveillance system that is to be designed.

8. The designer builds a mental model of the system or equipment he is developing. The model consists of his prior design knowledge and his analysis of all the preceding elements and the way he integrates them. The design model is progressively constructed as the system is being developed, is idiosyncratic to the individual designer and to his immediate design problem, and may be more or less clear and specific. The function of the mental model is to help the designer review at each stage what he must do to solve the design problem and what he has accomplished so far. It is a sort of *outline* of what he must do, phrased in terms of questions he must answer. He knows what the elements of the model are based on his past system development experience, and this guides him to the information he must secure and the analyses he must perform. At the start of design, his model of the system is largely empty (except for his stock of past design solutions), and he progressively fills in the model through the development of the design. The model then is a guide which directs the designer's actions with regard to immediate design.

It is important to note that all of the preceding elements interact. Moreover, design may proceed at any of four levels: the total system, individual subsystems, individual equipment, and individual modules (major subequipment). The *logical* progress of design is from molar to molecular, but in actuality the designer, at least early in design, may function at all four levels concurrently. When the design problem involves a major system, it is usually subdivided and each designer receives a small "piece of the action," but then the problem of integrating the pieces arises.

The following discussion is organized around the four states of problem solution mentioned previously, but we emphasize the mechanisms which produce design outputs. Table 1 presents an outline of a model of the design process. The design process is oriented around the effort to answer a number of generic questions. The designer utilizes certain mechanisms to secure the answers to these questions; it is these mechanisms in which we are most interested. Examples are provided from two ongoing Marine Corps projects, TOV and AROD. In combination, these two projects are an effort to develop teleoperated surveillance platforms for battlefield use. One is a land vehicle, and the other is an "eye in the sky."

C. Analysis of the Design Problem

The first step in analyzing the design problem is to decompose the system's[1] requirements into its component elements which were listed previously. This means determining what the system is supposed to do. However it is to be designed, the customer for the system intends that it will function in certain ways. For example, the AROD platform will rise in the air to X feet and then fly, under the control of a ground operator, a distance of Y feet; locate a specified target; return photograph-like stimuli describing the target; and return. Developing such a scenario (in much greater detail, of course) is a starting point in breaking down the system into its elements.

The starting point for the design analysis is therefore a *mission analysis* that has a centerpiece that is the development of a *scenario* or description of the sequential events in system operation. The scenario merely states how the system will be employed when it becomes operational. The scenario is initially developed to describe the total system and subsystems; later, as design becomes more molecular, it is expanded to describe equipment and modules within the equipment. At more molecular levels, the scenario describes not so much observable events as physical (e.g., electronic) processes. For example, in TOV, control of the multiple signals sent back over a fiber optic cable is hypothesized in terms of how a microprocessor distribution network functions.

The outputs of this analytic process are system *functions* to be performed and *parameters* of the problem. In some cases, functions and parameters are quite obvious (e.g., the function of an aircraft is to take off, fly, and land), but in

---

[1] The term "system" is intended to represent the total system, subsystems, equipment, and modules.

Table 1. A Conceptual Model of the Design Process.

	QUESTIONS ASKED		DESIGN MECHANISM
1.	**Analysis of the Design Problem**		
a.	What are the system, subsystem, equipment, and module supposed to do?	a.	Mission analysis; scenario development
b.	What are the functions and parameters of the system, subsystem, equipment, and module as related to the mission?	b.	Engineering knowledge; deductive logic
c.	What already established system elements are fixed? What is their effect on other system elements?	c.	Examination of system requirements
d.	What information about the system, etc. is known? What is unknown?	d.	Knowledge of the state of the art and of the test literature; gathering of additional information
e.	Do any aspects of the problem resemble those the designer has previously encountered? Are any previous design solutions applicable to the present problem?	e.	Designer experience
2.	**Generation of Alternative Solutions**		
a.	What are all the possible solutions that can be applied to this problem?	a.	Deductive logic: engineering knowledge; designer experience
3.	**Analysis of Alternative Solutions**		
a.	What criteria apply to this problem?	a.	Engineering knowledge; designer experience
b.	What constraints impact the alternatives?	b.	System requirements; information from management
c.	What alternative parameters can be traded off?	c.	Engineering knowledge
d.	What are the similarities/differences among alternatives?	d.	Engineering analysis
4.	**Selection of Preferred Solution**		
a.	What are the advantages/disadvantages of each alternative?	a.	Engineering knowledge; design and test data
b.	Which alternative is the best choice?	b.	Paired comparison of alternatives

other design problems functions and parameters are obscure and must be analyzed.

Where functions and parameters are unclear, the designer may make use of deductive logic, which one can term the "if this, then that" process. For example, if the AROD mission requires that the platform proceed to a location at which it will acquire a target, it must be navigated during its flight (navigation function). If the navigation is to be performed remotely (by an operator on the ground), the operator will have to receive remotely the same kind of information that would be received if he himself were actually flying and navigating. This presumes some sort of navigation sensor. Consequently, all the parameters that enter into such a sensor must be considered (e.g., field of view, resolution, etc.).

By the nature of the problem, certain system elements/aspects have been predetermined. It is intended that AROD be a remotely operated system; this means that all signals must be transmitted from the air to the ground. It is already established (for reasons that need not be discussed here) that these signals will pass over a fiber optic cable. It will, therefore, be necessary to provide some kind of cable status indicator for the remote operator.

Part of the design analysis, therefore, is for the designer to determine which system elements are fixed and with which he need not concern himself, and those which must be determined.

Problem analysis also seeks to determine what is known and unknown about the design parameters. For example, in AROD, the field of view of available sensors can be determined by referring to the manufacturers' literature, and minimum resolution required for target acquisition has been well researched. However, assuming that the flying platform on which the sensor will be mounted will experience a certain level of vibration, what will the effect of that vibration be on the operator's target acquisition performance? This is as yet unknown and must be researched. The unknowns in the design problem require the designer either to research the literature or to perform empirical tests.

By the conclusion of this analysis stage, the designer has constructed a preliminary mental model of the system to aid him in his search for a previous design solution. Analysis of the problem enables the designer to differentiate between two aspects of the problem. If some aspects of the problem are banal, they can probably be resolved by the application of past experience, whether this involves specific items of information or the designer's stock of past design solutions. If there are novel aspects to design for which information and past experience are unavailable, the engineer will have to use deductive logic, which involves making inferences from established fact.

The designer then asks whether the problem or some aspects of it are similar to ones that he has previously solved. To do this, the problem must be categorized in certain ways (a design taxonomy). Memory must then be searched to determine whether any previous problem solution or aspect of that solution matches any category of that taxonomy. If it does, he extrapolates from the previous solution.

D. Generation of Alternative Solutions

It is unlikely that the designer's previous solutions, even if applicable, will completely solve his present problem. At best, they may aid by suggesting alternative ways of solving the problem. The designer now proceeds to develop these alternative solutions. This is the most creative part of the design process. At this point, deductive logic and the designer's stock of experimental solutions interact. The goal is to provide *all* possible design solutions and then to select from among them. However, the previous design solutions tend to constrain the number of alternatives developed and narrow the designer's focus (meaning merely that most designers tend to be experience-limited).

E. Analysis of Alternative Solutions

The generation and analysis of alternatives tend to overlap because it is quite natural when conceiving of a candidate solution to start analyzing it almost immediately. Consequently, in practice, instead of analyzing all alternatives concurrently, the engineer tends to analyze them sequentially, which means that the designer may be biased toward alternatives conceptualized earlier or later in the sequence.

Evaluation of alternatives must be based upon criteria. The most important criteria are *performance capability* (will the proposed solution satisfy system requirements?); *reliability* (is the solution likely to stand up under adverse conditions?); *feasibility* (is the solution within the state-of-the-art technology and are there significant unknowns which may inhibit capability?); *cost* (is it excessive?); and *time* (can the development to the point of production be achieved in reasonable time?). Note that *human factors* is not on the designer's list of priorities, although concern for human factors will probably surface under the categories of capability or feasibility if there is any real question that the operator may not be able to operate the system.

Equally important are any *constraints* of cost or time which may bar a proposed solution, even though the solution, based on other criteria, is very promising.

As part of his examination of the alternatives, the designer may decide to trade off one value of a parameter for another. Thus, in AROD's sensor situation, a larger FOV can be traded off for less resolution.

The trading off of the system parameters is part of the designer's general examination of the similarities and differences among the alternatives. One alternative may be so slightly different from another that it is possible to eliminate it, thus reducing the difficulty of the designer's choices.

F. Selection of Preferred Solution

Ideally, the designer will systematically list the advantages and disadvantages of each alternative solution as a preliminary to selecting the best. The criteria developed in a previous stage will serve as the categories for this analysis. In specifying these advantages and disadvantages, the designer will make use of his accumulated knowledge, any information he has gathered during design, and any test data that are available. The selection of the preferred alternative can be performed systematically by means of a paired comparison of each alternative, using a quantitative procedure described by Meister (1971). This procedure involves the establishment of criteria *weights* which are multiplied against the results of a paired comparison of each alternative. In actual design, however, this analysis will probably not be quantitative. Most designers use a highly qualitative, almost intuitive, method of selecting a preferred alternative. It may involve a paired comparison of alternatives, but the basis for the comparison (criteria) may not be completely specified.

G. Summary

What are the implications of this model of design analysis for a design support system? Briefly stated, to be maximally useful, the design support system should incorporate the model. Phrased in this way, the principle is not very helpful to the potential system developer, but a major task in designing such a system will be deciding how to formulate the information the system provides. The system should not be a literal electronic transcription of a traditional technical manual or guide. It should be organized around the design process described previously in an effort to maximize its efficiency.

II. ACTUAL DESIGN PRACTICE

The design process described previously is a highly deliberate one, and it emphasizes logic, deduction, and the application of engineering knowledge. In actual practice, many (if not most) designers deviate from this process in ways that presumably reduce its effectiveness. We can adduce only partial evidence for the reduction of effectiveness because formal empirical studies of how engineers design are rare. The reader can refer to Meister (1971), but it requires actual immersion in design to see instances of the following:

1. Some of the analyses the designer performs are somewhat unconscious. He may be only partially aware that he is performing the analyses of the previous section because he tends not analyze his own processes. Consequently, some design steps may be skipped or truncated, certain questions may remain unanswered, and the engineer's biases are allowed a freer rein than would otherwise be the case.

2. The engineer is experience-oriented. He will, all other things being equal, try to repeat design approaches and solutions previously found effective. This tendency is exacerbated by the fact that most new

design is not revolutionary but builds incrementally on past technology. Revolutionary new design approaches like design of the atom bomb or development of the Apollo space module come along only once in a generation. There is nothing unusual or wrong with the experience orientation, but it may cause the designer to focus on particular design alternatives because they fit his experience and to ignore other possibilities because they do not.

3. The engineer is often intuitive in his thinking and fails to fully think through his design solutions.

4. The criteria which he uses to analyze candidate solutions usually/often do not include behavioral aspects which may result in selection of an inadequate alternative from a human factors standpoint.

5. The engineer's urge to come to grips with the hardware-/software-specific aspects of his design as quickly as possible often causes him to shortchange the analytic aspects that should precede hardware/software design.

6. When he lacks necessary information, he often does not know where to find it. Presumably this is where a design support system would be of value.

7. As a decision maker, the engineer is as likely as anyone else to deviate from an optimal (e.g., Bayesian) process of making decisions.

All of this does not mean that the engineer necessarily will develop an inadequate design (although this has happened, particularly from a behavioral standpoint), but the existence of such tendencies make a less effective design solution more probable.

III. THE SOCIOCULTURAL CONTEXT OF DESIGN

A. The Engineer's Attitude Toward Human Factors

It is impossible to understand design, and in particular how behavioral inputs are handled in design, without examining its sociocultural context. The role the engineer plays vis-a-vis his superiors (the management) and other contributors to the design process has significant implications for design efficiency (Perrow, 1983).

The sociocultural context of design is particularly important to its behavioral aspects because that context is often responsible for shortchanging behavioral design (although other specialties such as reliability or value engineering are also shortchanged, but perhaps not to the same extent).

We have talked as if the designer is one individual; in reality, he is and he is not. As systems become more sophisticated and design more specialized, the designer becomes part of an interactive group which, although the group reports to one man, functions in a very decentralized, almost competitive fashion.

One of the consequences of group design is that the design situation becomes much less structured. In very large projects, it is often difficult to determine who has the responsibility for what. If, as often happens, the behavioral input crosses traditional engineering specialties (e.g., electronics, hydraulics, controls), the human factors practitioner may have difficulty finding a focal designer on whom to press his inputs. In other words, it was much easier in the past when there was only one designer to convince. Under these circumstances, as one veteran human factors practitioner put it, "human engineering gets lost in the shuffle."

The interest that the human factors practitioner has in the design context is primarily to assure a more equitable consideration by the engineer of the behavioral aspects of design (henceforth referred to as behavioral design), in the anticipation that this will improve the efficiency of the total design. There is considerable empirical evidence (Meister & Farr, 1967) that designers have little or no interest in behavioral design. When behavioral information is provided to designers, they are often reluctant to use it. Of course, part of the problem is that they may not know *how* to use it, in which case human factors specialists have been deficient in communication. Consequently, an aid which the behavioral community endeavors to supply to the engineer must contain not only useful technical information, but must also possess characteristics that induce the engineer to use the technical material. It will not help the designer much if, as often happens, the aiding material (e.g., handbooks) is used only by the human factors practitioner. Consequently, the developers of a design support system must consider this aspect as a primary one, which is almost as important as the accuracy and relevancy of the technical material itself.

The reader may object: Suppose we provide the designer with material that is genuinely useful (at least potentially). Will he, as a rational being, use it, and will this use influence his behavior? Will he be able to recognize a potential benefit? Even in this case, the engineer may not recognize the material as being truly useful to him because the utility is not self-evident.

The problem of motivating the designer to consider behavioral factors in design more fully is a very difficult one for which there are no easy or immediate answers. Many engineers have a set of concepts that act to justify their lack of interest in behavioral design:

1. The designer has too much "on his plate" as it is without worrying about human factors. He is preoccupied with very serious problems that bear on whether he can make the system work at all, much less optimally.

A Cognitive Theory of Design

2. To consider behavioral design means that he must shuttle between the physical domain (in which he is reasonably comfortable) and the behavioral domain (about which he knows almost nothing). He is reluctant to engage in situations in which he is likely to fail.

3. Many designers feel that the effect of the human on equipment/system performance is minimal, and the system will overcome any behavioral deficiencies.

4. Whatever design deficiencies of a behavioral nature are not picked up as design proceeds (and the designer is certain they will be and that they will be rectified), the operator will be flexible enough to compensate for them.

5. There is nothing one can do to rectify the notorious incapability of the operator (a form of Murphy's Law); he will always "screw up," so it is useless to take precautions about this (i.e., to design specifically to avoid or eliminate behavioral defects).

6. There is in fact no problem with behavioral design because "good" design includes human factors engineering.

7. Even if behavioral design is inadequate, the application of human factors (as a discipline) will do no good because the discipline is too crude and cannot solve problems.

The fallacies are recognizable as prejudices, but they cannot be disdained merely as such because they will be operative to a greater or lesser extent whenever the designer seeks to use the support system.

There is no easy way to eliminate these prejudices, but this author feels that one promising way is to provide *quantitative evidence* that 1) error exists in system operation and affects that operation negatively, 2) error results from behavioral design inadequacies, and 3) eliminating these design inadequacies will improve overall system performance.

B. The Designer as Gatekeeper

In most design projects, one engineer is given primary responsibility for the design of the entire system or a subsystem. However, he functions as a leader of a team of designers and others like human factors practitioners with a relevant interest in that design. He generates design solutions on his own but also receives inputs and critiques from other team members.

As team leader, the designer must receive and pass upon the inputs of others. He acts almost as a goalie in a hockey game, batting away (rejecting) all inputs except those that, in his opinion, make a significant difference to the performance capability, reliability, cost, or development time of an equipment item.

It is important to realize that the more criteria the designer has to apply to these inputs, the more difficult his design decisions become. Consequently, he would prefer to ignore those inputs which, in his opinion, are least relevant to his primary problem. His goal is to reduce the number of inputs as much as possible without damaging the design or overlooking some pertinent aspects of the design.

The designer tends to divide design inputs into two categories—those that pertain to the primary hardware/software system being developed, and those auxiliary subsystems or system aspects that influence the primary system only marginally. The latter, among which the designer counts human factors, are considered of minimal importance and are often given the short shrift.

What this means practically is that whatever is provided (information, data, principles, etc.) in the design support system should be related as much as possible to the design of the primary system. A behavioral data input which has implications only for personnel selection or training would be considered by most designers as having little relevance to their primary design problem. Designer prejudices reinforce this attitude.

The engineer applies this input criterion on his own because it is not part of system requirements (one will not find it written down anywhere). However, he also takes his cue from his superiors. If they emphasize the importance of behavioral design, he will adopt the same philosophy.

The practical result of the designer's characteristic response to behavioral inputs is that behavioral material provided specifically for him usually tends not to be used by him. On the other hand, the human factors specialist, having usually been trained in a university tradition which emphasizes verbal indoctrination (the power of *logos*, the WORD, as it were), tends to feel that the kind of indifference to behavioral design displayed by the engineer can be overcome by verbal indoctrination. Such indoctrination requires, however, a high degree of creativity on the part of the human factors specialist.

IV. A BEHAVIORAL DESIGN DATA BASE

A. Required Characteristics

In this section, those characteristics that a behaviorally oriented design data base should have (that is, what we conceive a design support system to be) are described. We assume that the essence of any design support provided to engineers must be "hard" data; a design support system may contain other things (see below) but it is not viable without quantitative material. The inference one must draw is that the material provided in any design support system must have

A Cognitive Theory of Design 241

two characteristics: 1) it can be related easily to design of the primary system, and 2) its inputs should be in quantitative form. Arguably, one could supply the designer with general principles only. One might even recommend designs supposedly satisfying behavioral requirements (although we do not really know how to do this), but based on what we know about designers, they would be unlikely to accept such material without accompanying quantitative data.

Whatever the desired characteristics of a design support system are, some of the requirements specified will demand an intensive research effort before they can be incorporated into the support system.

The first question the developers of a behavioral data base must ask themselves is, what is the support system supposed to do? The reason for asking this question is that most of those who develop behavioral design aids have a motive that goes well beyond the provision of information. In most cases, not only do they wish the designer to make extensive use of the design aid, they want him to become more responsive to behavioral inputs in design and to be more approving of human factors as a discipline.

It is of course necessary for designers to use the design aid; if they do not, the effort has obviously failed. However, to induce the designer to use a support system is a difficult task and the only way to accomplish it successfully is to recognize that it is both necessary *and* difficult. A more positive attitude toward human factors may derive from use of the support system, but one cannot guarantee this and it should not be a primary goal.

If the support system is a data base, it should be useful in design at all system levels (i.e., system, subsystem, equipment, and module) because the designer will be functioning at all these levels. The nature of the data may or may not be the same at all these levels. The data supplied should of course be applicable to all system types.

The material provided should contain the following as a minimum:

1. Behavioral principles of design such as those found in MIL-STD-1472C (Department of Defense, 1981). The following is an example of such a principle. All controls should be located adjacent to the displays with which they will be used in operating the equipment. One would hope that principles are available that apply not only to molecular system units (like controls and displays) but also those that are meaningful at system and subsystem levels.

2. Associated with such principles should be human performance data, expressed in terms of probabilities of error or correct performance (one is, of course, the reciprocal of the other). For example, some estimate should be made of the heightened probability of error if functionally related controls and displays are widely separated; e.g., .0015 or 15 errors made in 10,000 opportunities to use these controls and displays. (This probability is, of course, completely hypothetical since no

empirical data exist relative to the behavioral principle in item 1 above.)

The reason for providing empirical error data is that the designer is unimpressed with any behavioral principle that lacks supporting data. One possible reason why designers pay comparatively little attention to human factors inputs in design is because it is merely our *assumption* that these principles are significant for system performance. Only if the designer can be shown the consequences (in terms of operator performance) of his ignoring these principles will he then consider them seriously. He does apply some human factors principles that appear rational to him (i.e., are commonsensical), but in any tradeoff in which he has to compromise, he will discard the behavioral principle because it has no *demonstrable* negative impact. To the extent that the designer is impressed by quantitative data, it might be said that even the motivational aspects of human factors are dependent on hard data.

Veteran human factors practitioners assert that designers are eager to accept any set of numbers the practitioner will give them, however unknown their validity. This may be an exaggeration, but the point is that the basic problem is not so much the designers' reception of these data as the reluctance of the behavioral community to consider development of quantitative data bases.

3. Data on the behavioral maxima and minima of fundamental design parameters, e.g., the minimum field of view for adequate operator surveillance; minimum number of TV lines to resolve alphanumerics, etc.

4. Quantitative relationships and tradeoffs between design parameters in human performance terms; e.g., the tradeoff between field of view and resolution.

5. Procedures the designer should follow in developing his design. These procedures would be based on a conceptual structure describing the engineer's cognitive processes in designing, such as the theory presented in the initial section of this paper. Since such a theory represents how the designer *should* think while designing, the function of these procedures would be to guide the engineer to follow *correct* procedure.

Presumably the support system will be computerized in the form of a dialogue between the software program and the designer. Following the procedures referred to in item 5 above, the program will ask questions of the designer and require him to make choices based on menu-type checklists. Ideally, it should be possible for the engineer to enter his design requirement (at whatever system level he is functioning) and have the computer guide him to a detailed answer by a series of narrowing, focused steps.

Essential to the development of the support system is the anticipation of the kinds of behavioral problems the engineer typically encounters and the provision of information in a format responsive to these problems. This will require some prior research to determine the design problems that occur most frequently. To help this aspect, the system should be developed with the continuing aid and participation of designers themselves.

If we look at these required characteristics, we must recognize that data for items 2 and 4 (the quantitative data-heart of the support system) are almost completely lacking. This is not the arena in which to discuss human performance data bases and data sources (for these, see Meister, 1982), but the development of an adequate data base oriented around human performance is a task of some magnitude. The remainder of the material, some of it quantitative, much of it qualitative, will pose a much less difficult problem for the support system developers. Emphasis has been placed on numerical data in this description of required characteristics, but the designer is also responsive to graphics and, wherever possible, numbers should be presented in graphic form.

Should a support system be developed as conceptualized in this paper, its effects would go well beyond its immediate intended purpose. It would, for example, show that it is possible to develop useful data bases constructed of empirical human performance data, and that such a data base has significant utility. The lack of such data relevant to design and system operation is one of the fundamental weaknesses of human factors. Any effort to strengthen that weakness cannot but influence every other area of human factors.

B. Recommendations

An effective computerized design support system for engineers should combine the best aspects of the verbal indoctrination approach and human performance data. One must ask, however, whether the ones who support the notion of such a support system also support the extensive research needed to secure the performance data?

It is important to reiterate that there are prerequisites to such a system:

1. We still know very little about design dynamics and how these relate to behavioral inputs. It is extraordinary that so little has been done in this field despite the fact that design is a major area of concern for human factors.

2. We lack a respectable human performance data base related to design variables.

The goal of developing a design support system for engineers is admirable but years of neglect of vital prerequisites severely hamper us.

REFERENCES

Department of Defense. (1981). *Human engineering design criteria for military systems, equipment, and facilities* (MIL-STD-1472C). Washington, DC.

Lintz, L.M., Askren, W.B., & Lott, W.J. (1971). *System design trade studies: The engineering process and use of human resources data*, (Report No. AFHRL-TR-71-24). Wright-Patterson Air Force Base, OH: Air Force Human Resources Laboratory.

Meister, D. (1971). *Human factors: Theory and practice.* New York: John Wiley & Sons.

Meister, D. (1982). What and where are the data in human factors? *Proceedings of the Human Factors Society 26th Annual Meeting*, 722-727.

Meister, D., & Farr, D.E. (1967). The utilization of human factors information by designers. *Human Factors, 9*(1), 71-87.

Meister, D., Sullivan, D.J., & Askren, W.B. (1968). *The impact of manpower requirements and personnel resources data on system design* (Report No. AFHRL-TR-68-44). Wright-Patterson Air Force Base, OH: Air Force Human Resources Laboratory.

Meister, D., Sullivan, D.J., Finley, D.L., & Askren, W.B. (1969a). *The design engineer's concept of the relationship between system design characteristics and technical skill level* (Report No. AFHRL-TR-69-23). Wright-Patterson Air Force Base, OH: Air Force Human Resources Laboratory.

Meister, D., Sullivan, D.J., Finley, D.L., & Askren, W.B. (1969b). *The effect of amount and timing of human resources data on subsystem design* (Report No. AFHRL-TR-69-22). Wright-Patterson Air Force Base, OH: Air Force Human Resources Laboratory.

Nadler, G. (1985). Systems methodology and design. *IEEE Transactions on Systems, Man, and Cybernetics, SMC-15*(6), 685-697.

Perrow, C. (1983). The organizational context of human factors engineering. *Administrative Science Quarterly, 28*, 521-541.

Rouse, W.B. (1986). On the value of information in system design: A framework for understanding and aiding designers. *Information Processing and Management, 22*(2), 217-228.

Chapter 19

DESIGNING FOR USER ACCEPTANCE OF DESIGN AIDS

Nancy M. Morris

Search Technology, Inc.
Norcross, Georgia

ABSTRACT

The potential utility of computer-based aids for system design is noted, and the importance of user acceptance to overall success of such systems is discussed. Four interacting dimensions are identified as being relevant to user acceptance: perceived usefulness, perceived usability, perceived level of discretion, and perceived organizational/peer group attitudes. Suggestions for encouraging development of acceptable aids for system design are offered.

I. INTRODUCTION

Design of large-scale systems can be a complicated and apparently chaotic undertaking. Undoubtedly, this statement is echoed in other papers in this volume. However, in spite of the complexities involved, the facts that planes do fly and power plants do produce electricity are evidence that systems are designed with a fair amount of success. Occasional problems such as the disaster in Flixborough, England in 1974 ("The Flixborough Disaster," 1974) and the incident at Three Mile Island in 1979 (Rubinstein, 1979) indicate that, in spite of impressive success stories, there is still room for improvement.

In light of the complex nature of the enterprise, it is not surprising that system design is increasingly becoming a computer-based activity. In fact, use of computers in design has enabled the design of more complex systems than has been possible without them (e.g., VLSI circuits). As is the case with so many other domains, visions of the potential benefits of introducing computer-based assistance into the design process have prompted this trend. Thus far, this assistance has been in the form of tools useful for detailed design. Tools facilitating rapid prototyping and visualization of design products have become available, as have capabilities to perform rather procedural design tasks such as manufacturability and stress analyses.

There is also interest in providing the designer with computer-based support in the less procedural aspects of design. Projects seeking to support the designer in making decisions, judging between alternatives, and, more ambitiously, designing systems at the conceptual level are in developmental stages. Not only are functional capabilities of these "designer support systems" expanding, but new roles for computer-based support systems are foreseeable as well. In contrast to the rather passive tools available today, the possibility of a support system being more actively involved in the design process by offering guidance or serving as a "coach" appears both attractive and feasible. Clearly, implementation of these enhanced capabilities for designer support can have a fundamental impact on the way designers perform their jobs.

When designing any sort of computer-based aid, there are many issues which must be addressed. Typically, hardware and software capabilities and constraints dominate the list of issues considered. Unfortunately, questions relevant to user acceptance of technology are often overlooked, or at least their full importance to overall success is not acknowledged at appropriate points during the design of the aid. The potential consequences of such oversight include intentional or unintentional misuse of the aid and failure to use the aid at all (Frey and Kisner, 1982). Further, the possibility for negative consequences associated with lack of user acceptance appears greater as the potential impact of an aid upon the way a job is performed increases.

Some of the factors which may influence user acceptance are discussed in this paper in an effort to illustrate the multidimensional nature of acceptance and the importance of considering user acceptance in all phases of support system design and implementation. Following this discussion, possible approaches to enhancing user acceptance are proposed. It is important to note that designers occupy two different roles in the different sections of this paper. In the section entitled, "Factors That Influence Acceptance," designers are viewed as the *end users of computer-based aids*. Thus, the discussion deals with factors which are expected to influence designers' acceptance of a support system as an aid for design of other systems. The section entitled, "Fostering Acceptance," considers the *designers of computer-based aids*. The focus in this section is on steps that can be taken to encourage those designers to develop aids that will be acceptable to the designers who are the intended end users.

II. FACTORS THAT INFLUENCE ACCEPTANCE

For present purposes, the following working definition is proposed: A computer-based aid is accepted to the extent that the end users willingly use it for its intended purpose. Acceptance of an aid is the result of an explicit or implicit tradeoff on the part of the user. If an aid is accepted, then the perceived benefits of using it outweigh the perceived costs.

The most obvious costs are the time and effort required to use the aid. Other costs are also possible. For example, the user might feel that using the computer-based aid reduces valued interaction with peers or discretion in the way his/her job is performed. Hopefully, a benefit of using the aid would be a perceived improvement in the performance of one's job. There may also be "hidden" benefits accompanying the use of an aid, such as increased job responsibility or compliance with the wishes of management.

Note that it is the user's *perceptions* of the aid that determine acceptance. Although it is sometimes reasonable to expect correspondence between the user's perceptions and objective "reality," such correspondence should not be taken for granted. Given this view of acceptance as a result of a cost-benefit tradeoff, the following dimensions are identified as relevant to user acceptance of an aid. For clarity, each dimension is first discussed in isolation, as if the effects upon acceptance were independent of values along other dimensions. Note, however, that strong interactions are expected and are discussed later.

A. Relevant Dimensions

1. Perceived Usefulness

Stated rather simply, if an aid is perceived as being useful, then it will be used. Although "usefulness" may be interpreted in many ways, perhaps the primary connotation is in terms of enhanced job performance. If quality of performance is the only consideration and use of the aid leads to better job performance than is achievable without the aid, then the aid will be used. However, there are other conditions which must be met in order for this to be true. The prospective user must perceive a need for and desire improvement in his/her own performance, and must further perceive that improvement is possible with the help of the aid.

If the aid provides a desired capability that is clearly beyond the human's ability to provide (such as application of a complex algorithm to a set of data), it is likely that these conditions will be met. However, meeting these conditions becomes less likely if 1) the functions performed by the aid are not perceived as necessary or relevant to the task at hand, or 2) the user has known the aid to make an error. Even a relatively low likelihood of error on the part of the aid may be sufficient cause for rejection (Zakay, 1982).

Chances that the user will perceive a need for the aid are also less likely if the aid performs functions which can be performed by the human. Laboratory evidence has shown that people may overestimate the quality of their own performance, even in well-defined situations, and thus may feel less need for computer-based assistance than they should (Morris & Rouse, 1986). If it is difficult to recognize a need for help in situations in which characteristics of "good" performance can be clearly specified, imagine the potential difficulty of realizing that an aid can enhance the often nebulous activities associated with design.

2. Perceived Usability

"Usable" is most commonly interpreted as "easy to use." All other attributes being equal, an aid which is easy to use will be used more readily than one which is not. If performance of a task with an aid is easier than performing the same task without it, then the aid will be used. Although intuitively obvious, the deceptive simplicity of these statements is realized when one considers the range of factors which may contribute to perceived ease of use.

For example, this perception may be influenced by the physical effort required to use the aid, as affected by such factors as the distance to the computer or terminal, the number of actions involved in accessing the aid (such as inserting diskettes), and the number of keystrokes required to accomplish a task. It may be pointed out here that, regardless of how little effort is required to use an aid, it will inherently be greater than that required to make a decision in an ad hoc manner (Phelps, 1986).

Mental effort required may be expected to contribute heavily to perceived ease of use, and may be influenced by a variety of obvious and more subtle factors. Some of these factors are aid response time, the hierarchical organization of menus, the size and mnemonic value of available command sets, and the appropriateness of the form and content of information provided.

Individual differences may contribute to the perceived magnitude of physical effort (e.g., entering a large number of keystrokes may not seem effortful to a skilled typist), and may influence perceived mental effort in a variety of ways. For example, impulsive persons may find a long aid response time more frustrating than do reflective persons. Features which make the aid appealing and usable to novices may appear tedious to experienced users. Whereas some persons may find verbal information useful, others may find a graphic representation of the same information more helpful. Studies of individual differences provide numerous examples such as these (e.g., Benbasat & Taylor, 1982; MacLean, 1986).

There is another aspect of usability which may have a substantial effect upon user acceptance. This is the level of enjoyment the user may derive from using the aid. Some aids are simply fun to use. (This is the case with many of the visualization aids available today.) The aid that is fun to use may not only be accepted, but may also cause designers to alter the way they perform their jobs so that they may use the aid more often.

3. Perceived Level of Discretion

The perceived level of discretion left to the user may also influence aid acceptance. The user-aid relationship is analogous in many ways to a manager-subordinate relationship. The manager may delegate some aspects of job performance willingly, but may find it highly desirable to maintain direct control over other aspects.

For an aid to be accepted it should perform those tasks which the user is willing to delegate, but it must not intrude upon those areas the user wishes to control directly (Rouse & Morris, 1985). Automation of tasks that users have reserved for themselves probably poses a greater threat to user acceptance than does failure to provide a support function that users desire. Generally, humans are more willing to delegate to a computer those tasks which appear to be well-defined and proceduralized (i.e., require little discretion) than they are tasks which appear to require judgment and planning. Additionally, attempts to convince users that judgment tasks can be proceduralized and thus delegated to a computer have been met with resistance (Adelman, Donnell, Phelps, & Patterson, 1982; Mackie, 1980; Price, 1985; Teates, 1986). Relative to this dimension, it seems possible that how users feel about an aid is not so important as how using the aid makes them feel about themselves and the contributions they make to overall job performance.

4. Perceived Organizational/Peer Group Attitudes

Finally, user acceptance of an aid may be affected by perceived organizational/peer group attitudes. This refers rather broadly to the wealth of perceived forces in the environment within which the user operates. Use of an aid may be endorsed by others in the environment, or it may be scorned. If the intended user perceives that independent thinking is valued, then use of a highly prescriptive aid may be unlikely. On the other hand, if "following the book" is the norm, a prescriptive aid may be more acceptable. Use of the aid may be avoided because it reduces valued interaction with colleagues; alternatively, consulting the aid may be viewed as a group process and an opportunity to gather with friends. Again, an aid may simply be fun to use. There is a large body of literature in the areas of management, social psychology, and industrial/organizational psychology that is potentially relevant to this topic (Tucker, 1981). Of the four dimensions discussed, it is the most context-specific.

B. Interaction of Dimensions

As noted earlier, these dimensions do not operate independently in their effects on user acceptance. Rather, the extent to which a user accepts an aid is dependent upon the interaction of all of these dimensions. Although it is possible to imagine plausible interactions between each of these dimensions, only a few are discussed here for illustration.

Consider the possible interaction between perceived usability and usefulness. For example, suppose the designer is provided with an aid in the form of a decision support system. If the intended user does not perceive that decisions are enhanced by using the aid, chances are that it will not be used. However, if the aid is extremely easy to use, it may be consulted as a second opinion for decisions after they are made. On the other hand, if the user's perception is that the aid is difficult to use, then a substantial performance improvement may be required to overcome this problem. Further, it must be apparent that the increment in performance is sufficiently valuable to the individual or organization to warrant the effort invested.

As a second example, the following relationships between perceived usefulness and desired discretion might be expected. It has been stated that an aid will be used if performance is enhanced by using it. However, if the aid performs a function that users would prefer to reserve for themselves, then the expected performance increment may have to be much greater than if the aid performs nonreserved functions. Additionally, users must feel that their own performance is definitely not as good as desired.

Finally, consider the possible interaction between perceived usability and organization/peer group attitudes toward the aid. If the aid is difficult to use, then the chances that it will be used are lower. However, a perception that use of the aid is valued by the environment may be sufficient to overcome the negative effects of difficulty.

Upon reading these examples, one might come to the conclusion that weaknesses in some areas might be compensated for by strengths in others. This conclusion is only partially true. It is suspected that there are minimum thresholds for acceptance along each of these dimensions. Failure to meet one of these thresholds may result in rejection of the aid, regardless of how well the aid fares along the other dimensions. For example, the aid that requires an inordinate amount of physical or mental effort will not be used, even if it offers the potential for large improvements in performance.

III. FOSTERING ACCEPTANCE

Consider now the people who will design these aids for designers. What can be done to assist them in developing aids which are acceptable to the intended end users? Identification of the above dimensions as being relevant to acceptance of an aid serves to highlight sources of potential problems. However, this analysis does not indicate what should be done to enhance acceptance. More must be done to encourage the development of acceptable aids. The following approaches are recommended.

A. Provide Guidelines for User Acceptance

If we consider the dimensions presented and attempt to identify their implications for aid design and implementation, a number of questions come to mind. For example, what can be done to influence users' perceptions of an aid along each of these dimensions? What is the relative importance of each of these dimensions to overall acceptance? To what extent are the answers to these questions context-specific?

We must seek to provide answers to these questions to the extent possible if we are to help designers. If we consider case studies from a variety of domains in which computer technology has been introduced into existing problem areas, it is possible to identify factors which have been associated with successful (i.e., accepted) and unsuccessful implementations. Results of these studies can be translated into guidelines for the design and implementation of design aids.

Some of these guidelines can be fairly specific, such as "provide an undo function" (Branscomb & Thomas, 1983) or "implement new systems during training if possible" (Phelps, 1986). Undoubtedly specific statements such as these can be very useful and should be heeded when they apply. However, in light of the multidimensional nature of acceptance, it is unlikely that very many specific, unconditional guidelines can apply across a wide range of situations. Further, such statements do not provide much insight into what the aid should be like or how it should be introduced.

It is more feasible to provide rather broad, general guidelines. Consider the following statement by Rouse and Morris (1985) concerning implementation of aids for system operation. "Assure that personnel understand both the abilities and limitations of the increased automation, know how to monitor and intervene appropriately, and retain clear feelings of still being responsible for system operation." Statements such as this synthesize a great deal of information in a concise form and have the virtue of being applicable to a wide variety of situations. However, translation of general guidelines into an appropriate course of action is not always straightforward. For example, how does one assure that personnel understand the abilities and limitations of automation? Evidence from other domains indicates that merely telling people about them may not be sufficient, as people may have difficulty putting information obtained via verbal instruction to use (Morris & Rouse, 1985).

In addition to examining case studies, research aimed at reducing deficiencies in our state of knowledge should be designed and conducted. Although some insights may be gained in the laboratory, the utility of simple laboratory experiments in answering questions about user acceptance will be limited. Researchers must explore the way "real people" (as opposed to, for example, college sophomores) perform tasks which are realistic to them. If experimental tasks are carefully designed and appropriate subject populations are selected, then it may be possible to address some of the questions related to ease of use and performance improvement in the laboratory. Investigation of other issues which are highly context-dependent, such as job discretion and peer attitudes, will require more field observation.

B. Encourage Designers to Be More Evaluation-Oriented

Although design guidelines should be helpful in enhancing aid acceptability, the multidimensional and often context-dependent nature of acceptance insures that it will never be possible to encapsulate the requirements for an acceptable aid in lists of specific "dos and don'ts" (Gould & Lewis, 1985). At best, it is reasonable to hope for a few specific statements, and a larger set of broader statements reflecting general areas of concern. Thus, these "guidelines" can never become "standards."

Designers cannot and should not wait upon standardization of user acceptance issues. Rather, they should be made aware of the relevant issues and should be encouraged to assume more responsibility for evaluation of their own work relative to its acceptability to specific users. Effective design for acceptance is virtually impossible without interaction with the intended user from the time of conceptual design until aid implementation is complete. Some authors have offered more detailed suggestions as to how the intended user may be included in aid design (e.g., Branscomb & Thomas, 1983; Carroll & Rosson, in press; Gould & Lewis, 1985), and this advice should be considered.

C. Encourage Designers to Value User Acceptance

If the above advice is to be followed, however, another more fundamental stumbling block must be overcome. Designers must be made to feel that devoting effort to enhancing user acceptance is worthwhile. Consider the design and implementation of decision support systems. Many designers seem to believe that decision support systems can be developed from a rational analysis of how decisions should be made and that involving the user in the design process complicates matters unnecessarily. If user acceptance is considered at all, it may be assumed that acceptance will be assured by providing people with an aid that is capable (in principle, at least) of improving performance.

In light of the strong interactions that might be expected between perceived performance and other dimensions discussed, the validity of this assumption is rather doubtful. The low success rate of previous attempts at implementing decision support systems attests to this as well. In spite of this negative evidence, however, significant forces from the environment may be required to change prevailing attitudes.

Unfortunately, the environment in which designers operate often does nothing to promote consideration of end user acceptance, and may even discourage it. This is due to the fact that the purse strings are usually controlled by people other than those who will actually use the aid. It is these third parties that designers must satisfy first if they are to sell their products. If the people procuring the aid do not demand aids that are acceptable to end users, then there is no impetus for designers to worry about acceptance. Designers may even develop systems they realize will *not* be acceptable to users but will be acceptable to buyers.

The further removed the buyers are from the intended end users, the less likely it is that they will be in touch with what the users need and want, and the less likely that the resulting aid will be acceptable to users. Regrettably, this is often true of the way in which systems are procured in large organizations, particularly in the case of the federal government. Thus, encouraging the development of systems which are acceptable to intended users may require a reexamination of organizational policies.

IV. CONCLUSION

The acceptability of an aid to the intended end users is of utmost importance to good aid design. Lack of acceptance on the part of users will undermine the effectiveness of an aid, regardless of how clever it is from a design standpoint. Designers of aids must come to realize that design features which are intriguing to them may not be of interest at all to end users. In other words, well-designed features do not always lead to perceived benefits.

In some cases, the user may have little choice other than to accept an aid, as performance of a necessary task may be impossible without it. For example, air traffic controllers must use available traffic monitoring systems to accomplish their jobs. However, if designers are viewed as end users and the task to be performed is system design, the activity involved is highly discretionary. Since designers may choose not to use an aid at all, the acceptability of the aid to designers is crucial to aid success.

As was noted earlier, designing for user acceptability is hampered by prevailing attitudes. The common tendency is to focus on software and hardware issues associated with making the aid operational. Consideration of issues related to user acceptance is postponed until "later." There are at least two problems with this approach. First, it is difficult to know what functions the intended users would like the aid to perform unless they are consulted. Thus, decisions as to aid functions which are made early in the design process may be inappropriate. Second, issues which are delayed until "later" often do not receive sufficient attention or support, and solutions to such issues may appear to be afterthoughts.

This attitude that it is unnecessary to consider user acceptance may be particularly strong in designers who are developing systems for other designers. Software designers of an aid may assume that they know what other designers want, even though these other designers may produce very different systems (e.g., aircraft). It is less likely that designers will assume that they know what other types of users (e.g., pilots) want.

Encouraging the development of user-acceptable aids will require more than reciting the litany of "consider user acceptance." Aid designers need clear statements of issues which should be considered and how the problem of user acceptance should be approached. Ways to provide these are suggested in this paper.

It is appropriate here to reemphasize the importance of guidance from the top. As noted, system developers will probably seek to satisfy the purchasers of the system. If the persons requesting an aid do not provide impetus to designing for user acceptability, then chances are that acceptability issues will not be stressed at the design level. Perhaps one way to encourage accountability for user acceptance might be to emphasize in the initial product request that evaluation of the usability and acceptability of the aid by the intended users is required. If this step is taken, however, purchasers should be willing to involve designers and users in specification of the aid, so that lessons learned by interacting with the intended users may be incorporated into the design.

REFERENCES

Adelman, L., Donnell, M.L., Phelps, R.H., & Patterson, J.F. (1982). An iterative Bayesian decision aid: Toward improving the user-aid and user-organization interfaces. *IEEE Transactions on Systems, Man, and Cybernetics, SMC-12*, 733-743.

Benbasat, I., & Taylor, R.N. (1982). Behavioral aspects of information processing for the design of management information systems. *IEEE Transactions on Systems, Man, and Cybernetics, SMC-12*, 439-450.

Branscomb, L.M., & Thomas, J.C. (1983). Ease of use: A system design challenge. *Proceedings of IFIP Conference.*

Carroll, J.M., & Rosson, M.B. (in press). Usability specifications as a tool in iterative development. In H.R. Hartson (Ed.), *Advances in human-computer interaction.* Norwood, NJ: Ablex Publishing.

Frey, P.R., & Kisner, R.A. (1982). *A survey of methods for improving operator acceptance of computerized aids* (Report No. ONRL/TM-8236). Oak Ridge, TN: Oak Ridge National Laboratory.

Gould, J.D., & Lewis, C. (1985). Designing for usability: Key principles and what designers think. *Communications of the ACM, 28*, 300-311.

Mackie, R.R. (1980). Design criteria for decision aids—The user's perspective. *Proceedings of the Human Factors Society 24th Annual Meeting*, pp. 80-84.

MacLean, R.F. (1986). User-system interface requirements for decision support systems. In S.J. Andriole (Ed.), *Microcomputer decision support systems: Design, implementation, and evaluation.* Wellesley, MA: QED Information Services.

Morris, N.M., & Rouse, W.B. (1985). The effects of type of knowledge upon human problem solving in a process control task. *IEEE Transactions on Systems, Man, and Cybernetics, SMC-15*, 698-707.

Morris, N.M., & Rouse, W.B. (1986). *Adaptive aiding for human-computer control: Experimental studies of dynamic task allocation* (Report in press). Norcross, GA: Search Technology.

Phelps, R.H. (1986). Decision aids for military intelligence analysis: Description, evaluation, and implementation. In S.J. Andriole (Ed.), *Microcomputer decision support systems: Design, implementation, and evaluation.* Wellesley, MA: QED Information Sciences.

Price, H.E. (1985). The allocation of functions in systems. *Human Factors, 27*, 33-45.

Rouse, W.B., & Morris, N.M. (1985). Understanding and avoiding potential problems in implementing automation. *Proceedings of the 1985 International Conference on Systems, Man, and Cybernetics*, pp. 787-791.

Rubinstein, E. (Ed.). (1979). Three Mile Island and the future of nuclear power (Special issue). *IEEE Spectrum, 16*(11).

Teates, H.B. (1986). The role of decision support systems in command and control. In S.J. Andriole (Ed.), *Microcomputer decision support systems: Design, implementation, and evaluation.* Wellesley, MA: QED Information Sciences.

The Flixborough disaster: Report of the court of inquiry. (1974). London: Her Majesty's Stationery Office.

Tucker, J.H. (1981). Implementation of decision support systems. *Proceedings of the International Conference on Cybernetics and Society*, pp. 654-659.

Zakay, D. (1982). Reliability of information as a potential threat to the acceptability of decision support systems. *IEEE Transactions on Systems, Man, and Cybernetics, SMC-12*, 518-520.

Chapter 20

ENGINEERING DESIGN SUPPORT SYSTEMS

Robin Popplestone, Tim Smithers, Jonathan Corney,
Anastasia Koutsou, Karl Millington, and Gideon Sahar

Department of Artificial Intelligence
University of Edinburgh
Edinburgh, England

I. THE NATURE OF DESIGN

Engineering design may be regarded as the passage from a specification of a requirement to a specification of an artifact that will satisfy that requirement. We shall refer to this artifact as the Designed Artifact (DA). If a computational system is to be used to advantage in the process, it is necessary that both the specification of the requirement and that of the DA be in a formalism comprehensible to the system. We shall use the term Design Support System (DSS) for such a computational support for the human designer. In this paper, we shall draw extensively on experience gained in our work on implementing the Edinburgh Designer System, which forms part of the Alvey Large Scale Demonstrator Project, "Design to Product" (Alvey Programme Annual Report, 1984; Department of Trade and Industry, 1982).

A. The Form of Transactions with a Design Support System

We will use the term "design description document" to refer to a record of all statements, textual or graphical, which are made about the possible form of the DA, and of all facts inferred by the computational system from these statements. Design is an exploratory activity, so we need to support people in changing their minds. Hence, the design description document may contain statements that do not form part of the final design.

Communication about a design may be conducted textually or graphically. Much of the conceptual content of a design will not be initially represented as conventional engineering drawings, but may be represented as sketches. The purpose of sketches is not to convey precise geometrical information, but to establish a basic vocabulary of entities and their relationships. Thus, a sketch should be considered as a statement in a graphical language. The form that such a graphical language might take if it is to be machine-interpretable is an issue for consideration. One might imagine it as being something rather like the MacPaint program, with a capability of labeling each graphical entity as it is created as representing an engineering entity.

As the capability of creating synthetic three-dimensional "worlds" is developed, these worlds may also serve as a means of communicating design intentions. Just as it is necessary and nontrivial to attach a meaning to a graphical act, so it will be necessary (and nontrivial) to attach a meaning to a manipulative act in such a synthetic three-dimensional world.

B. Design and Manufacture

A design must take account not only of the functional requirements for the DA to work and to work reliably, but it must also take account of the need to manufacture the DA economically and to maintain it in the field. However, as we shall see below, considerable revision to current approaches using computing systems in support of the human designer needs to be made if the full potential of such systems is to be realized.

1. The Conventional Organization of Engineering Design and Manufacture

The procedures and operations carried out in the design and manufacture of engineered products are traditionally organized in a sequential way. After the need and initial specification for a product have been identified, the process starts with its conceptual or schematic design, followed by its detailed design. Separate stages follow for manufacturing planning, manufacturing, and testing before the product is packaged and sent out into the field. It will be subject to regular maintenance and occasional repair procedures in the field.

Throughout this sequential process, knowledge about the particular product is being generated, some of which is passed on for use in the next or subsequent stages. For example, in the design of a round shaft element subjected to a torsional load which has a change in diameter from $D1$ to $D2$ part way along its length, and has a shoulder fillet radius of r, the values of $D1$, $D2$, and r will be related by a stress concentration factor, K. In other words, for given values of $D1$, $D2$, and K, r will have a specified value determined by the designer at design time. This knowledge about the size of the shoulder fillet radius will be passed on to the production engineer in the form of an engineering drawing of the component on which the dimensions $D1$, $D2$, and r will be specified. What will not be passed on, however, is the fact that the value of r depends on the value of $D1$ and $D2$ in a particular way, and in this example, in a possibly critical way. Thus, unless the production engineer and the operator of the machine on which the component is turned, or the part programmer for the numerically controlled machine, also know about the stress concentration factor relating r to $D1$ and $D2$, they may be tempted to change the value of r to one more convenient to the manufacturing plan (for example, to minimize tool changes).

When all the designing, manufacturing, and testing have been completed and the product is ready for dispatch, a great deal of knowledge will have been generated about the particular product's function, shape, manufacture, and test performance. The problem is that some of this knowledge is lost between the different stages, and what does survive is distributed throughout the organization. This loss and distribution of knowledge also means that some of it has to be recreated at different times and in different forms. This leads to inconsistency and the inability to transfer knowledge in ways other than those allowed by the sequential organization of the different stages.

2. CAD/CAM and Its Current Developments

The introduction of computer-aided drafting/design and computer-aided manufacturing systems into traditional engineering design and manufacturing procedures has not changed the fundamentally sequential process described in the preceding section. What they have done is improved the efficiency and speed with which certain stages in the sequence are performed. By their support of standard parts and part program libraries, they have improved the knowledge distribution problem in certain areas, but because they still embody the sequential knowledge transfer concept, they do not improve the knowledge loss and regeneration problem. The introduction of new technologies into this sequential and compartmentalized approach has thus lead to the creation of "islands of automation" (Mather, 1984).

The present developments in the CAD/CAM and CIM areas can be seen as attempts to link existing islands of computer-based aids or automation; for example, computer-aided drafting to NC cutter path generation; NC cutter path generation to NC part program preparation and planning; and part programming to production control. However, these islands are still part of the traditional sequence of the design and manufacturing stages, and joining them with common data bases and computer-controlled data or knowledge flow does not remove the inefficiencies inherent within this type of organization. The feedforward and feedback problems that occur in the sequential organization of engineering design and manufacture also are not removed, and neither are the data consistency and management problems. This failure to achieve real integration using current CAD/CAM and CIM systems means that the integrated interdisciplinary functioning and continuity of operations required to improve the efficiency and flexibility of a manufacturing company is not realized (Jefferies, 1984).

C. The Integrated Information Approach to Design

It is necessary, therefore, to avoid the losses in the product knowledge generation cycle, and to solve the knowledge consistency and management problems inherent in existing design and manufacturing methodologies. It is proposed that these aims will be met by an integrated and unified product description whose consistency will be maintained and managed by the DSS, and which will contain all the knowledge generated about a particular product at any time.

Thus, it is the commonality of the knowledge required and generated during a product's design, manufacture, and service life which is seen as the underlying, unifying, and integrating factor, not the separate implementation of some of the different stages which occur during design and manufacture as in the philosophy behind current CAD/CAM and CIM developments. The DSS has therefore to be seen not as one end of a sequence of design, planning, and manufacturing stages, but at the center of the whole product generation activity. This central position occupied by the DSS means the introduction of flexibility into the traditional sequential ordering of the various stages, thus more easily allowing aspects within them to influence aspects within other stages. In other words, the product design activity is placed at the center of the product generation process where each kind of activity (designing for function, manufacture, and serviceability) can more easily and directly influence the others in an integrated and consistent manner.

1. Artificial Intelligence Techniques

The concept of a unified and integrated knowledge base at the heart of the design and manufacturing activity provides an ideal application for some of the techniques which have come out of Artificial Intelligence (AI) research. The support of the various design activities is also seen as an important application of AI research. Of particular importance to the creation of a DSS is the AI work done in the areas of knowledge representation; algebraic equation solving; geometric and spatial relationship reasoning; and automatic robot and manufacturing planning, design support, and verification.

II. MANAGEMENT AND TRANSFORMATION OF KNOWLEDGE

A DSS can be conceived of as a sort of animated text book cum catalog collection, and the sort of knowledge present in it can be inferred by perusal of such documents. The knowledge will express facts about engineering entities;

for example, electric motors or aluminum alloys. Thus, it may contain information about the yield stress of a given alloy or an electric motor:

$$T_{out,m1} = k_{m1} * i_{m1}$$

(i.e., that the output torque is a constant times the input current). Thus, the DSS will have 1) a repository of long term "knowledge" of the branch of engineering supported by the DSS, which we will refer to as the "encyclopedia"; and 2) a means of creating a design description document which contains information relevant to a particular design activity. This document contains both information derived from assumptions the user makes about what is present in the DA, and deductions made from these assumptions by various "engines" which are information transformers.

A. The Form of the Encyclopedia

The forms of knowledge in the encyclopedia in the DSS can be classified, not necessarily mutually exclusively, as 1) English, 2) Tabular, 3) Graphical, 4) Mathematical, and 5) Heuristic.

1. Knowledge Expressed in English

While the interpretation of free English is an aim of AI systems, in our opinion the best use of English in a DSS in the medium term is to assist in the communication with the human user. That is to say, English text serves to clarify the meaning of more formal knowledge. As such, it should be embedded in the DSS in a way that formally relates it to other knowledge. If, for example, a user needs to know the interpretation of the symbol σ_{s1}, the DSS can respond, "It is shear stress in shaft s_1," by making reference to a formal association between the declaration of the quantity σ_{s1} and a descriptive phrase of English appearing as a comment on the declaration.

2. Knowledge Expressed in Tabular Form

Tables are used to express knowledge that is either arbitrary or empirical. An example of arbitrary knowledge is that about standard screw thread dimensions. An example of empirical knowledge is that about strengths of materials. Tabular knowledge can be quite well encoded in a form similar to that used in relational data bases, but interpolation of engineering data is commonly required. For example, Table 1 is an extract from a table found in an engineering textbook (Hall, Holowenko, & Laughlin, 1980).

A table of this form is not immediately usable in a DSS. First, the column headers are commonly descriptive English phrases. It is necessary to provide a more formal structure than this, which allows a correspondence to be made

Table 1. Form Factors y—for Use in Lewis Strength Equation. (From Schaum, with permission from McGraw-Hill Book Company.)

Number of teeth	14 1/2° Full-Depth Involute or Composite	20° Full-Depth Involute	20° Stub Involute
12	0.067	0.078	0.099
13	0.071	0.083	0.103
14	0.075	0.088	0.108
15	0.078	0.092	0.111
16	0.081	0.094	0.115
17	0.084	0.096	0.117
18	0.086	0.098	0.120
19	0.088	0.100	0.123
20	0.090	0.102	0.125
21	0.092	0.104	0.127
23	0.094	0.106	0.130
25	0.097	0.108	0.133

between quantities referred to in tables and those occurring in the equations and inequalities occurring elsewhere in the encyclopedia entry on gear teeth. Second, the tabular form may not directly correspond with a relational data base form, where the column headers refer to the values of certain quantities. Third, beyond the capabilities of most relational data base systems is the requirement to interpolate between values in the table; for example, to arrive at a suitable value of the form factor for 22 teeth, when only values for 21 and 23 teeth are listed.

The above table might need to be recast as Table 2.

Table 2. Form Factors for a Relational Data Base.

n	ϕ	frm	y
log	exact	exact	lin
12	14.5	fdi	0.067
12	20	fdi	0.078
12	20	si	0.099
13	14.5	fdi	0.071
13	20	fdi	0.083
13	20	si	0.103
14	14.5	fdi	0.075
.	.	.	.
.	.	.	.

Here n is the number of teeth, ϕ is the pressure angle, *frm* is a discrete parameter representing the tooth form (fdi = full depth involute form and si = stub involute form), and y is the form factor. The second row indicates whether interpolation is allowed on the parameter concerned, and if so, a scaling function is given to determine a metric on the space. For example, the difference between gears with 120 and 130 teeth is about the same as that between gears with 12 and 13 teeth, so a logarithmic scaling is indicated.

It should be noted that there is a close correspondence between data presented in the form of graphs and that presented in tabular form. Graphs, if based on empirical data, need to be input to most computer systems in tabular form. Output of tables and other constraints should be possible both in graphical and tabular form.

Another extension of the normal capabilities of relational data bases is the requirement that some tables have symbolic entries which are to have appropriate interpretations put upon them. Such entries are commonly found, for example, in textbook entries dealing with beam theory, where the moment of inertia for a given beam cross section is expressed symbolically in terms of the parameters defining that cross section.

3. *Knowledge Expressed in Graphical Form*

The important point here is to treat graphics as a language so that any combination of graphics primitives expresses the existence of an engineering entity or relationships between such entities. Precise quantitative data should not be conveyed graphically, although there is significance to be attached to the relative sizes of the graphical representations of entities.

4. *Knowledge Expressed in Mathematical Form*

Mathematical knowledge is not to be confused with procedural knowledge, which we treat under the heading of "transformation of information." Mathematical knowledge is intended for symbolic manipulation by the DSS and is consequently represented in a form governed by a formal syntax. The Edinburgh Designer System (Popplestone, 1985) uses Prolog terms (Clocksin & Mellish, 1981) for this purpose, but equally acceptable might be something derived from the input form of Macsyma (Bogen et al., 1983) or REDUCE (Hearn, 1984). The external form of LISP S-expressions is not sufficiently user-friendly to be satisfactory as a medium for presenting mathematical knowledge to the user; thus, some syntactic gloss, as provided by Macsyma or REDUCE, is needed for a LISP-based system.

Mathematical knowledge in mechanical engineering is more concerned with spatial considerations than would be the case for electrical engineering. At least the following areas should be covered:

a. Algebra of the reals, including elementary transcendental functions,

b. Algebra of the complex numbers,
c. Algebra of vectors and matrices over the above,
d. Algebra of shapes—constructive solid geometry,
e. Algebra of sets and sequences,
f. Differential and integral calculus, and
g. Temporal logic (for plan formation).

Pragmatically, it will probably be necessary to have both a "teletype form" and a "typeset form" for mathematical expressions. For example, it is easy with current equipment to prepare text such as:

sigma(1..n, i->tau $ i).

but this may be presented as:

$$\sum_{i=1}^{n} \tau_i .$$

5. Heuristic Knowledge

Heuristic knowledge is intended to help the DSS find a good or acceptable design. It may take the form of mathematical constraints which are flagged as being advisory rather than mandatory (e.g., "the face width of a gear tooth should be from three to five times the circular pitch" can be expressed as two inequalities). It may indicate the focus of attention that best serves to determine a design; for example: "When designing a gear box, first determine the major parameters of the gears, then select trial bearing positions, and work out bearing forces. Choose shaft diameters and compute shaft loads and deflections, select the bearings themselves" and so on. However, while it is fashionable to praise such "scripts," they do tend to obscure the possibility of the application of general principles. Thus, in the above example, assuming that the gears are run under normal conditions, the shaft loads *depend* on the gear forces and thus on the gear parameters, but not conversely. This fact should be evident from an examination of the mathematical knowledge employed.

B. The Organization of the Encyclopedia

There are difficulties in representing knowledge using "frame" - based knowledge systems. An excellent critique of such systems is to be found in Brachman (1983). Following Barrow (1983), we shall use the term "module" for engineering entities with a concrete referent. Examples of the engineering entities which are modules are energy converters, motors, shafts, keyways, bearings, oilways, and tapped holes. The Edinburgh Designer System makes use of a strict taxonomy of modules, described below, as the primary organizational structure of the encyclopedia. The use of this strict taxonomy should avoid the problems which are discussed by Brachman (1983). However, engineering knowledge encompasses other entities (vectors, for instance) which are not modules in the above sense.

By a module taxonomy we mean a classification of entities such that every entity class, apart from a unique maximum class, is a subset of exactly one class at the next higher level of the taxonomy. Thus, for example, the class *"spur gear"* is contained in the class *"gear."* The taxonomy is strict in the sense that any property of a superclass is of necessity a property of the subclass. If, for example, the angular velocity of a gear in radians per second is represented by ω, then the angular velocity of spur gears must also be represented by ω.

Such a taxonomic classification may be criticized as being too restrictive. For example, the class *deep groove ball bearing* might well be considered to be a subclass both of the class *revolute bearing* and of the class *rolling element bearing*. That is to say, the bearing may be classified both by the kind of motion it contains and the kind of technology employed.

There would seem to be no objection in principle to having a non-tree classification scheme of this sort, provided that inheritance rules are obeyed down all paths. The requirement for a tree taxonomy in biology derives from the need to represent the accepted state of the evolutionary tree at a given moment in time, and even then it ignores the possibility of the permanent incorporation of material in the genome carried by vectors like bacteriophages and molecular biologists.

While any organization of the encyclopedia would seem to require a taxonomic component, there are other relationships which are of importance. For example, one module will often have others as parts, components, or elements. If we use the word "component" to mean a distinct sub-body of a module, then we may use the word "element" to mean modules which are logically part of a larger module without being components. For example, a valve block may have elements which are oilways and other elements which are mounting holes.

Whatever the organization of the encyclopedia, it is important to present a graphical representation of this knowledge structure to the user of the DSS. This enables the user to readily locate entries concerning engineering entities relevant to his needs; for example, graphic representation is as a tree structure.

1. Problems Encountered in a Taxonomic Organization

A problem with using the taxonomic form is that of the possibility of the development of modules which combine the functions of two other modules. For example, there exists electric motors with built-in shaft encoders. These are neither motors nor encoders since they have more properties than either, but if one is building a system which requires both a motor and a shaft encoder, it is important to know about the existence of such combined modules. This difficulty can be dealt with, while preserving strict inheritance, by providing reference to conceptually related items which serve as user guides rather than as strict taxonomic links.

A related issue can be the treatment of entities which can serve more than one purpose. For example, a screw can be used both for the transmission of power and as a fastener, and it should be found in the encyclopedia under both headings. However, the theory required is different. For example, as a screw fastener, the power capacity of the screw will not be relevant. This difficulty can be surmounted in general by having an entity definition which encodes the common properties of the entities in question, and then by making this an *element* of the various manifestations. Thus, there would be an encyclopedia entry for screws, which would have basic facts about such things as the relationship between angular velocity, pitch, and linear velocity, as well as separate entries for screws as fasteners and power screws, which make reference to the general screw as an element.

Yet another difficulty that arises in the universal classification of engineering knowledge is the existence of multiple sets of standards, particularly with respect to the units used, but also with respect to the existence of various codes of practice. For example, are bevel gears designed to the British Standard BS545 logically identical to bevel gears designed to the corresponding American standard? Clearly making multiple entries in the encyclopedia would be a pragmatic solution to this problem, but a more elegant solution to this problem has yet to be proposed.

2. Designs Described in Terms of Modules

Using the DSS, a design is specified in terms of *modules* and in terms of how these modules combine to form supermodules. Such a combination serves to provide the same information as the *ports* used by Barrow. Modules have *parameters*, which are symbols denoting quantities whose values are determined during the design process, and *variables*, which are symbols denoting quantities which vary during the operation of the module. For example, a DC motor has its maximum power output as one of its parameters and the rotational speed of its shaft as one of its variables.

At the lowest level of the module taxonomy, the modules will be actual engineering components. Most of the knowledge for these modules will, therefore, be that presently provided in manufacturers' component catalogs. The taxonomy also provides the means by which standardization in the use of certain types of components can be introduced into the design and manufacturing system.

The use of functional unit modules to describe a design or set of designs should not be seen as implying "modular design" since the modules referred to in this paper are basic engineering functional entities. In order for the DSS to contain definitions of complex higher level modules and not coerce a modular form of design, the concept of *modular identification* has to be introduced. This allows the designer to decide that two modules which were originally considered to be distinct are in fact identical. For example, an electric motor always has a frame which supports its working parts, and a transmission, likewise, needs a frame. In an optimized design, the frame of the motor and transmission may be identified as being the same.

While Prolog terms are an acceptable medium for encoding much mathematics, the same is not true of the use of Prolog clauses for encoding the larger structures of the encyclopedia. We believe that in any DSS, there is a need to employ a specifically designed formalism in which to encode encyclopedic knowledge. Thus, in the case of the Edinburgh Designer System, taxonomy is encoded in an external form and automatically translated into the internal Prolog form.

C. Transformation of Information in the Design Support System

The purpose of the DSS is to assist in the activity of mechanical engineering design and manufacturing planning, which is seen as being a part of the design process. It is thus a means of exploring the space of possible designs via the consideration of different design assumptions, and providing the means to assess and analyze the suitability of a particular design or set of designs. It will also provide the mechanism for maintaining a consistent representation of the design or designs being worked on.

A number of subsystems are needed within the DSS to provide the different types of functionality needed to support engineering design. They come under the broad headings of knowledge representation, design space exploration and design description, inference engines, specialists, and user interface. These are presented in more detail in the following sections.

1. The Design Description Document and the Assumption Based Truth Maintenance System

The term, "design description document," is used to refer to the body of knowledge about a product which is built up using the DSS. It is where the description of a design or set of designs is maintained in terms of instances of engineering entities, including functional unit modules.

The activity of design involves the exploration of the space of possible designs. This exploratory nature of design is treated by regarding any statement made by the designer to the system as an *assumption*, which may or may not form part of the final design. For example, a designer may wish to consider the possibility of using an electric motor or a hydraulic motor in a given product, or at a more detailed level, the possibility of using a journal bearing versus a rolling element bearing. While some of the design work associated with a particular assumption may be lost if an alternative assumption is finally chosen, it is important to avoid losing more than is necessary since some if it may be used again when making further assumptions.

Thus, there is no requirement for the assumptions to be consistent. For example, the assumption that the parameter p of a module has the value of 10 units might coexist with the assumption that p has the value of 20 units. However, a final design is characterized by a set of assumptions which must not be consistent. Therefore, the two values for p quoted above can only form part of alternative possible designs.

Maintaining "true" knowledge in a system where facts are being continually changed and added is one problem frequently faced in AI research, and there have been a number of approaches developed to deal with it. The system being developed for the Edinburgh Designer System to deal with the assumptions made by designers during the design of a product is called an Assumption Based Truth Maintenance System (ABTMS). The ABTMS is based on the work of deKleer (1984). As well as assumptions about the existence of particular modules, the designer will also make assumptions about the values of module parameters, the way in which modules combine to form supermodules, the spatial relationships between their features, and the grouping of modules to form components.

While a small design exercise could be carried out with no additional structure on the ABTMS, serious design will require the provision of a means to *focus* the design space exploration. This is done by setting a limit on the set of assumptions and consequences used to infer new consequences, thus preventing the generation of possible consequences that are not of immediate relevance to the design. For example, in the case of a gear box design, the initial focus will be on the gear teeth, followed by the location of the bearings on the shafts; the loading of these bearings; the analysis of the shafts for stiffness and strength; the detailing of bearings, housings, and gear wheels; etc.

Conceptually, the design description document is at the center of the Design to Product system where it holds the knowledge about a product generated and required during its design, manufacture, and service life.

D. Engines

A number of "engines" within the DSS are required to derive or infer consequences of the assumptions made by the designer. They will work in a forward chaining mode, although it will be necessary to restrict their operation in order to avoid an undesirable proliferation of consequences.

1. The Algebraic Engine

The Algebraic Manipulation and Solution Engine is used to manipulate symbolic constraints and solve symbolic equations as required by engineering

Engineering Design Support Systems

design. It will also be used to follow paths in the relational structure of the design mediated by constraint equations, or by other types of symbolic representations, and to infer new relationships.

We have already referred to the possibility of using REDUCE or Macsyma to provide a representational basis for design, and, naturally, they could well be considered as also providing a possible basis for the Algebraic Manipulation and Solution Engine. Macsyma and REDUCE, for example, would provide a valuable capability in the area of symbolic integration.

2. The Spatial Relationship Reasoning Engine

This engine serves to perform inferences about the relationships between features of bodies and, consequently, to deduce facts about the location of bodies. For example, we may say that a bearing seating bs_1 of a shaft *fits* the inner ring of a bearing bg_1, so that the constraint bs_1 *fits* $inner_{bg_1}$ holds. Such relationships are basically kinematic and can be handled algebraically (Brooks, 1982; Kapur, Mundy, Musser, & Narendran, 1985; Popplestone, Ambler, & Bellos, 1980) or by systematically reducing a more geometrical representation (Corner, Ambler, & Popplestone, 1983). If an algebraic modus operandi is chosen, then this engine can be regarded as an extension of the Algebraic Engine. Earlier experience at Edinburgh indicates that there are considerable advantages of efficiency to be gained if a geometric approach is used at least to reduce complexity arising from the existence of multiple, rather simple, relationships. This engine will usually work in association with the Geometric Modeling Engine.

In the Edinburgh Designer System, the Spatial Relationship Reasoning Engine is based on the RAPT Inference Engine. It forms part of the RAPT Robot Programming System described by Corner, Ambler, and Popplestone (1983) and is used for inferring module locations from defined spatial relationships.

3. The Geometric Modeling Engine

The use of a solid geometry modeler is required so that space occupancy and intersection questions can be answered—something which is not possible with "wire-frame" modelers. The Geometric Modeling Engine thus provides the three-dimensional geometry and spatial occupancy reasoning capability required to support mechanical engineering design; for example: Do these two bodies interpenetrate? There is room for debate as to the relative merits of surface-based modeling versus constructive solid geometry (CSG) modelers of the sort described by Requicha and Tilove (1978). CSG modelers represent shapes as a boolean combination of relocated primitive shapes, often referred to as CSG trees. The authors advocate the use of CSG modelers for this purpose since they fit well with the concept of a module taxonomy. The shape of a composite module may be derived from the shape of its elements by applying a well-defined set of rules.

An additional modeling capability which is clearly important to design is the ability to model bodies in motion. Such a capability is embodied in the ROBMOD polyhedral modeler developed by Cameron (1984) and already in use in the Edinburgh Designer System.

4. Specialists

Apart from the support for designing the functional and geometric aspects of mechanical parts, there is also a need to support the "design for" type of activity. Examples include design for assembly, design for machining, design for maintenance, design for test and inspection, and design for cost. These specialized aspects of design will be supported partly by including relevant knowledge in the encyclopedia and partly by rule-based subsystems which are referred to as Specialists.

5. The User Interface

To build a user interface for a DSS is a large design problem in itself. One thing is clear, however, and that is that the continuing development of workstations with large high resolution screens supporting multiwindowing and non-keyboard interaction devices will be of particular importance.

In the future, the use of "synthetic worlds," made possible by helmet displays with tactile feedback, opens the possibility of entirely new methods of communication between man and machine about design. There is no need for such worlds to obey normal physical laws, and indeed it can be advantageous for nonphysical transformations to occur. For example, when a module is chosen from a high level in the taxonomy, it should be represented by what is in effect a "mock-up." For instance, a motor might be represented by two coaxial cylinders, one representing the body and one representing the output shaft. One can imagine that descending the taxonomy to levels of greater detail could be accomplished by equipping such mock-ups with buttons, which will cause a local representation of the taxonomy to appear magically on the mock-up itself. The user of the system will then be able to choose the more detailed module he requires by pressing a button on the taxonomy, whereupon the mock-up will transform itself into a representation of the new module.

There has been, and continues to be, a good deal of work on how man-machine interfaces should look and behave. A particular problem faced when designing a new tool as powerful as a DSS is that you cannot easily get a true picture of how engineers will use it before it is built. One approach currently being developed to deal with this problem is called "user modeling." It involves the interface to the system maintaining a "model" or "models" of the people who use the system to adjust its response to a particular user or type of user.

6. Manufacturing Planning and Scheduling

In the traditional sequential and departmentalized organization of engineering design and manufacture, planning and scheduling is often restricted to a slot between design and manufacture. In reality, however, the need to plan and schedule a product's manufacture starts from the moment designing begins and finishes when the last manufacturing or testing operation has been completed. It, therefore, needs to be supported at all stages of the product's design and manufacture. In a design-centered product generation system in which the DSS is surrounded by, and interacts with, all the other components of the factory system, the scope of planning and scheduling can, with benefit, be extended to a much more diverse and varied set of problems.

E. Implementation

A DSS of the type discussed in this paper forms the Edinburgh part of the Design to Product Demonstrator Project. It is being implemented in the POPLOG programming environment and is run under the UNIX operating system on a network of Sun Microsystem computers. The POPLOG (Barrett, Ramsay, & Sloman, 1985; Hardy, 1984) programming system provides the POP-11, Prolog, and Common LISP programming languages, which can be freely mixed. It also provides an integrated screen editor which allows incremental compilation of code modules into an existing system.

In terms of hardware, it is already clear that before a Design to Product system can realistically be implemented in a factory environment, new and more powerful hardware systems will be required. Of particular importance is the current research into relational data base hardware and parallel processor architectures for applicative languages.

ACKNOWLEDGMENTS

The work being carried out in the Department of Artificial Intelligence at Edinburgh University on the Design to Product Project comes under the Alvey Programme of Research, and is funded by H.M. Government through the Science and Engineering Research Council. Additional funding for other parts of the project is being provided by GEC plc, Lucas CAV Limited, and the Department of Trade and Industry. We express our thanks to Sue Renton, Anne Elder, and Chris Malcolm for their help and advice while writing this paper.

REFERENCES

Alvey programme annual report (1984). Published by The Institution of Electrical Engineers.

Barrett, R., Ramsay, A., & Sloman, A. (1985). *POP-11, a practical language for artificial intelligence.* Ellis Horwood.

Barrow, H.G. (1983). Proving the correctness of digital hardware designs. *Proceedings of the National Conference on Artificial Intelligence.* Washington, DC.

Bogen, R., Golden, J., Genesereth, M., Pavelle, R., Wester, M., Fateman, R., & Doohovskoy, A. (1983). *Macsyma reference manual.* Massachusetts Institute of Technology, The Mathlab Group, Laboratory for Computer Science.

Brachman, R.J. (1983, Fall). "I lied about the trees" or, defaults and definitions in knowledge representation. *A.I. Magazine,* pp. 80-93.

Brooks, R.A. (1982). Symbolic error analysis and robot planning. *International Journal of Robotics Research, 1,* 4.

Clocksin, W., & Mellish, C. (1981). *Programming in Prolog.* New York: Springer Verlag.

Cameron, S. (1984). *Modeling solids in motion.* Ph.D. Thesis, Edinburgh University, England.

Corner, D.F., Ambler, A.P., & Popplestone, R.J. (1983). Reasoning about the spatial relationships derived from a RAPT program for describing assembly by robot. *Proceedings of the 8th International Joint Conference on Artificial Intelligence.* Karlsruhe, Germany.

deKleer, J. (1984). Choices without backtracking. *Proceedings of the Conference of the American Association for Artificial Intelligence.*

Department of Trade and Industry. (1982). *A programme for advanced information technology, the report of the Alvey committee.* HM Stationary Office.

Hall, A.S., Holowenko, M.S., & Laughlin, M.S.M.E. (1980). *Schaum's outline of theory and problems of machine design—SI (metric) edition.* New York: McGraw-Hill.

Hardy, S. (1984). A new software environment for list-processing and logic programming. In O.Shea and M.Eisenstadt (Eds.), *Artificial Intelligence, Tools, Techniques, and Applications.* New York: Harper and Row.

Hearn, A.C. (Ed.). (1984). *REDUCE users manual version 3.2.* Santa Monica: The Rand Corporation.

Jefferies, R.S. (1984). The integrated engineer and his role in flexible automation systems. *International Conference on the Development of Flexible Automation Systems* (Conference Publication No. 237). Institution of Electrical Engineers.

Kapur, D., Mundy, J., Musser, D., & Narendran, P. (1985). Reasoning about three-dimensional space. *Proceedings of the IEEE International Conference on Robotics and Automation.*

Mather, D.V. (1984). Computer integrated manufacture. *International Conference on the Development of Flexible Automation Systems* (Conference Publication No. 237). Institution of Electrical Engineers.

Popplestone, R.J. (1985). An integrated design system for engineering. *Proceedings of the Third International Symposium of Robotics Research.*

Popplestone, R.J., Ambler, A.P., & Bellos, I. (1980). An interpreter for a language for describing assemblies, *Artificial Intelligence, 14*(1), 79-107.

Requicha, A.A.G., & Tilove, R.B. (1978). Mathematical foundations of constructive solid geometry: General topology of closed regular sets, (TM27a). University of Rochester, USA, Production Automation Project.

Chapter 21

DESIGNERS, DECISION MAKING, AND DECISION SUPPORT

William B. Rouse

Search Technology, Inc.
Norcross, Georgia

ABSTRACT

Designers are defined as those individuals who make judgments and choices for the purpose of intentionally influencing the function and form of the eventual design product. The types of individual who satisfy this definition are considered and the nature of their design decisions are discussed. It is concluded that information utilization is the primary behavior that is common to all of the types of decision maker involved in the design process. Approaches to supporting information utilization are considered in terms of assisting designers to overcome the limitations that potentially affect their use of information.

I. INTRODUCTION

Few people would argue with the assertion that in order to develop appropriate decision support systems, one should understand the nature of the decision making that is to be supported. This assertion has certainly been upheld by the R&D community's experiences in developing support systems for aircraft pilots, electronic troubleshooters, and nuclear power plant operators. In fact, the lack of success of many of these support systems often can be traced to an inadequate understanding of end users' needs.

Thus, most practitioners would agree that development of a support system for designers should be based on an understanding of the nature of design. An important concern, however, is whether or not these practitioners have the requisite understanding of design. Two particular types of misunderstanding are ubiquitous.

Some developers of support systems tend to view designers as another type of operator or maintainer, and assume that the factors of importance are similar across these domains. However, design is primarily a cognitive rather than psychomotor activity (Rouse, 1986a). While ambient lighting, seat geometry, etc. should not be ignored, these types of issue are certainly not central to supporting designers.

The second type of misunderstanding is common among engineers (and nonengineers) who think they know all about design. Based on an undergraduate engineering background and perspective, as well as little, if any, actual design experience, these individuals perceive design to be an orderly, top-down process that systematically moves from objectives, functions, etc. to a final detailed design. While these individuals recognize that the process may stray from this normative path during the design of any particular system, these variations are viewed as aberrations that eventually will be eradicated with appropriate methodologies and technologies. In fact, this view is very naive—one purpose of this paper is to provide a more realistic view of design.

In order to provide this perspective, the discussion first focuses on identifying the "designers" in the process of synthesizing a system and, in turn, describing the roles of these individuals in this process. The discussion then shifts to the nature of decision making throughout the design process. Finally, approaches to supporting this decision making are considered.

II. THE ELUSIVE DESIGNER

It seems reasonable to define designers as those individuals who make judgments and choices for the purpose of intentionally influencing the function and form of the eventual design product. The specification of intentional influence is important in that it excludes, for example, the possibility of consumers being defined as designers of products via their purchasing behaviors relative to these products. This definition does, however, include many types of individual beyond those whose job title is "designer."

The range of individuals who intentionally influence design products depends on the type of product. In some cases (e.g., architecture), the end users and designers may interact directly and regularly—although it is somewhat simplistic, one can view the users and designers in this situation as constituting the design team. In this way, the users are also designers, at least according to the definition proposed above.

For complex systems such as automobiles, aircraft, and power plants, the relationships among users and designers are much more indirect and complicated. Further, when the government is involved in terms of various regulatory activities (e.g., safety and environmental standards, cost accounting procedures, and other procurement regulations), the relationships among users and designers are very complex. In fact, it can reasonably be argued that the number of third party participants far outnumbers the users and designers.

Alternatively, one can view most of these participants as "designers" according to the proposed definition. This perspective is reflected in Figure 1 which depicts the primary relationships among individuals who influence design products that are procured by the government. Clearly, the relationship between government and industry is much more complicated than an "official" view (described below) would lead one to conclude.

The official view is that government users (e.g., pilots and their command structures) provide requirements to government technical personnel who develop a design specification or statement of work. Government procurement personnel then advertise the potential procurement in the *Commerce Business Daily*, soliciting proposals from industry. Contractor procurement personnel "notice" the announcement and then pass it on to contractor technical personnel who prepare a proposal that is submitted by the contractor procurement personnel to the government procurement personnel. Government technical personnel then determine the set of proposals that are technically satisfactory, and government procurement personnel then select the least expensive proposal from this list.

If the process actually worked this way, the design products resulting would, at best, be marginally acceptable and relatively inexpensive. More likely, however, there would be endless iterations of unreasonable requirements, impossible proposals, and wasted investments of government funds. Greater interaction among the participants is needed if these problems are to be avoided. The need for this interaction is clearly an underlying premise in the recent report of the Packard Commission (1986).

In fact, such interaction is inherent. The structure shown in Figure 1 illustrates this interaction. The triangle or "truss" that unifies this structure depicts the interaction among users and technical personnel, which is central to synthesizing reasonable requirements—an appropriate blend of what is needed and what can be done. The "windows" of this structure represent the processes whereby marketing personnel learn of needs as well as introduce ideas and capabilities. Finally, the "foundation" of this structure depicts the business aspects of procuring and developing design products.

All of the types of individual represented in Figure 1 are designers in that they promote particular needs, advocate specific solutions, or intentionally affect the feasibility of alternatives via government or industry practices or procedures. Thus, the development of a complex system involves many "designers" who interact in many complicated ways. From this perspective, it is quite unreasonable to discuss supporting *the* designer as if any particular type of

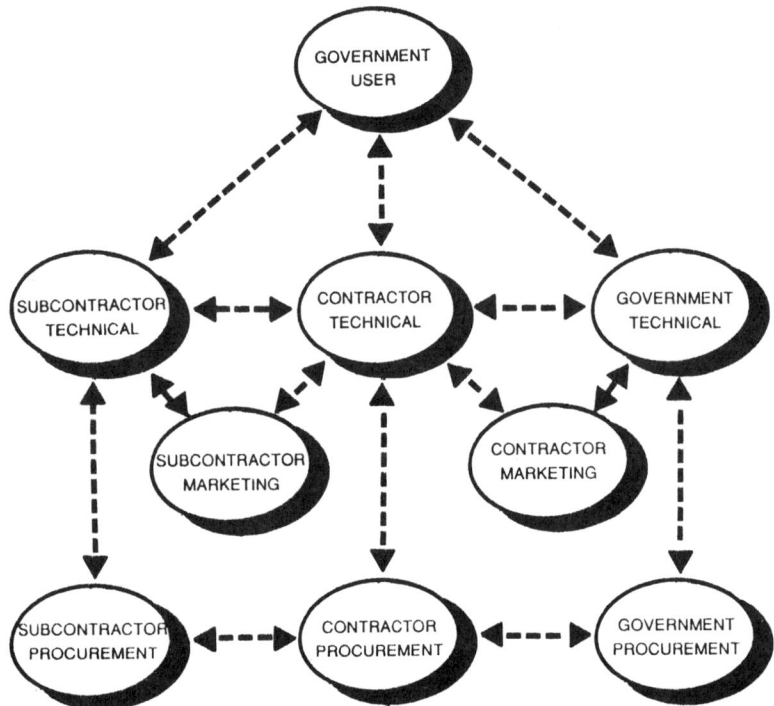

Figure 1. Types of Individual Influencing Design Products Either by Requesting, Suggesting, or Constraining the Form and/or Function of the Product.

individual, seated perhaps at a high-tech engineering workstation, is responsible for design decision making in general.

III. THE PERVASIVENESS OF DECISION MAKING

As argued earlier, the development of a decision support system should be based on an understanding of the nature of the decision making of concern. This understanding should include knowledge of the decision makers and decision situations involved. The earlier discussion related to the types of decision maker; this section begins by considering the decision situations of interest.

Relative to the life cycle of a complex system (illustrated in Figure 2), design decision making does *not* take place solely at one point or period of time. R&D decisions affect what technology is available. Procurement and sales decisions affect how technology is employed. Finally, operational experiences, including training, affect the definition of requirements and identification of opportunities as well as all of the above decisions.

Designers, Decision Making, and Decision Support

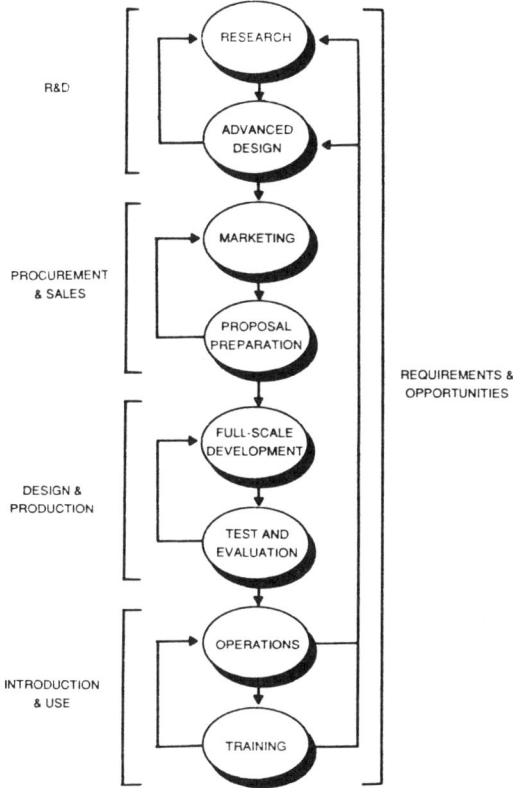

Figure 2. Life Cycle of Complex Systems Illustrating the Many Points in the Process Where Design Decisions Emerge.

As systems evolve, via engineering change requests and orders (ECRs and ECOs) as well as new versions of a particular system concept (e.g., F-15A through F-15E), many iterations through the cycle in Figure 2 may result. Consequently, a very large number of design decisions are made over a long period of time. The many people involved are likely to be widely distributed, both geographically and organizationally. As a result of this temporal, spatial, and organizational dispersion, decision making criteria tend to be quite diverse relative to tradeoffs such as cost versus performance and short-term versus long-term considerations.

Beyond diversity of criteria, the process also is oriented toward multiple goals. In one mode, the process can be objectives-driven in the sense of pursuing remedies of deficiencies or approaches to meeting new requirements. On the other hand, the process can be technology-driven in terms of trying to exploit technology opportunities appropriately. Realistically, of course, the process tends to be in both modes simultaneously.

Thus, many types of people, for many different reasons and with a variety of criteria, make judgments and choices that influence the design of complex systems—design decision making is ubiquitous. As a result of the distributed nature of design decision making, it is very difficult to identify individuals and times associated with particular design decisions. Instead, decisions tend to emerge from ongoing interactions of the variety of individuals who have a stake in the consequences of the decisions (Connolly, 1983; Wise, 1979).

IV. APPROACHES TO DECISION SUPPORT

The nature of decision making as described above presents developers of design support systems with a dilemma. While the success of a decision support system depends on understanding the decision making to be supported, design is such that particular individuals and situations often cannot be associated with the decisions of concern. In other words, if decisions are not always "made" but rather tend to emerge, it is not clear who is to be supported and in what ways.

This state of affairs presents considerable difficulty for the "systems approach" to decision support. In particular, it is virtually impossible to prescribe a normative path through the time-varying "decision space" associated with the individual, organizational, temporal, and spatial dimensions of design decision making. Thus, it is quite unreasonable to argue that sufficient research will eventually lead to a comprehensive methodology around which can be built a monolithic computer-based design support system.

This conclusion does *not*, however, imply eventual failure of computer-aided design tools. Certainly, geometric modeling, finite-element analysis, circuit layout, and drafting have benefited tremendously from the development of computer-based approaches. However, design decision making at the levels represented in Figures 1 and 2 is not amenable to the same type of algorithmic support.

Nevertheless, despite the individual, organizational, temporal, and spatial diversity at these levels, there is at least one very important common thread—information utilization. All types of designer (Figure 1) and stages of design (Figure 2) involve identification, access, manipulation, and management of various types of information as a basis for decision making. Understanding and enhancing this utilization of information appears to be a primary way in which the efficiency and effectiveness of design can be improved.

Two broad classes of information are of particular interest. One class is independent of the system being designed. This class includes technical information in general, government and industry technical standards and guidelines, government procurement requirements and constraints, and company practices and procedures. The second class of information is specific to the particular system being designed and relates to the current state of the design and the "audit trail" of how this state emerged.

Clearly, these two classes of information include a vast amount of time-varying information in numerous formats, and sometimes dubious quality. It appears that a major weakness of many large-scale design efforts is an inability to maintain up-to-date and accurate data bases of this information, both within and across design projects. Computer-based information systems have the potential for substantially improving this situation across all of the individual, organizational, temporal, and spatial dimensions of this problem.

Figure 3 provides a unified view of the types of support that a computer-based information system might provide in order to assist designers. This framework is organized in terms of approaches to help designers overcome the limitations that might affect their information utilization. This scheme of organization serves to emphasize two important distinctions. First, information utilization may be degraded if designers are 1) not aware of information, 2) aware but cannot access the information, and 3) aware and can access the information but judge it not to be worth the effort.

For the first two limitations, awareness and access, it is fairly easy to conceptualize alternative solutions. The technology required and creation of appropriate data bases may be prohibitively expensive relative to anticipated benefits, but the problems are not conceptually overwhelming. However, the third limitation, perceived usefulness, definitely is a very difficult issue.

In a recent comparison of perspectives of those who generate technical information (e.g., researchers) and those who potentially utilize technical information (e.g., practitioners), it was concluded that the use of technical information is often discretionary and, hence, relevance and accessibility are not always sufficient to justify utilization (Rouse, 1985). This is due to the fact that

Figure 3. Framework for Design Support System Illustrating How Designers' Limitations in Utilizing Information Can Be Supported.

TYPE OF SUPPORT	DESIGNERS' LIMITATIONS		
	DIDN'T	COULDN'T	WOULDN'T
INFORMATION RETRIEVAL	DIDN'T KNOW ABOUT INFORMATION	COULDN'T RETRIEVE INFORMATION	WOULDN'T ATTEMPT RETRIEVAL
INFORMATION MANAGEMENT	DIDN'T REALIZE IMPLICATIONS	COULDN'T DETERMINE IMPLICATIONS	WOULDN'T ATTEMPT DETERMINATION
INFORMATION TRANSFORMATION	DIDN'T UNDERSTAND CALCULATION	COULDN'T PERFORM CALCULATION	WOULDN'T ATTEMPT CALCULATION
PURPOSE OF SUPPORT	ENLIGHTENMENT	ENABLEMENT	ENCOURAGEMENT

practitioners seldom have the luxury of optimizing the design of a system independent of the cost of designing it. Further, it is not often clear that considering state-of-the-art research results will make any difference in the "bottom line" of the design and production process (Allen, 1977).

In order to overcome this limitation, research results have to be integrated and practical implications determined. Unfortunately, such integrative efforts do not receive sufficient attention and support (Branscomb, 1983). The DoD-sponsored information analysis centers (IACs) have the potential of helping to solve this problem—however, the task is immense, particularly if the results are to impact conceptual design where research data are less important than the hypotheses that were investigated and the principles that emerged.

The second important distinction depicted in Figure 3 is among types of support. Information retrieval is the most straightforward since research and development in this area have been pursued for many years. Information management has also long been recognized as important and the popularity of various microcomputer-based spreadsheet and data base packages has served to emphasize the general desire for support in terms of information management.

Information transformation is a more subtle type of support. Computer-based approaches to geometric modeling, finite-element analysis, and circuit layout represent successful applications. All of these applications represent manipulations that users know how to do manually, but cannot perform because of the complexity of the problems of concern. In other words, in terms of Figure 3, the computer provides "enablement" and "encouragement."

It is rather difficult to think of successful applications of "enlightenment" where the support involves performing transformations that are not understood by the user. The problem here appears to be user acceptance in that designers are unwilling to rely upon, for example, off-the-shelf mission analysis models or human performance models whose internal workings they do not understand. From this perspective, it would seem that "mega model" approaches to design support, or comprehensive "expert systems" approaches, are unlikely to succeed. Instead, modeling methods that help designers to develop their own models, as well as very focused expert systems, are more likely to be appropriate types of "enlightenment."

The foregoing discussion of Figure 3 has served to illustrate the potential of alternative types of support for the purpose of helping designers to overcome several limitations. Successful examples of some of the alternatives are easy to identify. Other alternatives are more problematic. Fortunately, it is not necessary (and perhaps not even desirable) that any particular design support system encompass all of the alternatives within this framework.

V. CONCLUSION

This paper has emphasized the complexity of design decision making in terms of the individual, organizational, temporal, and spatial dimensions of the process, as well as alternative approaches to design support. Although very little attention was devoted to specific aspects of individual design decision making, these aspects are important and are discussed elsewhere (Rouse, 1986a, 1986b). These considerations were avoided in this paper to allow single-minded pursuit of one theme. It is naive and unreasonable to pursue comprehensive, algorithmic approaches to design support—incremental approaches with particular emphasis on supporting information utilization are much more likely to improve the efficiency and effectiveness of system design.

REFERENCES

Allen, T.J. (1977). *Managing the flow of technology: Technology transfer and the dissemination of technological information within the R&D organization.* Cambridge, MA: MIT Press.

Branscomb, L.M. (1983). Improving R&D productivity: The federal role. *Science, 222,* 133-135.

Connolly, T. (1983). *Scientists, engineers, and organizations.* Monterey, CA: Brooks/Cole.

Packard, D. (Ed.). (1986). *Interim report by the President's blue ribbon commission on defense management.* U.S. Government Printing Office.

Rouse, W.B. (1985). On better mousetraps and basic research: Getting the applied world to the laboratory door. *IEEE Transactions on Systems, Man, and Cybernetics, SMC-15*(1), 2-8.

Rouse, W.B. (1986a). On the value of information in system design: A framework for understanding and aiding designers. *Information Processing and Management, 22*(2), 217-228.

Rouse, W.B. (1986b). A note on the nature of creativity in engineering: Implications for supporting system design. *Information Processing and Management, 22*(4), 279-285.

Wise, J.A. (1979). Cognitive bases for an expanded decision theory. *Proceedings of the 1979 IEEE International Conference on Cybernetics and Society* (pp. 336-339). Denver, CO.

Chapter 22

KNOWLEDGE, SKILLS, AND INFORMATION REQUIREMENTS FOR SYSTEMS DESIGN

Andrew P. Sage

School of Information Technology and Engineering
George Mason University
Fairfax, Virginia

ABSTRACT

This paper discusses issues relevant to the design of systems, in particular computer-aided systems for knowledge-based support of humans in purposeful activities that involve judgment and choice. We begin with a discussion of the design process itself and a methodology for systems design. This leads to a consideration of the nature of design and requirements for successful design of systems. Among these requirements are knowledge representations and information requirements for design. These are discussed in the final sections of the paper.

I. INTRODUCTION

Design is the creative process through which products, processes, or systems presumed to be responsive to client needs and requirements are conceptualized or specified. There are four primary ingredients in this not uncommon definition. The first is that *design results in specifications or architecture* for a product, process, or system. The second is that *design is a creative process*. The third important idea is that *design activity is conceptual in nature*. Finally there is the very important, and sadly often elusive, notion that *a successful design must be responsive to client needs and requirements.* Good design practice requires that the designer be responsive to each of these four ingredients for a quality design effort. The latter necessary ingredient results in the mandate to obtain, from the client for a design effort, a set of needs and requirements for the product, process, or system that is to result from the design effort. This *information* requirement serves as the input to the design process. Design is creative, and it is a process that is conceptual in nature. The result of this creative and conceptual process is *information* concerning the specifications or architecture for the product, process, or service that will ultimately be manufactured, implemented, installed, or brought to fruition in some other way.

Information is the key word in this systems model of the design process; it is an essential feature in the input to the design process, the design process itself, and the output of the design process. Design, then, is an information technology and a fundamental goal of systems engineering. These are the perspectives that we will take in this paper.

We first discuss systems design methodology as a needed element in order for design to be considered as a process. This will lead to a consideration of the nature of design. Associated with this will be purposeful methodological approaches and resulting formalisms for design. These lead to theoretical and conceptual frameworks that are inherent in the nature of design as a process.

It will be clear from our discussion of the nature of design that there are a variety of knowledge frameworks or perspectives necessary for design. We will be concerned with characterization and understanding of the thought process of designers in organizing information about design, including the acquisition, representation, and use of this information. We will pay particular attention to the requirements for successful knowledge-based support to designers, and aids that support design processes. This concern naturally turns to important issues relative to the knowledge base for design, and how this knowledge base might best be employed and exercised in a support system that assists in decision and design processes. To accomplish this, both analytical and perceptual capabilities are required. Also needed is a knowledge base and model base for support that allows a purposeful interplay of these characteristics of successful design.

Clients and designers have needs and requirements that should be satisfied by the results of a successful design process. Information requirements determination is, therefore, a multifaceted need for successful systems design. In a very real sense, this is the most important need for successful design as it is highly likely that functionality of the resulting design will only be noticed by its absence, unless there exists appropriate information concerning that which is to be designed.

II. SYSTEM DESIGN METHODOLOGY

There are many ways in which we can characterize design. We could describe design as an activity involving iterative *hypothesis generation* and *test* of alternatives or concepts. The hypothesis step involves primarily inductive skills, generally based upon experience, that enable generation of design alternatives. These are evaluated through the primarily deductive activity of evaluation or testing. An initial hypothesis is often not acceptable. When it is rejected, generally through evaluation, iteration through the hypothesis generation step enables modification of the identified alternative, and reevaluation such as to, hopefully, lead to a successful design alternative. When a design is evaluated as successful, design activity, at least the particular phase of design being undertaken, ceases. Effort then turns to implementation of the acceptable design or initiation of the next phase in a design effort.

This design methodology is somewhat incomplete in that it does not formally recognize the stimulus that leads to hypothesis generation. There are at least two types of hypotheses generated in the typical design effort. Hypotheses concerning client needs, or concerning enemy intent if a military conflict situation is envisioned, represents the first type. The other type of hypothesis relates to the design alternatives or options that are available for choice. Uncertainties and information imprecision abound relative to each of these hypotheses. The first is information input imprecision. This creates the need for hypothesis generation regarding client needs. The second is consequence of option implementation imprecision. This creates the need for generation of hypotheses relative to design alternatives. In a similar way, we might talk about stimuli that lead to hypotheses about client needs, and to hypotheses about options and their impact. The first stimulus is generally the result of requirements identification or specifications for the design effort. The second is the result of imperfect knowledge of the environment, and the effect of a given design option.

More often than not, the hypothesis generation and option evaluation efforts are first conducted in a preliminary way in order to obtain several concepts that might work. Several potential option alternatives are identified and the resulting options are subjected to at least a preliminary evaluation in order to eliminate clearly unacceptable alternatives. The surviving alternatives are then subjected to more detailed design efforts, and more complete architectures or specifications are obtained. The result of this is a system that can be subjected to detailed design, testing, and at least preliminary operational implementation. Once this has occurred, operational evaluation and test of the implemented system, product, or process can occur. The system design may be modified as a result of this evaluation and this will, hopefully, lead to an ultimately improved system and operational implementation. What is envisioned here is a series of steps of the design process, such as shown in Figure 1, and a series of phases within which the steps of Figure 1 are iterated. The process is highly iterative and interactive.

This leads us to a system design methodology that consists of five phases which are sequenced in an iterative manner (Sage, 1981a):

1. Requirements Specifications Identification

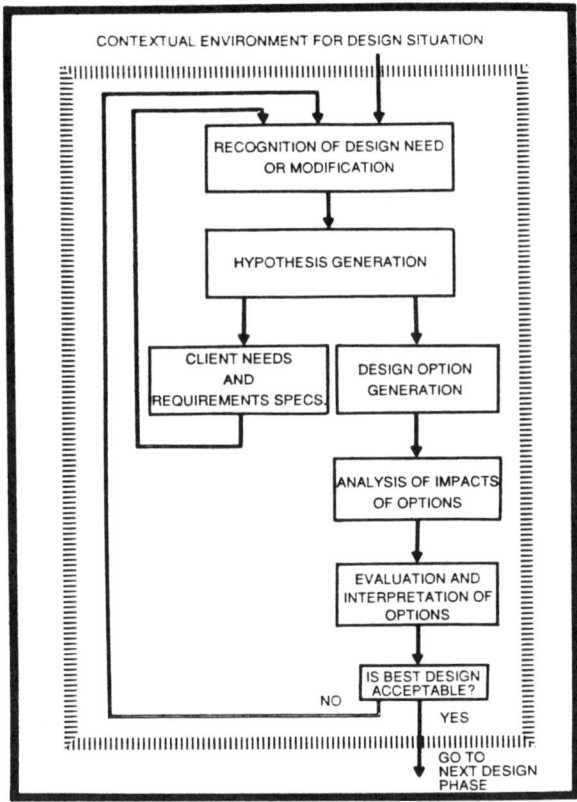

Figure 1. Conceptual Model of Design Steps.

2. Preliminary Conceptual Design
3. Detailed Design, Testing, and Implementation
4. Evaluation
5. Operational Deployment

There are many descriptions of systems design methodology and associated frameworks (Nadler, 1985; Sage, 1977), and we outline only one of them here.

The requirements specification phase of our system design methodology has as its goal the identification of client needs, activities, and objectives to be achieved by implementation of the resulting design as a product, process, or system. The effort in this phase should result in the identification and description of preliminary conceptual design considerations that are appropriate for the next phase. It is important to note that it is necessary to translate operational deployment needs into requirements specifications in order that these needs be addressed by the system design efforts. Thus, we see that information requirements specifications are affected by, and affect, each of the other design phases of the systemic framework for design.

As a result of the requirements specifications phase, there should exist a clear definition of design issues such that it becomes possible to make a decision concerning whether to undertake preliminary conceptual design. If the result of the requirements specifications effort indicates that client needs can be satisfied in a functionally satisfactory manner, then documentation is typically prepared concerning specifications for the preliminary conceptual design phase. Initial specifications for the following three phases of effort are typically also prepared and a concept design team selected to implement the next phase of the design effort.

Preliminary conceptual design typically includes and results in specification of the content and associated architecture and general algorithms for the product, process, or system that should result from this effort. The primary goal of this phase is to develop conceptualization of a prototype that is responsive to the requirements identified in the previous phase. Preliminary concept design according to the requirements specifications should be obtained. Rapid prototyping of the conceptual design is clearly desirable for many applications.

The desired product of this phase of design activity is a set of detailed design and architectural specifications that should result in a useful product, process or system, and prototype design. There should exist a sufficiently high degree of user confidence that a useful product will result from detailed design, or the entire design effort should be redone or possibly abandoned. Another product of this phase is a refined set of specifications for the evaluation and operational deployment phases of the design process.

A product, process, or system is produced in the third phase of design. This is not the final design, but rather the result of implementation of the prototype design that resulted from the conceptual design effort of the last phase. User guides for the product should be produced such that realistic operational test and evaluation can be conducted.

Evaluation of the design and the resulting product, process, or system is achieved in the fourth phase of the design process. Preliminary evaluation criteria are obtained as a part of requirements specifications and modified during the following two phases of the design effort. The evaluation effort must be adapted to other phases of the design effort such that it becomes an integral and functional part of the overall design process. Generally, the critical issues for evaluation are adaptations of the elements present in the requirements specifications phase of the design process. A set of specific evaluation test requirements and tests are evolved from the objectives and needs determined in requirements specifications. These should be such that each objective measure and critical evaluation issue component can be measured from at least one evaluation test instrument.

If it is determined that the product, process, or system cannot meet user needs, the systemic design process reverts iteratively to an earlier phase and effort continues. An important by-product of evaluation is determination of ultimate performance limitations for an operationally realizable system and identification of those protocols and procedures for use of the result of the design effort that enable maximum user satisfaction. Often, operational evaluation is the only realistic way to establish meaningful information concerning functional effectiveness of the result of a design effort. Successful evaluation is dependent upon explicit development of a plan for evaluation developed before initiation of the evaluation effort.

The last phase of a system design effort concerns final acceptance and operational deployment. This description of design methodology contains a strong process flavor. For our purposes, a process is the integration of a methodology with the behavioral concerns of human judgment in a realistic operational environment. Our description of systems design in this section has emphasized the methodological concerns and, perhaps to a lesser extent, operational environment concerns. We now turn our attention to some of the human and behavioral concerns in design.

III. THE NATURE OF DESIGN

Regardless of the way in which the design process is characterized, all characterizations will necessarily involve the following (Sage, 1977, 1981a, 1981b, 1982):

1. *Formulation of the design problem* - in which the needs and objectives of a client group are identified, and potentially acceptable design alternatives, or options, are identified or generated;
2. *Analysis of the alternative designs* - in which the impacts of the identified design options are evaluated; and
3. *Interpretation and selection* - in which the design options are compared by means of an evaluation of the impacts of the design alternatives. The needs and objectives of the client group are used as a basis for evaluation. The most acceptable alternative is selected for implementation or further study in a subsequent phase of design.

Without question, this is a formal rational model of the way in which design is accomplished. Even within this formal framework, there is the need for much iteration from one step back to an earlier step when it is discovered that improvements in the results of an earlier step are needed in order to obtain a quality result at a later step of the design effort. Also, this description does not emphasize the key role of information and information requirements determination, which is concentrated in the formulation step but which exists throughout all steps of the design process (Sage, Galing, & Lagomasino, 1983).

More importantly, this morphological framework, in terms of phases of the design process and steps within these phases, does not emphasize the different types of information and different types of support that are needed within each step at the various phases. During the issue formulation step, the support that is needed tends to be of an affective, perceptive, or Gestalt nature. Intuition-based experiential wisdom will play a most important role in this. During the analysis step, the typical need is for quantitative and algorithmic support, perhaps through the use of the formal models of operations research and management science. In the interpretation step, the needed effort shifts to a blend of the perceptive and the analytical.

Even when these realities are associated with the morphological framework, it still represents an incomplete view of the way in which people do, could, or should accomplish design, planning, or other problem solving activities. The most that can be argued is that this framework is correct in an "as if" manner. There are a variety of ways that might be used to describe how people seek information with which to describe their perceptions of the world around them, and then use this information in an effort to accomplish an identified task. Among the characterizations that have been proposed are the inquiry system models of Churchman (1971); the multiple perspectives models of Linstone (1984); the decision framework and organizational choice models of Diesing (1962), Simon (1979, 1983), Allison (1971), Steinbruner (1974), March and Olsen (1979), and others. Summary discussions of many of these models, which have particular relevance to purposeful design activities, are contained in Sage (1981b) and in several chapters of Nystrom and Starbuck (1981). The central message of essentially all of these studies is that the inquiry system that one adopts and the type of rationality employed is very much a function of the background and experiential familiarity that is brought by the problem solver, or the organization involved in problem solving or design, to the issue under consideration. Included also within the elements affecting design or problem solving behavior are such "stress"-related elements as time and resource availability. The stress-based model of Janis and Mann (1977) is one particularly useful model that explicitly incorporates these considerations into a descriptive model of judgment and choice. This model suggests that judgment "style" will range from unconflicted adherence to a present pattern or change to a new but familiar option, to hypervigilance or panic, to "decidophobia," to the vigilant information processing of the economically rational actor. The dominant variables influencing this choice of style are experiential familiarity with the task, the significance of the decision, and the time allowed to exercise judgment and choice.

The central result of the majority of these studies is that there is no single unique judgment and decision style that is adopted independently of the contingency task structural variables of issue, environment, and problem solver familiarity with task and environment. Rather, the style of decision behavior adopted is a function of these three ingredients. This research appears to have strong descriptive, normative, and prescriptive implications for design and computerized support systems that aid system designers. We will first look at knowledge perspectives for design in terms of potential ways in which we might represent human knowledge in the knowledge base of a support system, and then examine ways in which humans represent knowledge as a function of the contingency task structure.

IV. KNOWLEDGE PERSPECTIVES FOR DESIGN

Critical to success in design, and especially in the design of computerized support systems to aid designers, is an understanding of the way in which humans formulate issues, identify possible alternative courses of action, analyze the impacts of these alternatives, and integrate these impacts in accordance with a value system. It is now generally recognized that experiential familiarity with the perceived task at hand is the dominant influence upon the way in which task requirements are cognized and the way in which problem solving activities proceed. It is convenient to first discuss this in terms of methods and frameworks that have been suggested for machine representation of information and knowledge, and then in terms of ways in which humans represent knowledge.

A. Machine Representation of Knowledge

There are several different knowledge representation frameworks. These include logic, procedural representations, semantic networks, production systems, direct (analogical) representations, semantic primitives, cognitive maps, frames and scripts, and schemata. Each of these may be used to describe the four different types of factual knowledge elements that may be captured in a knowledge base that is used for design purposes: objects, events, performance, and metaknowledge. These representations may be used to describe knowledge retrieval, reasoning, and acquisition of new knowledge, and to relate this new knowledge to already known knowledge. New knowledge can cause any of several changes in a knowledge base. It can cause expansion of the knowledge base to encompass new knowledge. It can cause contraction of the knowledge base because of rejection of old knowledge as currently inappropriate.

Knowledge can be represented from any of several perspectives, as we have indicated. Also, there are several frameworks that can be used to represent knowledge. In the means-ends representation of problem solving (Newell & Simon, 1972), there exists:

1. *problem space*;
2. set of all possible *problem states* within the problem space;
3. one state which is known as the *initial problem state*;
4. one, or possibly more than one, state which is known as the *goal state*;
5. a set of *conditional operators* that will transform one problem state into another problem state, all within the problem space;
6. an *error comparator* which computes the difference between two states, typically between the goal state and the present state;

Knowledge, Skills, and Information Requirements

7. a *controller* which applies a control that is a function of the error difference that is detected; and
8. a set of *path constraints* that must be satisfied in order for a problem solution to be admissible.

Within the problem space, problem solving using means-ends analysis is comprised of four generic activities:

1. comparison of the current state with the goal state;
2. choice of a control that will reduce this difference;
3. application of this control if it is admissible and, if this is not the case, determination of a suitable subproblem and applications of means-ends analysis to the subproblem; and
4. resumption of effort on the original problem or task when the subproblem has been solved.

The purpose of a particular knowledge representation is to enable ultimate human use of the knowledge for purposes of problem solving and design. Typical tasks include retrieving factual information that is judged relevant to the task at hand from the knowledge base, reasoning about these facts in the search for a resolution of the task requirements, and acquiring more knowledge. The appropriateness of a particular representation may be judged, in large part, in accordance with how effectively it enables humans to achieve these three tasks.

There are many mechanisms that can be used to represent the organization of knowledge. The most common, and in many ways the simplest, representation is that in which knowledge is structured as a set of facts, with each fact related to one or more facts in a causal type of inferential relationship. The basic idea behind a production system is that there exists a set of productions, or rules, in the form of various condition-action pairs, generally in the form of "IF THEN" combinations. Initially, these were exclusively explicit rules, although there is much current interest in incorporating fuzziness and imprecision into production rule concepts. Generally, production rules are heuristic in nature, and may be appropriate or inappropriate to the task at hand. The normative goal of an expert system is to use "good" heuristics, of course.

The basic concept of having nodes that represent elements in the universe and links that represent the contextual relations between these elements is a very appealing form of knowledge representation known as a semantic network. A semantic network is intended to represent concepts expressed by natural language words and phrases as interconnected nodes which are connected by a particular set of arcs called semantic relations. A semantic network is a convenient computational representation, readily implementable on computers, in which to represent knowledge which is expressed in natural language. The extent to which this type of knowledge representation can be effectively utilized for inference processes depends on the existence, or lack thereof, of a well structured and sophisticated set of rules. The antecedent-consequent type of rule is commonly used. Structurally, the left-hand side of the rule, or antecedent, represents the set of assumptions or conditions appropriate for use of the rule. The right-hand or consequent side of the rule represents the set of end results. Conjunctions, disjunctions, and negations can occur on either side of the typical relational rule. The contemporary frame and script concept, in which the structure and framework within which new information is interpreted in terms of the concepts that have been acquired through previous experience, has evolved from semantic network concepts. Again we see a strong contextual dependency, or expectation-driven processing, of scripts and frames such that one looks for things based on the context that one believes exists.

A well-designed semantic network is able to represent physical characteristics of actions and the purposeful aspects of these actions. A semantic network representational scheme must, as a consequence of this requirement, include the goals or objectives that are obtained as a result of actions, those scripts that describe scenarios in simple stereotyped situations, the plans that allow for flexible description of potentially adaptable action-impact pairs, and the themes which allow for such environmental descriptions as the occupations and aspirations of actors involved in the issue under consideration. In this way, a semantic network may be a descriptive as well as a prescriptive mechanism for aiding the designer through knowledge support.

There are bound to be difficulties when the size of the semantic network, in terms of the associated data base, becomes sufficiently large such that it can represent a nontrivial amount of knowledge. The computational difficulties in processing the network and the cognitive difficulties in coping with the associated complexity may be overwhelming. This leads to the need for network aggregation in order to obtain "summary" representations that are efficient and effective. These aggregate networks are called frames. Thus, the concept of a frame potentially eliminates many of the defects of semantic networks.

There are also other difficulties associated with the semantic network concept. There are, for example, difficulties associated with representing time because structural models are basically static devices. There are difficulties associated with maintaining a distinction between "facts" and "values," and in incorporating concepts of uncertainty, fuzziness, and imprecision. Situations often arise, for example, in which reasoning must be accomplished using incomplete, inconsistent, and perhaps even contradictory data. These may arise when collecting data from information summaries or from imperfect and/or distributed sources. Much research is being done in semantic networks to improve their usefulness as representational schemes for general knowledge (Findler, 1979).

A much simpler network-based representation is one in which knowledge is structured as a set of concepts, with each concept related to one or more facts by a single causal type of relationship. Needed to accomplish a structural model of this sort is a theory of psycho-logic (Axelrod, 1972, 1976); pulsed digraphs

(Roberts, 1976); or structural models in which relations may have enhancing (+1), inhibiting (-1), or neutral (0) causal influences on other relations. A number of applications of the resulting cognitive maps, as the resulting structural models are often called, have been reported. In its more general form, a cognitive map is the result of an attempt to capture an individual's view of the world with respect to a particular issue. An alternate approach is that of interpretive structural modeling (Sage, 1977; Warfield, 1976), which is a two-level logic representation that may be used to obtain cognitive map-like representations.

Unlike semantic networks, which are typically multirelational structures and as such require a sophisticated set of production rules or grammars for representation and interpretation, a cognitive map is generally based on a single specific contextual relation, such that any given element will have enhancing, inhibiting, or neutral causal impacts on each element in the cognitive map. Thus, the representation and analysis of a cognitive map usually will be simpler than is the case for semantic networks. The simple contextuality of the cognitive map may make it difficult for such a map to replicate a complex belief structure. However, the elements which represent concepts in a cognitive map are variables that can take on different values. The linkage, or contextual relation, among concepts may represent causal assertions and perceptions concerning how one concept variable affects another concept variable. Since the dynamics of the reasoning mechanism in a cognitive map are imbedded in structural considerations, it is a simple matter to simulate the reasoning process of a person if we can assume that the cognitive map has been faithfully constructed.

Minsky (1975) presents a rather different approach to semantic networks that eliminates some of their defects. He advocates the use of local procedures within a "frame" in order to represent structured knowledge. A frame is, as we have noted, a chunk of knowledge for representing a stereotyped situation. Attached to each frame are several kinds of information. Some of this information concerns how to use the frame and some of the information may concern what can happen next. A frame can then be represented as a hierarchical network of nodes and relations. The top-level element in the frame will represent facts that are always assumed to be true about the generic situation at hand. The lower levels of the frame will have many empty terminal slots that must be filled by the specific context-dependent information about the frame and the situation at hand.

The concept of scripts and frames is related to that of schema representations of knowledge. Fundamentally, schemata are conceptual structures that explain the relationship between information acquired, task variables, and decision making performance. Their use is suggested whenever decisions are made based on what has worked before in previously encountered situations that are believed to be similar. There are three layers to a schema: a feature slot layer, a constraint layer, and an inference and action layer. Slots contain features of situations that are believed relevant to a given issue. Features may be spatial relationships or other characteristics of a situation. When enough slots are filled, a schema becomes activated and may be used to understand a decision situation. The constraint layer defines the degree to which features with different characteristics fill a slot. The inference and action layer contains inferences about decision situations that are associated with a schema, and alternative courses of action or design options that are believed appropriate to the particular decision situation that is associated with a schema.

Schemata have two essential functions. They enable situations, objects, and concepts to be recognized and classified. They specify actions and inferences that are believed appropriate to entities that are associated with a schema. Schema theory argues the prevalence of decisions based on recognizing that a situation corresponds to a particular schema. It does not assert, however, that all decisions are of this type. When situations are ambiguous and correspond to several schemata, or when the lack of experiential familiarity with a situation does not suggest any schemata, then alternative methods are used to make a decision.

Analogies and analogous inference play a very important role in human judgment. Philosophers of science often claim that reasoning by analogy is the basis for hypothesis formation and identification in science. Often analogous reasoning is used when there exists uncertainty and imprecision associated with the judgment task at hand. Learning is a process whereby we are able to do things more efficiently or effectively, or more efficiently and effectively the next time that we do them. Thus, learning can involve rote memory and direct implementation of instruction. It can involve reasoning by analogy. Learning also occurs through discovery and observation of a concrete or formal nature.

When there is a lack of explicit, certain, and concrete knowledge about a specific issue, reasoning by analogy will often be used. In such cases, one searches for similarities between the extant task situation and a previously experienced and familiar situation. When the situations are sufficiently analogous so that one can see "parallels" between elements in one situation and elements in the other, reasoning by analogy becomes possible.

Silverman (1983, 1985), in an extensive review of research concerning analogous inference in systems management, identifies a taxonomy that facilitates the development of procedures and protocols for the identification and correction of pitfalls in this form of reasoning. Sternberg (1977a, 1977b, 1981) and Gick and Holyoak (1980) also present descriptive theories of analogous reasoning and experimentally identify sources of errors that may occur during each of the operations constituting these paradigms for analogous reasoning. The basic operations in Sternberg's model are encoding, inference, mapping, application, and response. This research provides a great deal of descriptive information that has motivated the development of normative theories of concept formation and issue formulation.

Carbonell (1983, 1978) has modeled analogous reasoning using means-ends analysis. He uses this model to integrate skill refinement with knowledge acquisition as an effective procedure for reliable and effective problem solution. His conceptualization of analogous reasoning as a blend of holistic and "wholistic" thought allows representation, of at least the holistic portions, of analogous reasoning through means-ends analysis.

Nakamura and Iwai (1982), and Nakamura, Sage, and Iwai (1983) have developed a questioning-answering system for information retrieval in which the system relates its own associative knowledge with the user's knowledge in an analogous fashion. The system's associative knowledge is represented as a semantic network and an information metric is introduced to measure topological distances that represent analogous reasoning similarities. The question-answering dialogue is such as to encourage user responses which enable identification of analogous situations through direction of the questioning for maximum similarity determination. This representation has direct implications for potential effective and efficient use of knowledge in design.

B. Human Representation of Knowledge

As we have noted, there are a variety of knowledge representation techniques including rules, formal logic, semantic networks, cognitive maps, schemata, frames, and scripts. There also exists a number of taxonomies of the ways in which humans, in an unaided fashion, approach problem solving tasks including design. One particularly useful taxonomy (Rasmussen 1983, 1985, 1986) conceptualizes three distinct types of problem solving behavior or reasoning: 1) Formal Knowledge-based Behavior, 2) Rule-based Behavior, and 3) Skill-based Behavior.

The choice of which type of reasoning to employ is made by the problem solver on the basis of experiential familiarity with the task at hand and the environment into which this task is imbedded.

Often, there is a mismatch between the problem solving behavior that a particular human will use in a given situation, and the behavior that a machine might be programmed to emulate. Expert systems often attempt, for example, to use rule-based representations of skill-based knowledge, or alternately to use rules when an experienced expert would use a more skill-based representation of knowledge. This knowledge may initially be expressed in any of several affective or intuitive forms that have been learned by the expert on the basis of much relevant experience. Although experts may well have learned how to be an expert through formal means that require an absorbed and alert monitoring of the task components, they may no longer be formally aware of all of their monitoring actions and may, therefore, give incomplete information concerning their activities. This simply reflects the view that expertise develops through perceptual learning *and* through the acquisition of a larger number of rules. Experts are not just faster and more accurate with their use of a larger number of rules; they can perceive situations, similarities, and task objectives with considerable clarity. Dreyfus (1982, 1984) and Klein (1980) have stated very useful models of human problem solving activity based on these observations.

Expert systems may be used in environments in which there exists a requirement for advice before the expert system is able to conduct an exhaustive "search" of its knowledge base. In this case, it becomes especially important to be able to generate a reasonably good solution based on a limited search and then proceed in the time remaining to improve, to the extent possible, upon initial recommendations. This casts another requirement for the efficient and effective structuring and use of metaknowledge.

There are various kinds of knowledge that comprise expertise. Among these are facts about the specific domain of the decision maker or problem solver and others to be aided by the expert system; fixed rules and procedures that will always be valid; problem situations in which there is imprecision and uncertainty and in which heuristic rules need to be applied with caution; skill-based perceptions based on experiential familiarity with the task at hand; and formal theories of the domain that need be used when experiential familiarity with the task at hand and the environment into which it is imbedded is insufficient to enable wise use of skill- and rule-based knowledge. In accordance with these thoughts, it appears reasonable to define *learning* as the way in which task performance is enhanced through an improvement in skills; an increase in the ability to acquire, organize, and extend rules of action; or an increase in formal reasoning ability. This complements our earlier discussion of learning which, in general, leads to intelligent behavior. *Intelligence* is the ability to adjust to an environment and adapt to a changing environment to achieve objectives.

It is especially necessary that a knowledge representation scheme be capable of coping with the types of expertise and the reasoning perspectives that can be expected to exist among the users of the knowledge-based system. In this way, we will be better able to accomplish needed activities that involve learning and discovery such as the diagnosis of faulty theories, the proper assimilation of new knowledge, and useful frameworks in which to pose questions, so that they are understandable and interpretable in the way in which the questioner (should have) intended. Clearly, all of this has major implications for subjects such as decision making in general and in important subareas such as design. A particularly important role for a knowledge support system is to assist the user in minimizing errors between perceived knowledge level relative to a particular task and the user's actual knowledge level. When both perceived and actual knowledge level are at the level of "master," for example, skill-based knowledge is generally appropriate for judgment and choice tasks. When the perceived and actual knowledge level is that of "novice," the knowledge support system user is generally aware of the need for support. When the perceived knowledge level is that of "master" and the actual knowledge is that of "novice," perhaps due to an unrecognized change in environment, it is very likely that acts of judgment and choice will be associated with self-deception. The task for a knowledge support system, in this regard, would be to alert the decision maker to the potential difficulties of skill-based behavior in an appropriate manner. A very important role for a knowledge support system, therefore, is that of alerting the decision maker or designer to the information requirements appropriate for the decision maker, the task at hand, and the environment surrounding the task and task requirements.

V. INFORMATION REQUIREMENTS FOR DESIGN

Judgments concerning design, at least prudent judgments, are seldom made without information. Information is often defined as data of value for decision making. Activities associated with acquisition, representation, storage, transmission, and use of pertinent data are generally associated with information processing. The task of information requirements determination is associated with each of these.

There are many ways in which we can characterize information. Among the attributes we might use are accuracy, precision, completeness, sufficiency, understandability, relevancy, reliability, redundancy, verifiability, consistency, freedom from bias, frequency of use, age, timeliness, and uncertainty. It is also possible to define information at several levels. At the *technical level*, information and associated measures are concerned with transmission quality over a channel. At the *semantic level*, concern is with the meaning and efficiency of messages. At the *pragmatic level*, information is valued in terms of effectiveness in accomplishing an intended purpose. From the viewpoint of design, we are clearly concerned more with pragmatic and semantic issues than we are with technical level issues. At the pragmatic and semantic levels, our concerns with information for design purposes are five-fold:

1. Information should be presented in very clear and very familiar ways to enable rapid comprehension;
2. Information should improve the precision of understanding of the task situation;
3. Information that contains an advice or decision recommendation component should contain an explication facility that enables the user to determine how and why results and advice are obtained;
4. Information needs should be based upon identification of the information requirements for the particular situation; and
5. Information presentation and all other associated management control aspects of the design process should be such that the designer or decision maker, rather than a computerized support system, guides the process of judgment and choice.

Clearly, this relates directly to the concept of *value of information* and indicates that this concept is very dependent upon the contingency task structure. The mix of task at hand, the environment into which the task is imbedded, and the problem solver's familiarity with these interact to determine both the perceptions and the intentions of the problem solver.

The concept of economic efficiency value of information (Sage 1977) is important and of value for design purposes. Effectiveness measures are of even more importance, however. Yovits, Foulk, and Rose (1981) have expanded upon the classical value of the information concept so that it includes effectiveness and efficiency measures. The value of a new item of information is dependent upon the difference in decision maker effectiveness before and after the information is received. There have been a number of other approaches to the value of information concepts and to information requirements analysis; many of these are described by Sage, Galing, and Lagomasino (1983).

There are at least three human limitations associated with information processing that may significantly affect the requirements for information that are likely to be identified: limited information processing capability, potential bias effects in the selection and use of information, and limited knowledge of appropriate problem solving behavior that results in incorrect assessment of the contingency task structure associated with a given issue.

The methodology of Davis (1982) is especially appropriate in that it is sufficiently comprehensive to address each of these concerns. Davis suggests four strategies for determination of information requirements. The first of these is simply to ask people for their requirements. The appropriateness and completeness of the information determined by this approach will depend upon the extent to which the people questioned can define and structure their problem space and can compensate for their biases. This approach can be further subdivided into the use of interacting and nominal group approaches to inquiry. The second strategy is to elicit information requirements from existing systems that are similar in nature and purpose to the issue or design task under consideration. Properly executed anchoring and adjustment strategies, or analogous reasoning strategies, are potentially useful here since a starting point can be determined from the existing system and extrapolation of this can be made. Examining existing plans or reports represents one approach of identifying information requirements from an existing or conceptualized system or previous design. The third approach consists of synthesizing information requirements from characteristics of the utilizing system or individuals. This permits definition of an analytical structure for the problem space and will often allow use of formal analytical procedures, such as input-output analysis or decision analysis. The fourth strategy consists of determining information requirements through experimentation on an actual system, perhaps one specifically constructed for this purpose. Generally, the first approaches are more simple and economic than the latter approaches. The approaches can certainly be combined, and it would be desirable to be able to determine which of the approaches is best in a given situation.

Information obtained from any, or a combination, of these approaches must be capable of representation in a support system to aid the designer. A purpose of a design support system should be to determine possibilities of insufficient and/or inappropriate information; that is, information that is sufficiently imperfect so that the likelihood of an acceptable design is low. The support system should then be able to determine the nature of the missing or otherwise imperfect information and suggest steps to the user to remedy this deficiency. This suggests that a support system should be capable of detection, diagnosis, and correction of faults in a set of information obtained for an issue. The type of information that the user of a particular support system will wish to, or should, use is very much dependent upon the contingency task structure (Sage & Lagomasino, 1984). Another information-associated requirement for a support system is that of identifying the information, judgment, and choice perspectives that a designer will and should wish to use, and to be capable of coping with requirements for design support from these multiple perspectives. Satisfactorily coping with these needs should result in truly innovative and used support systems for design processes.

ACKNOWLEDGMENT

This research was sponsored by the Army Research Institute for the Behavioral Sciences under Contract MDA-903-82-C-0124.

REFERENCES

Allison, G. (1971). *Essence of decision*. Little Brown & Co.
Axelrod, R.M. (1972). *Framework for a general theory of cognition and choice*. University of California at Berkeley: Institute of International Studies.
Axelrod, R.M. (Ed.). (1976). *Structure of decision: The cognitive maps of political elites*. Princeton University Press.
Carbonell, J.G., Jr. (1978). Politics: Automated ideological reasoning. *Cognitive Science, 2*, 27-51.
Carbonell, J.G., Jr. (1983). Learning by analogy: Formulating and generalizing plans from past experience. In R.S. Michalski, et al. *Machine learning: An artificial intelligence approach* (pp. 137-162). Tioga.
Churchman, C.W. (1971). *Design of inquiring systems*. New York: Basic Books.
Davis, G.B. (1982). Strategies for information requirements determination. *IBM Systems Journal, 21*(1), 4-30.
Diesing, P. (1962). *Reason in society*. Urbana, IL: University of Illinois.
Dreyfus, S.E. (1982). Formal models versus human situational understanding: Inherent limitations in the modeling of business expertise. *Office, Technology, and People, 1*, 133-165.

Dreyfus, S.E. (1984). The risks and benefits of risk-benefit analysis. *OMEGA, 12*(4), 335-340.
Findler, N.V. (Ed.). (1979). *Associative networks: The representation and use of knowledge by computers.* Academic Press.
Gick, M.L., & Holyoak, K.J. (1980). Analogical problem solving. *Cognitive Psychology, 12*, 306-355.
Janis, I.L., & Mann, L. (1977). *Decision making: A psychological analysis of conflict, choice, and commitment.* New York: Free Press.
Klein, G.A. (1980). Automated aids for the proficient decision maker. *Proceedings of the 1980 Conference on Systems, Man, and Cybernetics,* (pp. 301-304). Boston, MA.
Linstone, H.A. (1984). *Multiple perspectives for decision making: Bridging the gap between analysis and action.* New York: Elsevier North-Holland.
March, J.G., & Olsen, R. (1979). *Ambiguity and choice in organization.* Universities for Press.
Minsky, M. (1975). A framework for representing knowledge. In P.A. Winston (Ed.), *The psychology of computer vision.* New York: McGraw Hill.
Nadler, G. (1985). Systems methodology and design. *IEEE Transactions on Systems, Man, and Cybernetics, SMC-15*(6), 685-697.
Nakamura, K., & Iwai, S. (1982). Topological fuzzy sets as a quantitative description of analogical inference and its application to questioning-answering systems for information retrieval. *IEEE Transactions on Systems, Man, and Cybernetics, SMC-12*(2), 193-204.
Nakamura, K., Sage, A.P., & Iwai, S. (1983). An intelligent database interface using psychological similarity between data. *IEEE Transactions on Systems, Man, and Cybernetics, SMC-13*(4).
Newell, A., & Simon, H.A. (1972). *Human problem solving.* Prentice-Hall.
Nystrom, P.C., & Starbuck, W.H. (1981). *Handbook of organizational design* (Vols. 1-2). Oxford University Press.
Rasmussen, J. (1983). Skills, rules, knowledge; signals, signs, and symbols; and other distinctions in human performance models. *IEEE Transactions on Systems, Man, and Cybernetics, SMC-13*(3).
Rasmussen, J. (1985). The role of hierarchical knowledge representation in decision making and system management. *IEEE Transactions on Systems, Man, and Cybernetics, SMC-15*(2), 234-43.
Rasmussen, J. (1986). *On information processing and human-machine interaction: An approach to cognitive engineering.* North Holland.
Roberts, F.S. (1976). *Discrete mathematical models: With application to social, biological, and environmental problems.* Englewood Cliffs, NJ: Prentice Hall.
Sage, A.P. (1977). *Methodology for large scale systems.* McGraw Hill Book Company.
Sage, A.P. (1981a). A methodological framework for systemic design and evaluation of computer aids for planning and decision support. *Computers and Electrical Engineering, 8*(2), 87-102.

Sage, A.P. (1981b). Behavioral and organizational considerations in the design of information systems and processes for planning and decision support. *IEEE Transactions on Systems, Man, and Cybernetics, 11*(9) 640-678.

Sage, A.P. (1982). Methodological considerations in the design of large scale systems engineering processes. In Y.Y. Haimes (Ed.), *Large scale systems* (pp. 99-141). North Holland.

Sage, A.P., Galing, B., & Lagomasino, A. (1983). Methodologies for the determination of information requirements for decision support systems. *Large Scale Systems, 5*(2), 131-167.

Sage, A.P., & Lagomasino, A. (1984). Knowledge representation and man machine dialog. In W.B. Rouse (Ed.), *Advances in Man-Machine Systems Research*, (pp. 223-260), JAI Press.

Silverman, B.G. (1983). Analogy in systems management: A theoretical inquiry. *IEEE Transactions on Systems, Man, and Cybernetics, SMC-13*(6), 1049-1075.

Silverman, B.G. (1985). The use of analogs in the innovation process: A software engineering protocol. *IEEE Transactions on Systems, Man, and Cybernetics, 15*(1), 30-45.

Simon, H.A. (1979). Information processing models of cognition. *Annual Review of Psychology, 30*, 363-96.

Simon, H.A. (1983). Search and reasoning in problem solving. *Artificial Intelligence, 21*, 7-29.

Steinbruner, J.D. (1974). *The cybernetic theory of decision.* Princeton University Press.

Sternberg, R.J. (1977a). Component processes in analogical reasoning. *Psychological Review, 84*(4), 353-378.

Sternberg, R.J. (1977b). *Intelligence, information processing, and analogical reasoning: The componential analysis of human abilities.* Erlbaum.

Sternberg, R.J., & Turner, M.E. (1981). Components of syllogistic reasoning. *Acta Psycologica, 47*, 245-265.

Warfield, J.N. (1976). *Societal systems: Planning, policy, and complexity.* John Wiley & Sons.

Yovits, M.C., Foulk, C.R., & Rose, L.L. (1981). Information flow and analysis: Theory, simulation, and experiments. I, II, and III. *Journal of the American Society for Information Science, 32*, 187-202, 203-210, and 243-248.

Chapter 23

INTUITION BY DESIGN

J. MacGregor Smith

Department of Industrial Engineering
& Operations Research
University of Massachusetts
Amherst, Massachusetts

> I shall decompose the conjecture into lemmas which are not conjectures any longer but intuitions, that is, nondubious apprehensions of a pure and attentive mind which are born in the sole light of reason.
>
> Descartes, *Rules for the Direction of the Mind* (1628)

> Whatever the process and the means may be by which knowledge reaches its objects, there is one that reaches them directly, and forms the ultimate material of all through, viz. intuition (Anschauung).
>
> Kant, *Critique of Pure Reason* (1781)

ABSTRACT

Intuition is viewed as a fundamental part of the design process. The origins and development within western culture are briefly reviewed and its relationship to the design process is demonstrated. Central to this discussion is the question how should intuition be integrated into the design of a system for designing? In response to this question, a design methodology based on a dialectical process involving *generational* and *judgmental* intuitions is proposed. Finally, the origins of this dialectical design process and its natural evolution into a *conversational optimization* process are presented together with a small example illustrating the proposed design process.

I. INTRODUCTION

The motivation for this paper arises from a curious incident that occurred while I was studying for my doctorate in Industrial Engineering and Operations

Research at the University of Illinois at Urbana-Champaign. I was having coffee with a fellow student in the student union and a mathematician friend who happened to be a geometer (actually a topologist) asked me about my background. I told him that I had studied architecture and had actually taught it for three and one-half years before I decided to enter the field of operations research, and that I thought operations research embodied many of the tools, techniques, and methods by which design and planning decisions made in architecture could be augmented. He smirked and said, "Architects primarily make decisions based on *intuition*, don't they?" At the time, I was somewhat stunned by his remark and subsequently maneuvered the conversation onto another topic.

When this opportunity arose to write about the design process and its information needs for this volume, I remembered that conversation. I thought I would explore the notion of intuition in design and examine the implications of the mathematician's remark to see if he was right, uninformed, or just kidding. As we shall see, his remark has led me to some very interesting observations about the design process.

II. OBJECTIVES

To be a bit more specific, let me raise the following issues which will be addressed throughout the paper:

1. What do we mean by intuition?
2. What do we mean by design?
3. What is the relationship between intuition and the design process?
4. How should a system for design be designed?

III. INTUITION AND DESIGN

As the quotes from Descartes and Kant at the beginning of this paper imply, the notion of intuition has been examined by some of the greatest philosophers of western culture. Foremost among them are Descartes, Spinoza, Pascal, Kant, and Brentano (among others); many of the Gestalt field psychologists and behaviorists (Bastick, 1982)[1]; and to my surprise some of the greatest mathematicians of our age including Brouwer, Poincaire, Polya, and Weyl (unfortunately, however, my mathematician friend was not one of those whose name I encountered).

As a definition of intuition, the *Oxford English Dictionary* yields the following which I will paraphrase for the purpose of brevity:

---

[1] Bastick's work is the most recent and comprehensive survey of intuition currently available.

> *intuition* - the immediate apprehension of an object by the mind without the intervention of any reasoning process...direct or immediate insight.

From the above definition, let us identify some of its key characteristics. First of all, intuition implies immediate apprehension of knowledge. An object or phenomenon is perceived through its essence alone. This characteristic is fundamental to Spinoza's categorization of how we think and act (Spinoza, 1666/1955; Churchman, 1971).

Secondly, it is knowledge which is "holistic," i.e., since it is more than the sum of its parts and represents the emergent properties of the system of the object under study. A bicycle is a bicycle because it looks like a bicycle.

Thirdly, through intuition, a person knows that he or she knows the intuition to be true. It represents knowledge with certainty.

Finally, one of the characteristics that intuition connotes is irrationality (Maslow, 1957). This aspect of intuition is perhaps a sociocultural characteristic, principally the one to which I probably reacted when conversing with the mathematician. As we shall see, I do not believe that this connotation is valid or deserving of intuition. Rather, intuition is a fundamental part of the design process and should be made a visible aspect, not something one shuns.

While the above represent some of its characteristics, there exist certain types of intuition important to our discussion which naturally emerge within the context of this paper on the design process. There appear to be two general types of intuition: one which is generational and the other which is judgmental.

By generational, we refer to intuitive ideas which give birth to new design concepts; new theorems to be proved; and new inspirations for novels, algorithms, artwork, and so on. Most of the intuitionists of mathematics feel that once they know the theorems (i.e., intuitions), they are then carefully sifted, refined, and organized by the logic of the mathematician working with them (Kline, 1980). Part of the revolution and "loss of certainty" in mathematics can be attributed to the intuitionists' school of thought. Perhaps my mathematician friend was recoiling from an encounter with one of his intuitionist colleagues when he made his earlier remark to me.

The other type of intuition is judgmental in nature or concerned with making off-hand judgments about an object or system; e.g., "That so-and-so is more than the sum of his parts, he is an *idiot*" (Checkland, 1984); or this design really does not look like a "bicycle" (Rittel, 1972a, 1972b). These intuitive judgments are critical starting and ending points of the logical analysis of alternatives.

Both types of intuition which I have chosen to discuss exist within the design process; however, before we progress much further, let me digress with a definition of design so we can unite the two central concepts of the paper.

I will paraphrase the concept of design formulated by Rittel (1972a) which has always struck me as a quite general and useful definition of design:

design - an activity aimed at the production of a plan which if executed yields certain anticipated consequences and no undesirable, unanticipated consequences.

Thus, from the definition, design is an activity not an object. While the outcome of the design activity is a plan, it is the process which is fundamental and of crucial interest to this paper.

Second of all, design is not a linear process, it has many interrelated steps, has a great deal of feedback, and is extremely complex to untangle. The essential activities of design involve idea generation and evaluation—what Rouse (1986) calls *facts, fantasies, and feedback*.

Thirdly, design is a purposeful (teleological) endeavor which due to its inherent complexity, often referred to as its wickedness (Rittel & Webber, 1973; Smith, 1981), must involve a wide range of people with various backgrounds and world views so that the proper formulation of the problem, let alone the "correct" solution, may occur. In this regard, design is essentially information deficient. The most effective design system revolves around the design of the information system used for design (Churchman, 1968a, 1968b, 1971; Rittel, 1972a, 1972b; Rouse, 1986). Thus, the concern with information systems for design is a proper focus for this volume.

Finally, design is something everyone can do, it is not the province of any single discipline. Of course, some people may claim to be better designers than others, but this claim is not an inherent trait of the design process; it is only the (ex)claim of the designer.

IV. INTUITION BY DESIGN

The central issue of this paper concerns the relationship between intuition and design and what this relationship implies for a system of design. The key characteristics of intuition and design revolve around the information system for design. Intuition is knowledge, albeit a very special type of knowledge, but a type of central importance to the design activity. Intuitions represent holistic knowledge, *knowledge chunks*, which are essential if the designer is to progress through the design process. A parallel may be struck here between the design process and the field of stochastic processes in that the state changes in a design solution (design variables) correspond to the arrival of the intuitions or the renewal points along the time axis.

A renewal (counting) process $X_t, t \geq 0$ is a nonnegative, integer-valued stochastic process which registers the successive occurrences of events during the time interval $[0,t]$, where the time durations between consecutive events are positive, independent, and identically distributed random variables (Taylor & Karlin, 1984). The events for our concern here are the intuitions themselves. In this way, one could experimentally operationalize and quantify the concept of intuition by recording the state changes in a design solution, i.e., the times between arrivals of the generational X_t and judgmental Y_t intuitions. It would be interesting to see if the process were Markovian or could be approximated as such so that the changes in a design process would depend only on the previous solution state.

In fact, the two types of intuition I mentioned are the key links to the design activity in that, during the design process, generational and judgmental intuitions constantly are created, processed, evaluated, and renegotiated.

Well, you might say this is all well and good, but what does it imply for our system of design? This naturally leads to the following question: Can we design a system for designing without intuition?

I think the answer to this question is an emphatic *no!* However, most of the computer-aided design systems used in practice today with which I am familiar, viz. IBM's CADAM, Dassault Systems' CATIS, and Computer Vision's CADDS-4, have very little intuition designed within them and provide very little chance for a designer to enhance his or her intuition. These designer systems assume that the design is "done" and all that is needed is to draw it. Perhaps it is unfair of me to criticize these systems since they are only supposed to be design aids, yet I think it important to realize that a system for designing without intuition is doomed to fail.

Now, skeptics will raise the timely question: How are you to carry out this formidable task you have constructed for yourself? The key to incorporating intuition in a design system is to develop a design methodology—a process wherein intuition, essentially generational and judgmental intuitions, play a crucial role because, as I have argued, they must be a central part of the knowledge base of the design system. In the next section of the paper, I will outline this methodology which is based on a dialectical approach to design where intuition becomes a fundamental and visible part of the decision making process.

V. DIALECTICS OF DESIGN

As we have seen, the knowledge base, the information system, and the design process are all united through intuition because intuition represents "holistic" knowledge. Before I outline this design methodology, it is useful to examine the rise of operations research and systems analysis over the past three decades to see the genesis for this design methodology.

Since its inception in the 1950s, operations research or more generally, systems analysis, maintained the following characteristics or features:

- o Operations research was felt to be a discipline in which a certain set of tools and techniques (e.g., linear programming, queuing theory, etc.) were part of the kitbag of an operations research practitioner and, as such, these tools and techniques were sacrosanct.

o The operations researcher was felt to be an "expert" the "client" would trust to solve his or her problem.

o Since the expert designer knew all the answers, the values and value systems underlying the problem were easily captured by a singular objective function, e.g., profits or cost, which the client would readily produce on demand.

o Implementation of the solutions arrived at by operations researchers were felt to be the problem of the client and not the researcher.

As a diagrammatic way of illustrating the evolution of the design process to its present stage, circles a and b in Figure 1 represent the *first* and *second* generations of our situation with the designer and the client for whom the field exists.

During the first generation, Rittel (1972a, 1972b) argues that there was an *asymmetry* of ignorance for our design problem situation. This is due to the designer knowing more than the client about how to solve the client's problem.

As it turned out, this paradigm was strongly held until around the 1970s when it became clearer that this *asymmetry* of ignorance was a deficient paradigm of the problem world view, and rather the more appropriate paradigm was that there is a *symmetry* of ignorance as depicted in the right half of circle b in Figure 1. This latter world view is reflective of the failures of designers and the fact that the client knows as much as the designer about his or her problem, but perhaps less so of the technical or nonapparent solutions possible for attacking the problem. The designer is often *necessary* but certainly not *sufficient* for solving the problem.

As a forecast of what is likely to occur in the next generation, circle c in Figure 1 is offered as a scenario.

Figure 1. Three Generations of Systems Analysis.

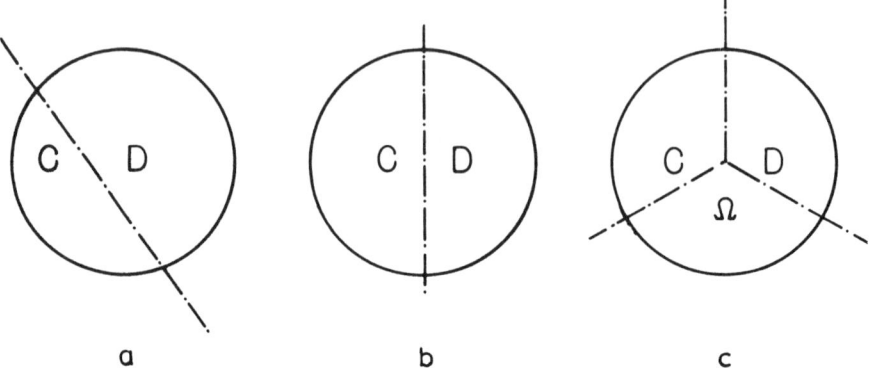

Intuition by Design

The upcoming generation is viewed with another component, that of the computer Ω where the design process integrates the C, Ω, *and* D $(C_\Omega D)$ in a system for design, i.e., a *trinity* of ignorance. That the design information system underlying the system for design should be "conversational" is a natural outgrowth of the previous discussion on the evolution of operations research and our discussion on the design process.

How is one to realize this conversational design process with the knowledge that intuition is crucial to the process?

As one approaches a design problem π, two intuitions should be developed in parallel, an intuition and one *counter* to the initial intuition, so that the antithetical intuition represents the polar opposite viewpoint to the original one. This dialectical paradigm as applied to design has been championed by a number of authors and has become one of the key concepts in recent operations research literature (Churchman, 1971, 1979; Checkland, 1984). For example, the new linear programming algorithm by Karmarkar is antithetical to the simplex method of George Dantzig in that the simplex technique is an extreme point method while the newer method is an interior point method.

Within the I_α and the I_ω, there may be a number of specific alternative courses of action characterized by the I_α or the I_ω. Developing the polar design concepts could also be computer generated, (Smith & Pelosi, 1984; Smith & Macleod, 1986), yet they should be clearly as opposite as possible in the spirit of the dialectical process. Figure 2 is an example of a computer generated set of

Figure 2. Computer Generated Solutions.

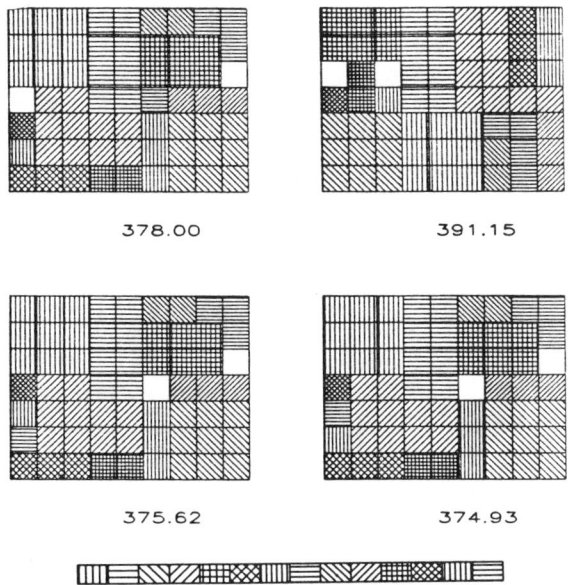

solutions via an automated layout generation program (Smith & Macleod, 1986). The solution in the upper right-hand corner is the optimal solution found with the computer program which maximizes a multi-attribute utility function. The other three solutions represent ones which are feasible but suboptimal. From the total set of solutions, two should be selected to continue the design process.

I_α or the I_ω may represent generic design concepts which embody specific design alternatives.   Through the evaluation process, a specific design alternative should emerge, but initially, no single alternative should dominate. Inspecting Figure 2, one immediately sees that while I_α should be the solution in the upper right-hand corner, the I_ω could be any of the other three and really should be the one most opposite to I_α from the suboptimal set.

Once the intuition and its counterpart are developed, the evaluative process should proceed to analyze each idea and its subset of realizable alternatives. At the initial and terminal stages of this evaluation process are the *judgmental* intuitions, Σ α and Σ ω (i.e., off-hand judgments), which guide the evaluation process.   The detailed, systematic evaluation process should be based on specific criteria unique to the problem and any formal mathematical analysis designed to integrate the partial scores on each of the individual attributes would be utilized as necessary. In this regard, formal decision analytic tools would be appropriate such as Multi-Attribute Utility Theory, Multi-Objective Programming, and related techniques (Keeney and Raiffa, 1976; Hwang and Masud, 1979).

After the detailed evaluation process, the two contrasting intuitions I_α or the I_ω would be either synthesized into I_Σ or else one alternative would dominate and be selected to continue the process. An example of this scenario with this methodology appears in the final section of the paper.

One might argue that to continue this process into the detailed part of the design process may be extremely costly or unnecessary. Certainly this may be the case, yet the more it is done even at the detailed level of design, the firmer the confidence in the solution.

VI. EXAMPLE OF A SCENARIO

The following example is a design problem with which I recently became involved. It concerns the redesign of an entrance to the Marcus Engineering Building of the University of Massachusetts. For many years, the main entrance has been subject to much ridicule. It is unsafe because the exterior stairs are cracked and spalling; the entrance area provides no milling space for those coming to and going from the auditorium; it provides little climate control; and generally maintains a poor entrance image in that there are no displays, exhibits, or adequate room for announcements of events. In general, it is a poor design solution.   The impetus for doing something about the problem arose as a

Intuition by Design

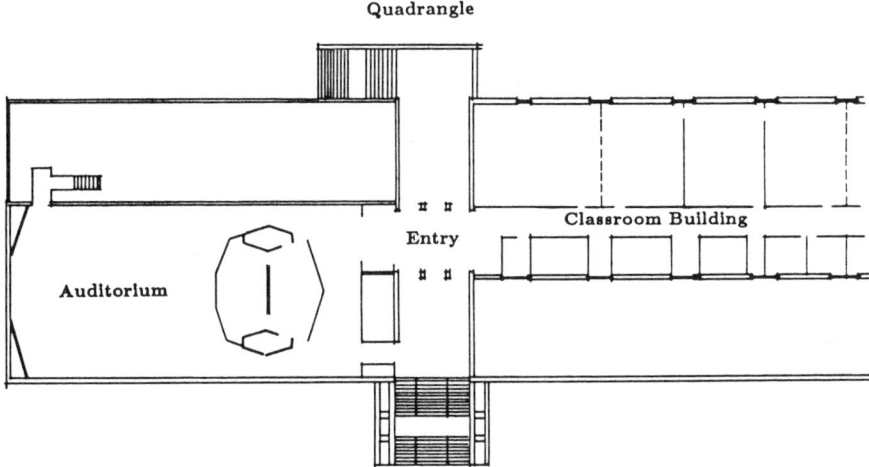

Figure 3. Existing Conditions for Entrance Problem.

memorial to the recent passing of a faculty member. Figure 3 represents a schematic drawing of the existing conditions.

When I began to examine the problem, it appeared to be a good test case for the design methodology. As I began to generate ideas for the entrance, I realized that one class of solutions could be characterized as making the entrance more symbolic, an allegory for the deceased faculty member, which would draw more attention to the entrance area as a unique space. In fact, this generic intuition I_α was a "point of attraction." At this juncture in the methodology, it was apparent that the polar opposite concept I_ω was that the entrance should be a "link"—a passageway interconnecting the auditorium and classroom buildings. These are clearly polar opposites and embody a range of specific design alternatives. Through the methodology, I was forced to develop an alternative concept, I_ω, one which really had not been considered.

When the design process for this problem was only just beginning, the dialectical methodology served to strengthen my intuitions and make certain that no alternative would go unnoticed. As the process unfolded, more specific design alternatives within each of the generic intuitions were being developed which were evaluated according to the attributes of structural stability, constructability, cost, image, safety, security, maintainability, flexibility, and accessibility. Finally, what resulted could be considered a "point-link" synthesis I_Σ of the two generic concepts, and this was entirely due to the dialectic (see Figure 4).

The one remaining issue which I have not directly addressed but has always lurked through the arguments in the paper is: Can we design a system for design with intuition?

Churchman (1971, 1979) has something to say about this, and it interestingly

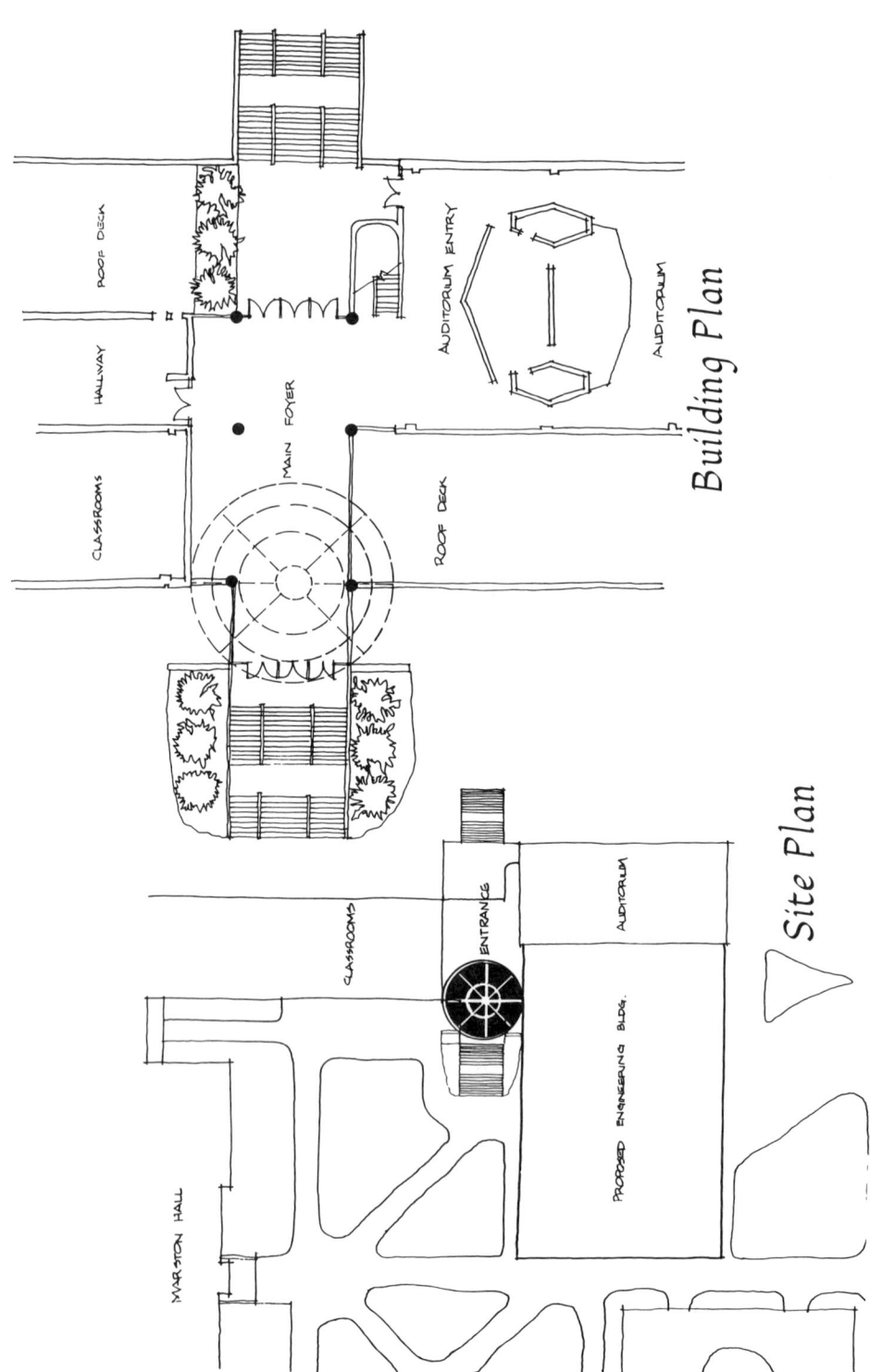

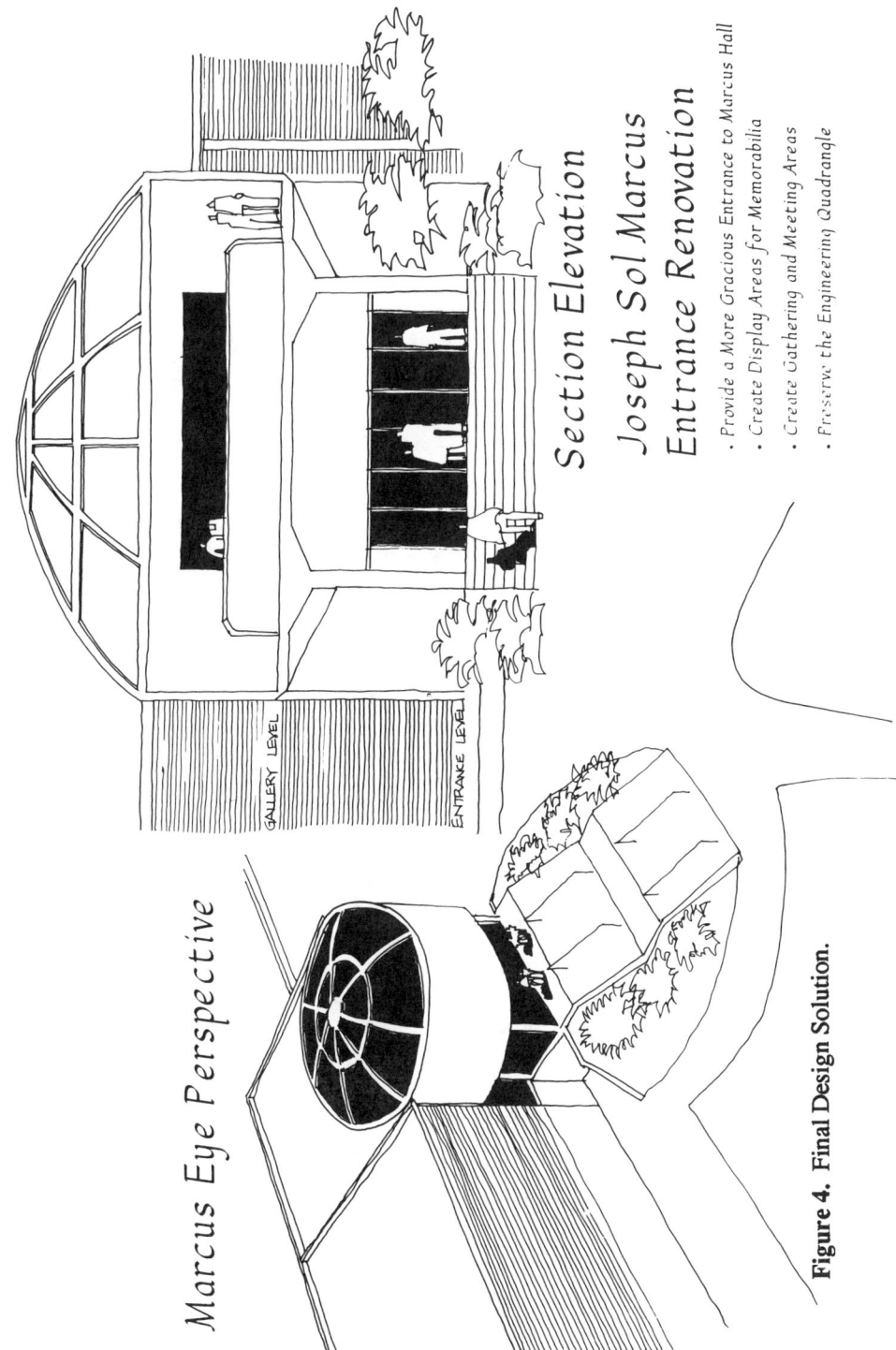

Figure 4. Final Design Solution.

enough ties its resolution to our understanding of aesthetics. While this may take us far afield into a metaphysical discussion relating intuition and aesthetics, it points out the fundamental difficulty of our problem to design a system for design; i.e., the aesthetics of the system itself are a crucial aspect of our design problem. Many systems approaches to problem solving, let alone design problems, often ignore aesthetics, yet in the context of our discussion, it is tantamount to ignoring the intuitions. After all, our design system should "look like a design system."

To summarize the paper, I would like to list some of the key attributes that the system for design should have based upon our discussion of intuition. The summary will take the form of a series of interrelated issues based on the concepts presented in the paper.

A. Generational Issues

o How should we effectively generate I_α and I_ω? Should it be a team approach where one group or individual develops I_α, the other I_ω?

o Who is to be involved in the generation process of the I_α and the I_ω?

o As the process of design becomes more computerized (my intuition tells me that this is going to happen), what is to be done with conference meetings, information conversations, telephone conversations, index cards, old photographs, dated record drawings, and so on? Is it reasonable to computerize all the information records? If not, how should we accommodate these other forms of information transfer?

o How are we to discriminate between I_α and I_ω? Figure 2 illustrated the problem with computer generation of solutions in that the computer often generates too many.

o Realistically speaking, how far along in the design process do we continue to develop alternative solutions I_α and I_ω in parallel? Could the dialectical process degenerate to the stage where no alternative emerges? This last issue naturally leads us into a discussion of the judgmental concerns.

B. Judgmental Intuitions

o Who should provide the initial off-hand judgments in a design process? C or D or both? How should this be orchestrated?

o How often should they occur?

- o How are deliberated judgments to be managed within the design process? Who generates the attributes, quantifies them, and measures their importance?

- o What underlying theory should guide the evaluation process: a value function or utility function approach? Is there perhaps another theory more appropriate than these existing paradigms, especially for the design process?

- o How can we easily assess the underlying value (utility) function of C and D or is this really impossible according to Arrow's Theorem (Keeney & Raiffa, 1976)?

- o Who is responsible for making the final overall off-hand judgment on the design solution?

VII. CONCLUSION

In this paper, we have discussed the notions of intuition and design and shown that there is a direct relationship between the design process and the knowledge created and evaluated through our intuitions. That this process is critical to the design of a "system for designing" has also been argued. While our excursion into intuition has been rather brief, hopefully you will also be smitten by its infinite possibilities.

ACKNOWLEDGMENTS

This research was partially supported by the National Science Foundation under Grant No. MSM-8417942.

REFERENCES

Bastick, A. (1982). *Intuition: How we think and act.* New York: Wiley.
Checkland, P. (1984). Rethinking a systems approach. In R. Tomlinson and I. Kiss (Eds.), *Rethinking the process of operational research and systems analysis* (chapter 4). Oxford: Pergamon Press.
Churchman, C.W. (1968a). *The challenge to reason.* New York: McGraw-Hill Book Company.
Churchman, C.W. (1968b). *The systems approach.* New York: Dell Publishing Company.
Churchman, C.W. (1971). *The design of inquiring systems.* New York: Basic Books.

Churchman, C.W. (1979). *The systems approach and its enemies.* New York: Basic Books, Inc.

Hwang, C.L., & Masud, A.S.M. (1979). *Multiple objective decision making methods and applications.* Berlin: Springer-Verlag.

Keeney, R.L., & Raiffa, H. (1976). *Decisions with multiple objectives: Preferences and value tradeoffs.* New York: John Wiley & Sons.

Kline, M. (1980). *Mathematics: The loss of certainty.* Oxford University Press.

Maslow, P. (1957). *Intuition versus intellect.* New York: The Life Science Press.

Rittel, H. (1972a). On the planning crisis: Systems analysis of the 'first and second generations'. *Bedrifts Okonomen* (8), 390-396.

Rittel, H. (1972b). Structure and usefulness of planning information systems. *Bedrifts Okonomen* (8), 398-401.

Rittel, H., & Webber, M. (1973). Dilemmas in a general theory of planning. *Policy Sciences, 4*, 155-167.

Rouse, W.B. (1986). On the value of information in system design: A framework for understanding and aiding designers. *Information Processing and Management, 22*(2), 217-228.

Smith, J. M. (1981). Wicked design problems and interactive optimization. *Proceedings of the International Conference on Cybernetics and Society* (pp. 218-224). Atlanta, GA.

Smith, J.M., & Macleod, R. (1986). A relaxed assignment algorithm for the quadratic assignment problem (submitted for publication). Paper presented at the *Mathematical Programming Symposium*, Boston, MA.

Smith, J. M. & Pelosi, R.S. (1984). Conversational optimization and facility layout planning. *Environment and Planning* (B11), pp. 63-86.

Spinoza, B. (1955). On the improvement of human understanding. In R.H.M. Elwes (Ed.), *The chief works of Benedict Spinoza.* New York: Dover Publications. (Spinoza's work published in 1666).

Taylor, H., & Karlin, S. (1984). *An introduction to stochastic modeling.* Academic Press, Inc.

Chapter 24

THE NATURE OF DESIGN (AND THE DESIGNER)

Edward J. Zagorski

School of Art and Design
University of Illinois at Urbana-Champaign
Champaign, Illinois

ABSTRACT

A series of student design problems that address the importance of the problem statement and given constraints is discussed. Emphasis is placed upon generating unique and creative solutions to a variety of design problems at a classroom level that can hopefully be paralleled at a professional level. Many of the problems are a discovery process of design principles found in nature or exercised in our man-made environment. The design instructor has to continually challenge the student designer, not only to move him through a series of self-revealing experiences but also to learn for himself how the student thinks and what he is thinking about. Avoiding the stereotype solution becomes a working challenge and goal. Human drives and motivation become apparent throughout the design program and a budding philosophical approach evolves.

I. INTRODUCTION

Over the past thirty years, I have been teaching (and practicing) industrial design at the University of Illinois and have been intrigued by the idea of teaching young people "how to design." In the process, I discovered that it was more important that I learn "how to *teach* how to design." Never having had formal courses in methodology or psychology, some truths of these two subjects were revealed to me through a series of experimental exercises that I devised for first and second year students enrolled in industrial design.

Perhaps the method could be called "design by challenge," for I discovered that in order to gain a maximum return of effort from the student, it was necessary to pique the imagination of the designer, to motivate, to enliven the competitive spirit, to raise awareness of the world around, and to openly introspect about shortcomings and strengths. The exercises which I explain in this paper ("position paper" seems to suggest a person with his heels dug into some dogma and refusing to budge) are quite simple, and the lessons learned are quite basic, but almost all of them, in perhaps a naive way, touch upon the issues in this volume on the psychology of system design and problem solving.

II. THE PROBLEM STATEMENT

Some of the most miserable and trite solutions that I have ever engendered from a group of students was to assign to them the problem of designing "a toy for a child, ages 5 to 12." Unwittingly, I had doomed every solution submitted. Their research and state-of-the-art search consisted of looking at and recording existing toys in department stores and catalogs, then modifying shapes and colors, substituting materials and processes, and ending up with warmed over push-pull, bland, and tired solutions.

The next time around, the toy problem was preceded by two preliminary problem statements. The first was to "design a device or a system that would render a random decision," and the second was "to observe a movement in nature, whether it be flora or fauna, subtle or violent, and to build a mechanical equivalent." Both preliminary exercises produced a variety of solutions, quite imaginative, and in the end were pushed to become toys or parts of toys. One solution for the first exercise consisted of a T-shaped pipe with a deflated balloon on each end of the tee. By blowing into the vertical arm, both balloons begin to inflate, one marked "yes" and the other "no" until eventually one would explode and give you the answer. Another solution utilized a plastic bowl with two outlets at the bottom marked "yes" and "no." The bowl is filled with water and a goldfish is added. Upon removal of the two plugs the water begins to recede and the final answer is left up to the goldfish who is forced to swim through one of the outlets. In other solutions, electronic circuitry was used as well as bouncing ping-pong balls, mousetraps, dodecahedron dies, and perhaps the most dramatic solution was provided by a young lady who fed her pet 6-foot boa snake two mice, one black and one white, to see which one would be taken first as a meal. Years later, a young man who is now a president of a large toy design office attributed this problem to his design of a toy called "Simon," utilizing an electronic chip for a random sequencing toy.

The second preliminary problem required a close observance of nature.

Some solutions included the mechanical hopping of a frog; the flip-flop actions of a click beetle; the undulating movement of a snake; the hopping of a kangaroo; the subtle movements of a field of wheat; objects bobbing on water; drifting snow and sand for the observance of form; a mechanical earthquake; and the erratic movements of a tadpole which consisted of gear movements, electrical circuitry, a piece of soft leather simulating the tadpole, and a heat sensitive lamp that provided the erratic power. One student brought in some spiders, observed as to how they dropped, and then built a mechanical device using a pendulum which opened and closed a release for a gravity bound weight which accurately duplicated the interrupted drop of a spider. The above problems also lead to a further study of bionics. A quote from Eric Hoffer's book, *The Ordeal of Change*, is in order here: "We are more ready to try the untried when what we do is inconsequential. Hence the remarkable fact that many inventions had their birth as toys."

Almost all problem statements at the student level are composed to be ambiguous, to be metaphorical; to allow for more than one right answer; to encourage soft and hard thinking; and to encourage fantasy, play, crossing boundaries of expertise, and breaking rules.

III. CONSTRAINTS

Too many constraints and the student feels stifled; too few and the student responds aimlessly. A machine was programmed with the letters of the alphabet and spaces in the proportions that they might appear on a printed page. Without any constraints they were projected at random on a screen producing absolute gibberish. When constraints were added a bit at a time (e.g., following specific vowels, a consonant that most often follows that vowel would appear), a semblance of partial words and sentences would appear on the screen. In the classroom, the freedom to design creatively would only occur when the right balance of constraints were introduced into the problem statement. These constraints usually involved materials, cost, function, size, time, manufacturing, space, etc.

A. Cost Constraint

In a given problem to design a microscope for a high-school environment that should cost less than $1.00, a student came up with a solution that cost $.10. It consisted of a scored sheet of cardboard that could be folded, punched out, and tabbed to form the stand and stages for the microscope—a drop of vegetable oil became the lens. It worked. In an unlimited cost version, another student designed a microscope out of cast acrylic designed in such a way as to take advantage of the "piping" characteristic of acrylic so that all the available light in the classroom became focused through the microscope under the stage, thereby eliminating the need of batteries and a bulb.

B. Media Constraints

It is often best to design with the material that will be used in the final prototype. However, since this is not always practical or possible (die castings, plastic moldings), substitute materials or a different media are often used. Curiously, it was discovered that 1) when a student worked at a drafting table using a T-square, triangles, ellipse guides, French curves, etc., his solutions reflected the tools that he used and his three-dimensional models had a two-dimensional appearance, and very few compound curves were introduced because they were too difficult to draw; 2) when a student worked in clay, his solutions rarely had a straight line and compound curves abounded. Charles Eames once said that "clay is a spineless material and it is extraordinarily easy to do something bad with it without half trying. The material puts up no resistance, and whatever discipline there is, the artist himself must be strong enough to provide. With granite (or wood) it is not easy to do something good, but it is extremely difficult to do something bad. Clay (and the airbrush) should be reserved for artists over fifty"; 3) today's favorite mock-up material in design offices is Fomcor. Fomcor has considerable limitations, especially in bending a radius. Fortunately, the resulting radii, although limited, are quite pleasing and acceptable and are found on every computer and machine housing, not as a predetermined form decision, but as a limitation of the mock-up material; 4) when asked to design a personal plastic sweep (ellipse guide), each student developed one unique to his personal preference for curves and radii that were reflected in his design solutions throughout the course. It was pointed out to the class that the Northwest Indians have done the same exercise for generations. Each Indian family passed down their own set of patterns to their children and each family exercises subtle differences in their designs.

C. Manufacturing Constraints and Communication

Each student is given a piece of basswood, 4 x 4 x 10 inches (the common denominator) and is allowed up to three cuts or passes on the band saw. He is not allowed to throw away any of the pieces but must instead rearrange the volume into a pleasing shape. After gluing the pieces together, he can spray-paint the final result. In addition to turning in his prototype, he must turn in directions or plans on how to go about making his design. This can be in the form of orthographic projections, isometric drawings, perspectives, or any conceivable form of communication he desires. Sometimes a movie is made showing each step of execution; or a comic strip format is used (surprisingly one that most students could follow easily); or a series of slides. Very often patterns or duplicate break-away models are turned in.

The Nature of Design (and the Designer) 323

In any event, the prototypes are hidden and the drawings are given to another class which is asked to make the prototype described thereon. If any question arises as to procedure or dimension, a written request has to be submitted asking for clarification. The designer has to clarify all requests and sometimes in the process generates a whole new set of questions. When all duplicate models are made, (theoretically executed by a manufacturing company), they are compared with the original prototype and the results are generally shocking and hilarious. The art of communication seems to be a lost and difficult one. When a student is asked to duplicate someone else's work, his craftsmanship acumen plummets to zero; radii become soft buttered blobs, dimensional tolerances are seldom held (and rarely given), and 180-degree errors are committed, producing mirror images of curves and forms. The duplicating student can become maddeningly dense when it comes to reading someone else's drawings, and there seems to be an underlying, subconscious desire to "screw it up."

The problem does not have to end at this point. The original prototypes are put on display by their makers using any means whatsoever to enhance it. This becomes a simulated trade exhibit and nondesign students are paraded in to select the models they would buy. Since there is no function associated with these models and the cost is the same for each one, the customer is purchasing or selecting pure aesthetics. After a number of "customers" make their selections, the models are arranged in winning order. The prototype makers are told that in one week another trade exhibit will be held. They then have the option of standing pat, or making slight modifications, or completely changing their design. Although it is obvious that the ones at the bottom will introduce some radical change, influenced to a large degree by the shapes of the leading consumer selections, the leaders might also modify their entry in the hopes of staying ahead of their redesigned competition.

D. Time Constraints and Mechanical Ingenuity

Given a steel ball five-eighths of an inch in diameter, design a machine that will keep this ball moving for precisely ten seconds, and at the end of this time it should strike and light a wooden match. The machine must be efficient and take up no more than a cubic foot of volume, and a timing device must be incorporated to indicate the elapsed seconds at any point of the cycle. The student is introduced to simple mechanics such as the lever, wheel, pulley, inclined plane, wedge, and screw. In addition, systems for storing energy are discussed such as tension, compression, counterpoise torque, gravity, balloons, chemicals, and magnets.

One of the more imaginative solutions involved a funnel supported on one side by a compressed spring and on the other side by a balloon filled with air. The steel ball, released by a pinball spring, whirled interminably in the funnel. At the same moment the ball was released, a water column began to descend down a small pipette marked off in seconds. After rounding a bend, the water began to rise, still under pressure, floating a small cork fixed with a pin, its point projecting upward directly beneath the balloon. At precisely ten seconds the balloon popped, upsetting the balanced funnel, dropping the steel ball directly onto a mouse trap arrangement holding a wooden match which struck an overhead piece of emery paper thereby igniting. Every student turned in a solution and no two had even a close resemblance to each other.

E. Space Constraints

Each student is given a glass gallon container with a neck opening of one inch. At the bottom of the container is a nut, a bolt, and a washer (unassembled). The students' first challenge is to design and make tools to assemble the three items through the neck of the bottle. The second challenge is to push ten Japanese cocktail umbrellas into the bottle, open them up, and assemble them in a formal arrangement using glue to make the structure permanent.

Ingenuity becomes evident here in designing tools that would delicately transfer forces of tension and compression at a distance through a restricted opening. The application of glue also required a rethinking of normal procedures. It was easy to parallel this problem with that of the surgeon performing a bypass, a scientist handling toxic material, or a car mechanic repairing a remote piece of machinery.

IV. RESEARCH

At the sophomore level, research into a problem is generally conducted at a very superficial level, generally relegated to looking at existing product solutions and varying the proportions somewhat. When asked to turn in the dimensions for a dining room chair, they simply measure an existing dining room chair. An experiment was tried where all the chairs and stools were removed beforehand from the classroom, and when the class convened, the students were divided into disparate, anthropometric teams: the fat with the skinny, girls with boys, the short with the tall, etc. Equipped with chalk, paper, and string, they were to turn in the dimensions of an ideal dining room chair based on the team's needs. This "raw" research took an hour and the remaining time was spent nailing together some plywood forms that reflected their dimensions. The five teams turned in five different sets of dimensions, all of them suitable for dining.

Anthropometric data, such as garnered from Henry Dreyfuss', *Measurement of Man*, can often be misleading. Very often data is obsolete. Information is often static dimensioning instead of dynamic dimensioning where the subject is in motion. Ethnic anthropometry is difficult to find. Usually, data is based upon perfect and normal subjects excluding the young, the old, and the handicapped. An ongoing study of a design for a delivery table resulted in a halt while raw data had to be collected on the anthropometric measurements of pregnant women because none existed. In cooperation with a local hospital and consenting pregnant women, a team of nurses and students visited Wright Patterson Air Force Base in Ohio to learn how to take proper measurements, and proceeded to record measurements of the subjects at the hospital over a period of one year.

V. INDIVIDUAL HUMAN DRIVES AND MOTIVATION

The student finds himself marooned on an unknown island and takes shelter in a nearby cave. Equipped with a box of dry matches, he lights a fire and discovers one-fourth pound of a strange material nearby. (He is actually given one-fourth pound of a wax-based clay.) A small pellet of this clay thrown into the fire turns into hardened steel, and he realizes that he has an opportunity to use this clay for some functional practical purpose. He is first asked to write a paper of his intentions—the problem statement—which he gives to the instructor, and then he proceeds to make whatever prototypes or devices he feels necessary. The personality and motivation of the designer becomes apparent in the results. Some eagerly accept the challenge of isolation, and, like Robinson Crusoe, decide to make the best of it. They design knife blades, adzes, awls, and fish hooks for building shelters, for trapping animals and fish, for making items for cooking, and in general all the tools for survival and an extended stay. Others want to find another island and plan on building a boat and leaving. Others cannot face the ordeal and design items for rescue: starting signal smoke fires, reflecting surfaces for signaling, designing a bludgeon to fend off intruders and perhaps even seek them out. One young lady writes that she will not make a knife because she could never kill an animal or a predator and instead makes a mirror and a comb to look presentable should someone else show up on the island. Another student designs arrow heads and a spear tip—he states in his paper that there is a tiger on the island and he plans to kill it. Still another student designs a small icon, a Buddha statue, and decides to pray for his rescue.

A. The Bomb

Inscribed with chalk, on two 4 x 8-foot tables, are twenty irregular and disproportionate shapes, each representing a municipality. Each student, upon entering the classroom, is given control over one of these areas and the contents of his pockets, in turn, become the capital goods of his municipality. This wealth may be in the form of a pocketknife, a paper clip, a rubber band, or a fingernail file. Upon a blackboard is a sketch of a partially complete system for constructing a "bomb," consisting of a paper missile fired from a tube-like rocket with the aid of a rubber band and a triggering pin. Above the table tops is suspended a flat disc which represents the moon. The following instructions are given: if a missile, when fired, falls upon someone's property, the one who did the firing will own that property, plus all the capital goods thereon. If a missile lands on the "moon," it gives the student the advantage of mobility; he can now move anywhere around the classroom to fire his missiles. End of ground rules. No one has stated that it is necessary to fire the missiles.

Almost immediately, the two tables cooperate but independent of each other. Their goal is to "get the other table." Alliances spring up between adjoining neighbors; some try mass producing and sharing the scarce supply of rubber bands and elastic. (The top of a pair of jockey shorts is sacrificed for the apparent war effort.) Those with knives cut paper easily and efficiently; those without use their fingernails and improvise straight-edges and are accused of making "dirty bombs" because of the resultant ragged edges. The "bombs" begin to rain down and neighbors begin to mistrust one another and protect their flanks by lobbing missiles on adjoining municipalities with a curt, "Sorry." No one tries for the moon because time is running out. When it is over, a discussion and analysis follow and the arguments drift to world politics. In the end, some simple truths are revealed.

B. Moses

As a group, the class was asked to recall the Ten Commandments and put them on the blackboard. Then, still acting as a group, they were asked to list them in the order of their importance. A majority vote would determine the listing. The most important commandment, and by a unanimous vote, was "Thou shalt not kill." Tied for eighth and ninth place was "Thou shall not commit adultery" and "Honor thy father and mother." The most disturbing listing was number two, that of "Thou shall not steal." When it was pointed out to them that a faculty member had to lock his car, his home, his office, his shop, or lose whatever the students could walk off with, the answer was that faculty members are part of the establishment and thereby fair game. Furthermore, this was not stealing. Stealing meant taking things from each other. Honor among thieves.

C. The World

Redesign the land masses of the world. The first three weeks are to be spent researching temperature zones, water currents, food-producing areas, mineral deposits, ethnic and language considerations, population distribution, etc. The finished results are to be presented with a series of acetate overlays depicting and defending the redesigned land masses. If the student designers had their way, we would all be living near an ocean shore or on an island, somewhere near the equator. In addition, we would be neatly compartmentalized as to color, religion, and language.

The Nature of Design (and the Designer)

VI. THE RELATIONSHIP OF PARTS

In an effort to get students to "see" the space surrounding objects, a simple expedient was employed. The student carved a lower case letter out of linoleum by removing the linoleum surrounding the letter, thereby concentrating on "meaningless" space. The exercise was extended further to observe the relationship between parts and to observe the function, especially aesthetically and psychologically, of the spaces between them. Carried away in a burst of enthusiasm, a poem of Lao-tse was held up as a bannerhead which read as follows:

> "The wheel's hub holds thirty spokes
> Utility depends on the hole through the hub.
> The potter's clay forms a vessel
> It is the space within that serves.
> A house is built with solid walls
> The nothingness of window and door alone renders
>   it usable,
> That which exists may be transformed
> What is non-existent has boundless uses."

We had finally found the fountainhead! We pointed to masters such as Josef Albers and others who showed us that one did not have to touch an object to change its appearance; simply change its environment, concentrate on its space and one could alter its color, size, and meaning. The students posted a slogan, "Think Negative!" Everything came crashing down the next day when a bright young lady addressed the class and said that Lao-tse expressed Eastern philosophy and thinking. The immaterial may indeed be more important to them, but in the U.S.A. a more appropriate poem expressing Western materialism would read as follows:

> "As you wander on through life, brother,
> Whatever be your goal,
> Keep your eye upon the doughnut,
> And not upon the hole."
>
>   Mr. Donut

VII. AVOIDING THE STEREOTYPE

In 1950, John Arnold ran a program in creative engineering at MIT before leaving to start up the design program at Stanford. Arnold announced to his students the discovery of an imaginary planet called Arcturus IV, where the gravity was 11 times greater than Earth's, the crops grew upside down, and the inhabitants had three eyes and fragile bones. Students were asked to design consumer goods for this world. Arcturan power tools, for example, ended up being cable-driven, their motors on the ground; a conventional drill weighing 4 pounds on Earth would have weighed 44 pounds on Arcturus IV, too heavy for a native's delicate arm to support. Other students designed egg-shaped cars and stereo viewers for the three-eyed heads.

In 1980, George Lynn ran a program in creative industrial design at Carleton University in Ottawa, Ontario in Canada where he passed out a book to his students entitled, "Gnomes" by Wil Huygen. They were to develop human factors and anthropometric charts about these people, and then design consumer products for their use. The gnome has the strength of 10 humans and averages 15 centimeters in height. Transportation vehicles, food preparation products, and laundry facilities such as washing machines and dryers were designed and built for these mythical people.

In 1961, I gave the students the following problem. Given a raw egg to represent an astronaut, the problem was to protect the egg from shock, water, and fire; these are typical blast-off and reentry problems. The students designed and built a capsule which was then launched from a student designed launching pad consisting of two bolted truck springs, cocked and held down until the proper countdown release moment. The capsule had to pass a preliminary fire test consisting of a welding torch held for ten seconds at a distance of eight inches from the capsule. After a 200-foot catapult into the University's reflecting pool, the capsule was opened and examined for cracks and moisture.

Using such cushioning devices as a wooden sphere filled with grease and a cocktail shaker lined with foam rubber and carpet padding, many landed their eggs in the water unbroken. One student used two halves of a coconut shell with sheet rubber stretched across the opening of the two halves. The egg was securely nestled in the middle and the device was compared favorably to a placenta's function in human birth. The champion of the day was a student who surpassed the requirements—with showmanship. His missile blew itself apart in midair with a timed firecracker, releasing a handful of tiny silk American flags, and a parachute that gently lowered into the pool a capsule containing the egg packed in gelatin and peat moss. When the capsule touched the water, the moisture melted an aspirin tablet, holding apart two electrical contacts and a tiny electric motor powered the craft ashore.

VIII. CONCLUSION

Life is a continuum, and it is difficult to formulate or state explicit conclusions on a continuing process that changes every second of our lives. Design is treated here as a life force; problem statements are in a continual state of flux and are usually formulated after open discussions with the current body of students. The marketplace changes; world politics change; and ethical and moral considerations change.

When a certain class of students seemed to lack responsibility for anyone but themselves, they were each given a hothouse plant to take care of for three months. Their grade would be based upon how healthy the plant appeared at the end of the term. This meant researching the plant species, adding nutrients and water, and coming in on weekends and their spring break to take care of it. All of them realized that life support systems require cooperation and attention to details. The rewards were a few blossoms and that tremendous inner satisfaction of sustaining living growth.

When another class complained about the grades they received, their next problem was graded with a complex coded group of symbols. (A kindergarten set of rubber stamps were used—combinations of ducks, pigs, pelicans, chickens, etc.) They broke the code quite quickly by evaluating the returned work themselves, as a group. They learned that they each had evaluative powers and did not have to rely on the teacher's judgment. Their complaints about grades was a form of carping. The analogies of taking care of a plant and evaluating their own work will hopefully be seen when they leave the University.

It was Eric Hoffer that stated, "The tendency to carry youthful characteristics into adult life, which renders man perpetually immature and unfinished, is at the root of his uniqueness in the universe, and is particularly pronounced in the creative individual. Youth has been called a perishable talent, but perhaps talent and originality are always aspects of youth, and the creative individual is an imperishable juvenile."

To nurture and encourage this uniqueness in young men and women has always been the goal in the many problems given in this design program.

AUTHOR INDEX

Adams, J. L., 212, 219
Adelman, L., 249, 254
Alavi, M., 150, 156
Allen, T. J., 3, 6, 87, 90, 91, 95, 282, 283
Allison, G., 291, 301
Alvey Programme Annual Report, 257, 271
Amabile, T. M., 212, 219
Ambler, A. P., 269, 272, 273
Andriole, S. J., 255
Arabian, J. M., 161, 172
Arnold, J., 327
Askren, W. B., 230, 244
Atzinger, E. M., 165, 173
Axelrod, R. M., 294, 301

Ballay, J. M., 65
Barrett, R., 271, 272
Barrow, H. G., 264, 266, 272
Bastick, A., 306, 317
Bean, T. T., 161, 173
Beavers, L., 43

Beitz, W., 85, 96, 222, 225, 226, 227
Bellos, I., 269, 273
Benbasat, I., 248, 254
Bennett, E., 110
Benzon, B., 128, 138, 142
Blanchard, G. S., 165, 173
Bloom, B. S., 104, 110
Bobrow, D. G., 212, 219
Boff, K. R., 3, 5, 6, 7, 8, 10, 13, 16, 43, 83,
    85, 87, 89, 93, 95, 96, 142, 147, 177,
    180, 223, 224, 227
Bogen, R., 263, 272
Bonder, S., 169, 172
Borning, A., 142, 143
Boulding, K. E., 83, 95
Brachman, R. J., 264, 265, 272
Branscomb, L. M., 251, 252, **254**, 282, 283
Brentano, F. C., 306
Brezovic, C. P., 178, 179, 181, 186
Brooks, R. A., 269, 272
Brouwer, L. E. J., 306

Calderwood, R., 180, 186
Calhoun, G. L., 3, 6, 8, 10, 16, 89, 95, 224, 227
Calluci, F., 40
Cameron, S., 270, 272
Capote, T., 35
Carbonell, J. G., Jr., 296, 301
Card, S. K., 212, 219
Caro, P. W., 223, 225, 227
Carr, T., 87, 95
Carroll, J. M., 252, 254
Checkland, P., 307, 311, 317
Childs, L., 98
Chubb, G. P., 116, 125
Churchman, C. W., 104, 110, 291, 301, 307, 308, 311, 313, 317
Clements, P. C., 147, 157
Clinton-Cirocco, A., 180, 186
Clocksin, W., 263, 272
Connolly, T., 280, 283
Cooper, R. S., 5, 43, 63
Coppola, A., 187
Corner, D. F., 269, 272
Corney, J., 257
Creighton, J. W., 96
Curran, A. R., 216, 220

Dantzig, G., 311
Davis, G. B., 300, 301
DePuy, W. E., 169, 172
Debons, A., 103, 106, 110
Degan, J., 110
deGroot, A. D., 176, 186
deKleer, J., 268, 272
Department of Defense, 241, 244
Department of Trade and Industry, 257, 272
Department of the Army, 165, 172
Descartes, R., 306
Diesing, P., 291, 301
Donnell, M. L., 249, 254
Doohovskoy, A., 263, 272
Doppelt, F. F., 1, 43
Dreyfus, S. E., 298, 301, 302
Dreyfuss, H., 324

Eames, C., 322
Eberts, C. G., 96
Eberts, R. E., 96
Eggleston, R. G., 113
Einstein, A., 34, 128, 131
Eisenstadt, M. 272
Elwes, R. H. M., 318
Essoglou, M. E., 90, 95

Farr, D. E., 229, 230, 238, 244
Fateman, R., 263, 272
Ferrell, W. R., 117, 125
Findler, N. V., 294, 302
Finley, D. L., 230, 244
Fiorello, M., 170, 173
Fleming, A., 128
Flixborough disaster, 245, 256
Ford, G., 98
Foster, J., 40
Foulk, C. R., 300, 303
Franco, M. M., 165, 173
Frey, P. R., 146, 156, 246, 254
Friedman, F. L., 165, 173
Furness, T. A., III, 31, 127, 140, 142

Galing, B., 290, 300, 303
General Accounting Office, 160, 172
Genesereth, M., 263, 272
Getty, W., 32
Gick, M. L., 296, 302
Gionfriddo, T., 140, 143
Golden, J., 263, 272
Gould, J. D., 252, 254

Hall, A. S., 261, 272
Hampden-Turner, C., 130, 142
Hardy, S., 271, 272
Hartel, C. R., 161, 172, 173
Hartson, H. R., 254
Havelock, R. G., 90, 96
Hearn, A. C., 263, 272
Hicks, D., 41
Hinrichsen, K., 140, 143
Hochberg, J., 133, 142

Author Index

Hoffer, E., 321, 329
Holowenko, M. S., 261, 272
Holyoak, K. J., 296, 302
Hunt, R. M., 145, 146, 156
Huygen, W., 328
Hwang, C. L., 312, 318

Iwai, S., 297, 302

Janis, I. L., 291, 302
Jefferies, R. S., 260, 272
Jefferson, T., 99
John, P. G., 182, 186
Johnson, E. M., 159
Jolly, J. A., 96
Jones, J. C., 66, 82, 212, 220
Jones, M. E., 13, 16, 93, 96

Kane, J. J., 161, 173
Kant, I, 306
Kaplan, J. D., 161, 172, 173
Kapur, D., 269, 272
Karlin, S., 308, 318
Karmarkar, K. R., 311
Kaufman, L., 142
Keeney, R. L., 312, 317, 318
Kerwin, W., 165, 173
Kidd, J. S., 170, 173
Kirchner-Dean, E., 161, 173
Kisner, R. A., 246, 254
Kiss, I., 317
Klein, G. A., 175, 177, 178, 179, 180, 181, 182, 186, 298, 302
Kline, M., 307, 318
Kocian, D. F., 140, 142
Koutsou, A., 257
Krueger, M., 140, 143
Kulp, B., 187

Lagomasino, A., 290, 300, 301, 303
Laird, M., 40
Lao-tse, 327
Larson, A., 106, 110
Lassiter, L. W., 199
Lauber, J. L., 167

Laughlin, M. S. M. E., 261, 272
Leifer, L. J., 211, 216, 220
Lenat, D. B., 119, 125, 142, 143
Lewis, C., 252, 254
Licklider, J. C. R., 106, 110
Lincoln, J., 3, 6, 8, 10, 16, 89, 95, 224, 227
Linstone, H. A., 291, 302
Lintz, L. M., 230, 244
Lott, W. J., 230, 244
Lowry, J. C., 170, 173
Lynn, G., 328

Mackie, R. R., 249, 254
MacLean, R. F., 248, 256
Macleod, R., 311, 312, 318
Maddox, M. E., 146, 156
Mann, L., 291, 302
March, J. G., 291, 302
Marcus, A., 161, 172, 173
Martin, E. A., 10, 16, 85, 95, 221, 223, 224, 227
Martino, J. P., 13, 16, 93, 96
Maslow, P., 307, 318
Masud, A. S. M., 312, 318
Mather, D. V., 259, 272
McDonald, D., 142, 143
McKim, R. H., 128, 131, 143, 212, 220
Meister, D., 91, 96, 229, 230, 235, 237, 242, 243
Mellish, C., 263, 272
Michalski, R. S., 301
Miller, J. G., 106, 110
Millington, K., 257
Mills, R. B., 140, 143
Minsky, M., 295, 302
Mirabella, A., 182, 186
Moran, T. P., 212, 219
Morris, N. M., 15, 17, 245, 248, 249, 251, 255
Morton, P., 107, 110
Mostow, J., 212, 220
Mullin, J. L., 165, 173
Mundy, J., 269, 272
Musser, D., 269, 272

Nadler, G., 231, 244, 288, 302
Nakamura, K., 297, 302
Narendran, P., 269, 272
Naval Research Advisory Committee, 160, 173
Newell, A., 71, 82, 212, 219, 292, 302
Nixon, R. M., 40
Norris, R., 27, 28, 30
Nystrom, P. C., 291, 302

Office of the Director of Defense Research and Engineering, 160, 173
Olsen, R., 291, 302
Organization for Economic Co-operation and Development, 110
Osborn, A., 212, 220

Packard, D., 40, 277, 283
Pahl, G., 85, 96, 222, 225, 226, 227
Parnas, D. L., 147, 157
Pascal, B., 306
Patterson, J. F., 249, 254
Pavelle, R., 263, 272
Pelosi, R. S., 311, 318
Perez, R. S., 182, 186
Perrow, C., 162, 173, 237, 244
Phelps, R. H., 248, 249, 251, 254, 255
Poincaire, J. H., 306
Polya, G., 306
Popplestone, R. J., 257, 263, 269, 272, 273
Price, H. E., 170, 173, 249, 255
Pritsker, A. A. B., 116, 125
Promisel, D. M., 161, 172, 173

Raiffa, H., 312, 317, 318
Ramsay, A., 271, 272
Rasmussen, J., 297, 302
Ray, M. S., 222, 224, 225, 226, 227
Reitman, W. R., 69, 82
Reitman, W., 13, 16, 93, 96
Requicha, A. A. G., 269, 273
Rhode, A. S., 165, 173

Rinderle, J. R., 212, 220
Rittel, H., 307, 308, 310, 318
Roberts, F. S., 295, 302
Rose, L. L., 300, 303
Rosson, M. B., 252, 254
Rouse, S. H., 15, 17
Rouse, W. B., 3, 5, 6, 7, 8, 12, 13, 15, 16, 17, 43, 88, 96, 146, 147, 156, 177, 180, 216, 220, 222, 223, 224, 227, 230, 243, 248, 249, 251, 256, 275, 276, 281, 283, 303, 308, 318
Rubinstein, E., 245, 255

Sage, A. P., 285, 287, 288, 290, 291, 295, 297, 300, 301, 302, 303
Sahar, G., 257
Sawyer, C. R., 170, 173
Schneiderman, B., 151, 157
Seifert, D. J., 116, 125
Seum, C. S., 116, 125
Shea, O., 272
Sheridan, T. B., 117, 125
Shorey, R. R., 194
Siddall, J. N., 212, 220
Sides, W. H., 146, 156
Silverman, B. G., 296, 303
Simon, H. A., 35, 71, 82, 291, 292, 302, 303
Sippl, C. J., 104, 110
Skinner, B. B., 165, 173
Sloman, A., 271, 272
Smith, J. M., 305, 308, 311, 312, 318
Smith, M. G., 170, 173
Smithers, T., 257
Smullin, L., 32
Speigel, J., 110
Spinoza, B., 306, 307, 318
Starbuck, W. H., 291, 302
Steinbruner, J. D., 291, 303
Sternberg, R. J., 296, 303
Suh, N. P., 212, 220
Sullivan, D. J., 230, 244
Sutherland, I., 134, 140, 143

Author Index

Taft, W. H., 41, 194
Taylor, C., 142, 143
Taylor, H., 308, 318
Taylor, R. N., 248, 254
Teates, H. B., 249, 256
Thomas, J. C., 251, 252, 254
Thomas, J., 142
Thurman, M., 162
Tilove, R. B., 269, 273
Tomlinson, R., 317
Topper, P. E., 165, 173
Tucker, J. H., 249, 245, 255
Turner, M. E., 296, 303

VanGundy, A. B., 212, 220

Warfield, J. N., 295, 303
Watson, J. D., 128, 141, 143

Webber, M., 308, 318
Weinberger, C., 40
Weischedel, R.M., 13, 16, 93, 96
Weitzenfeld, J., 177, 182, 186
West, H. M., III, 168, 173
Wester, M., 263, 272
Weyer, S., 142, 143
Weyl, H., 306
Whittenburg, J. A., 161, 173
Wilde, D. J., 212, 220
Winston, P. A., 302
Wise, J. A., 280, 283
Wortman, D. B., 116, 125

Yovits, M. C., 300, 303

Zakay, D., 247, 255

SUBJECT INDEX

AAMRL (*see* Armstrong Aerospace Medical Research Laboratory)
ABM (*see* anti-ballistic missile)
ABTMS (*see* assumption based truth maintenance system)
Academic-industrial research programs, 110
Academic institutions, 103, 105
Acceptance of support systems, 15
Access to technical data, 83, 84
Adaptive interfaces, 224
Advanced, forward swept wing, integrated technology aircraft, 32
Advanced Research Projects Agency, 19, 20, 22, 28, 30, 32, 33, 40, 43
Advanced tactical fighter, 182
Advanced technology demonstrations, 39
Advice, 13, 14
Aerodynamic flow, 36
Aerodynamics, 208
Aeronautical designs, 21
Aeronautical engineering, 21, 55

Aerospace industry, 46, 208
Aesthetics, 66, 69, 316, 323
Affirmative action, 59
AFWAL (*see* Air Force Wright Aeronautical Laboratories)
Aiding, 14, 114
Aircraft, 9, 30, 276
  cockpits, 45
  design, 31, 48
  pilots, 275
  sizing, 206
  structures, 206
Aircrew training device, 223
Air Force, 193
  Aeronautical Systems Division, 40
  Aerospace Medical Division, 2, 43
  Armstrong Aerospace Medical Research Laboratory, 140
  Chief of Staff, 188
  Human Resources Laboratory, 195
  Wright Aeronautical Laboratories, 191

Airplane loft lines, 200
Air traffic controllers, 253
Algebraic engine, 268
Algebraic equation solving, 260
Algorithmic support, 280, 291
Algorithms, 13, 37, 197
Alienation, 16
Alternatives
  evaluation of, 8
  generation of, 8
  selection of, 8
Aluminum alloys, 261
Alvey large scale demonstrator project, 257
Ambiguity, 80
AMD (*see* Air Force Aerospace Medical Division)
American Defense Preparedness Association, 194
Analogical matches, 9
Analogical reasoning, 182, 296, 300
Analogies, 52, 54, 55, 56, 119, 175, 179, 181, 296
Analysis, 8, 62, 148, 230, 290
Analytical behaviors, 48
Analytical decomposition, 47
Analytical strategies, 175
Analytical tools, 14
Anchoring and adjustment, 300
Animation, 214
Anthropometry, 133
  data, 324
  data bases, 141
  models, 169
Anti-ballistic missiles, 23
Apollo program, 21
Apollo space module, 237
Apple MacIntosh, 136
Applications-specific integrated circuits, 26
Architects, 44, 46, 53, 153
Architecture, 276, 306
Arcturus IV, 328
Armstrong Aerospace Medical Research Laboratory, 2
AROD (*see* eye in the sky)

ARPA (*see* Advanced Research Projects Agency)
ARPANET, 32
Array processors, 139
Arrow's theorem, 317
Artificial intelligence, 110, 206, 212, 218, 260 (*see also* machine intelligence)
Artificial intelligence workstations, 218
Artistic behaviors, 48
Artists, 8
ASD (*see* Aeronautical Systems Division)
ASIC (*see* applications-specific integrated circuits)
Associative knowledge, 297
Associative thinking, 129
Assumption based truth maintenance system, 268
Astronaut problems, 328
ATD (*see* aircrew training device)
Atom bombs, 237
Atomic fusion, 23
ATPG (*see* automatic test pattern generator)
Attitudes, 253
  toward risk, 37
Audit trail, 87, 280
Authority, 209
Automata theory, 110
Automation
  design support, 147
  design systems, 71
  information management, 93
  machinery, 196
  production machinery, 196
  target acquisition, 35
  test pattern generators, 190
Automobiles, 45, 276
Auxiliary subsystems, 240
Aviation industry, 169
Aviation metaphor, 211, 218
Axiomatic design, 212

Ballistic missile reentry discrimination, 23
Ballistic missiles, 21

Subject Index

Baselines, 52, 85, 88, 123, 124
Basic research, 37
Basketball teams, 116, 119
Bayesian decision making, 237
Bayesian theory, 185
Behavioral design, 238
Behavioral design data bases, 240
Behavioral research, 2
Behavioral science, 10
Benchmark, 124
Biases, 9, 299, 300
Binocular display electronics, 135
Binocular helmet-mounted displays, 140
Biomedical research, 2
Bionics, 321
Biotechnology, 3
BIT (*see* built-in test)
BLACK HAWK (UH-60A) helicopter, 161
Blue Ribbon Commission on Defense Management, 99
Boeing Aircraft Company, 31, 195
Boeing 757/767 aircraft, 169
Bolt Beranek and Newman, 30
Bookkeeping, 14
Bootstrapping, 21
Bottom-up composition, 47
Brain, 106, 108
Brainstorming, 44
British Standard BS545, 266
Buddha, 325
Built-in test, 189
Bureaucracy, 22, 38, 61
Bureaucratic processes, 210
Butterfly computer, 30

C³ (*see* command, control, and communications)
CAD (*see* computer-aided design)
CAD (*see* computer-aided drafting)
CADAM, 195, 217, 309
CADDS-4, 309
CAE (*see* computer-aided engineering)
California Institute of Technology, 29
CAM (*see* computer-aided manufacturing)
Capitol Hill, 38

Carleton University, 328
Car mechanics, 324
Carnegie-Mellon University, 30
Carpentry, 69
CAS (*see* computer-aided support)
Categorization, 108
CATIS, 309
CEDAR, 33
Central nervous system, 106
CEO (*see* chief executive officers)
Checklists, 164
Chess mastery, 176
Chief executive officers, 60, 115
CIM (*see* computer-integrated manufacturing)
Circuit boards, 27
  design engineers, 20
  layout, 280, 282
Classic designs, 51
Classification, 108
Class six computers, 21, 36
Clerk, 14
Clients (*see* users)
Client needs (*see* user needs)
CMOS (*see* complementary metal oxide semiconductor circuits)
Coach, 14, 15, 115, 246
Cognition, 113, 125
Cognitive complexity, 222
Cognitive design, 230
Cognitive functions, 203
Cognitive input/output port, 136
Cognitive maps, 292, 295
Cognitive ports, 139
Cognitive processes, 241
Cognitive style, 10, 223
Cognitive transactions, 114
Colleges, 105
Collimated pictures, 135
Color codes, 179
Combining information, 79
Command and control system, 33
Command, control, and communications, 103, 106, 140
Command languages, 11

Commerce Business Daily, 277
Common LISP, 271
Communications, 120, 194, 206, 209, 210, 221, 225, 226
  control system, 149
  satellite system, 25
  science, 106
Company practices and procedures, 280
Comparison-based prediction, 182
Competition advocates, 41
Competitiveness, 20
Complementary metal oxide semiconductor circuits, 28
Complexity, 57, 58, 59, 108, 159, 211, 213, 217, 282, 294, 308
Complex systems, 14, 62, 86
Component catalogs, 266
Components, 10
Computational power, 20
Computational tools, 14, 215
Computation techniques, 189
Computer-aided design, 13, 15, 54, 55, 56, 65, 69, 80, 101, 127, 132, 140, 142, 184, 188, 193, 195, 199, 201, 211, 215, 218, 309
Computer-aided design networks, 101
Computer-aided drafting, 259
Computer-aided engineering, 13, 15, 199, 201
Computer-aided engineering workstations, 189
Computer-aided logistics support, 194
Computer-aided manufacturing, 13, 15, 188, 193, 195, 199, 201, 218, 259
Computer-aided support, 188, 194, 195
Computer-based information systems, 281
Computer-driven instructor stations, 181
Computer-generated display systems, 146
Computer-generated imagery, 141
Computer-generated worlds, 136
Computer graphics, 19, 31
Computer-integrated manufacturing, 199, 201, 259
Computer networking, 19, 32
Computer programming, 69

Computer refresh rates, 179
Computer science, 34, 212
Computer Vision, 309
Computing algorithms, 139
Concept formation, 296
Conceptual design, 147, 288
Conceptual models of human-machine relationships, 116
Configuration control, 150
Congress, 38, 40, 58, 60, 98, 166
Congressional Commission on Government Procurement, 98
Consortia, 193
Constrained optimization, 52
Constraint-driven design, 147
Constraints, 230, 235, 246, 293, 321
Constructive solid geometry, 269
Consultants, 14, 15
Consumer products, 48, 328
Consumer Reports, 183
Control dynamics, 178
Control theoretic techniques, 55,
Conversational design, 311
Conversational optimization, 305
Cooperative efforts, 193
Cooperative systems, 122
Coordinate transformation matrices, 139
Coordination, 209
Corporate goals, 98
Corporate knowledge bases, 222
Corporate memory, 36, 221, 224
Corporate policies and procedures, 222
Correctness, 50, 53, 59, 60, 61
Corrosion prevention, 191
Cost accounting
  procedures, 277
  regulations, 12
Cost-benefit analysis, 55,
Cost constraints, 321
Cost of design, 29
Cray computer, 30
Creativity, 2, 7, 10, 23, 37, 39, 48, 51, 57, 59, 60, 66, 84, 88, 125, 129, 131, 151, 153, 154, 192, 207, 210, 212, 214, 240, 285, 319, 327

Subject Index

Crew stations, 9, 140, 142
Crew systems, 3
Criteria, 230, 237
Cross-disciplinary expertise, 211
Cross-disciplinary flow of information, 95
Cross-disciplinary information access, 83, 223, 226
Cross-disciplinary information utlization, 83
Cross-disciplinary interfaces, 221
CRT terminals, 11, 132
Cruise missile, 31
CSG (*see* constructive solid geometry)
Cue perception, 179
Cultural norms, 58

Daisy System Corporation, 26, 194
Dassault Systems, 309
Data, 13
  explosion, 200
  management, 110, 145, 154
Data bases, 36, 56, 133, 138, 142, 162, 187, 189, 191, 196, 199, 202, 259, 281
  development, 9
  maintenance, 9
  management, 197, 199, 201, 202, 205, 210
  packages, 282
  structures, 110
DBMS
  (*see* data base management)
DEC computer systems, 218
Decision analysis, 181, 218, 300
Decision making, 106, 212
  criteria, 279
Decision space, 280
Decision style, 291
Decision theory, 110
Deductive logic, 234
Deductive skills, 287
Defense budget, 38
Defense community, 160
Defense management, 37
Defense policy, 166
Defense systems, 1
Deliberated judgments, 317

Department of Defense, 12, 20, 21, 24, 30, 35, 37, 87, 178, 208, 282
Depth perception, 133
Deputy Secretary of Defense, 194
Design
  activities, 285
  aids, 27, 125, 181
  artifacts, 257
  assurance, 208
  automation, 26, 29, 32, 33, 154, 200
  by challenge, 320
  constraints, 169
  criteria, 66, 148, 178
  data, 54, 55, 101, 196
  data bases, 140
  decision making, 7, 177, 278, 280
  deficiencies, 160, 164, 239
  description documents, 267
  documentation, 147, 149
  education, 103, 105, 212, 216, 319
  effectiveness, 83, 84, 92, 95, 148, 280, 283
  environments, 211, 213, 215, 216
  guidelines, 146
  information systems, 62
  innovation, 61
  integration, 123, 156
  management, 105
  methodology, 286, 287, 305
  methods, 212
  packages, 26, 29, 32
  philosophy, 146
  problems, 53
  problem solving, 65, 72, 177
  process, 2, 4, 20, 59, 65, 69, 132, 147, 154
  process flexibility, 154
  requirements, 84, 119, 146, 242
  specifications, 88, 150, 151, 277, 285
  standards, 169
  style, 231
  support systems, 2, 5, 13, 51, 62, 92, 106
  support tools, 177
  taxonomy, 234

teachers, 70
team communication, 56
team interaction, 225
teams, 49, 86
technology, 213
theory, 212
tools, 21, 33, 54, 125, 132
to product, 257
to product demonstrator project, 271
tradeoffs, 7
trends, 20
workstations, 21, 26, 29, 31, 32, 210
Designers
  activities, 46
  as pilot, 214
  biases, 184
  desktop, 78
  intentions, 54, 139
  limitations, 275, 281
  preferences, 184
Designer's Associate, 19, 35, 92, 138, 141, 216
Designer-user relationships, 46
Detailed design, 147
Diagnostic test equipment, 9
Dialectical methodology, 313
Dialectical process, 305
Dialectics of design, 309
Dialogue structure, 15
Digital satellite communication systems, 20
Dimensional refinement, 75
Discipline-oriented instruction, 105
Discretionary information utilization, 281
DNA, 128
Documentation, 222, 226
Documentation costs, 150
DoD (*see* Department of Defense)
Domain expertise, 145
Domain knowledge bases, 148
Doppler imaging, 24
Drafting, 280
Drawings, 196
DSS (*see* design support systems)

Ease of access, 14
Easybanker, 66, 70
EATPUT model, 103, 106, 110
ECO (*see* engineering change orders)
Economics, 55, 205
ECR (*see* engineering change requests)
EDIF (*see* electronic design interchange format)
Edinburgh design system, 257, 263, 264, 267, 268, 270
Effects of organization, 11
Electrical engineering, 55
Electric motors, 261
Electromagnetic designs, 21
Electromechanical systems design, 211
Electron guns, 23, 32
Electronic countermeasures, 195
Electronic design, 205
Electronic design interchange format, 194
Electronic engineers, 21
Electronic flight control systems, 207
Electronic gates, 20
Electronic relationships, 33
Electronic troubleshooters, 275
Emergent decision making, 280
Emotion, 113, 125
Encyclopedia, 261, 264
Encyclopedic knowledge, 142
Engine components, 195
Engineering, 10, 50, 212
  change orders, 279
  change requests, 279
  design workstations, 187
  drawings, 257
  organizations, 209
  psychology, 53
  requirements, 67
  workstation, 201, 278
Engineer's biases, 236
Entomology, 141
Entrepreneur, 210
Environmental standards, 277
Equal opportunity, 59
Ergonomics, 66, 85, 168, 223

Subject Index

ESL, 30
Evaluation, 230
Evaluation criteria, 289
Event world analysis, 108
Executive expert systems, 191
Experience, 214
Experimental aircraft, 39
Expertise, 298, 310, 321
Expert systems, 13, 34, 36, 76, 142, 156, 191, 199, 203, 216, 282, 297
Explanation, 299
Explication (*see* explanation)
Exploratory development program, 37
External information, 78
Eye in the sky, 232
Eye position trackers, 145

F-16 aircraft, 182
Factories, 188
Facts, 13
Failure modes and effects analyses, 196
Familiarity matches, 180
Fantasies, 13, 321
Fatigue, 11
Fault
  correction, 301
  detection, 301
  diagnosis, 301
  (*see also* diagnostic test equipment)
Feasibility, 235
Federal government, 12, 98, 253
Federal pork barrel, 61
Feedback, 13
Field experiences, 182
Field of view, 235
Field theory, 141
Fighter aircraft, 32, 61
Fighter aircraft design, 35
Financial risks, 207
Finite-element analysis, 54, 280, 282
Fire ground commanders, 180
Fitzhugh Report, 40
Flexibility, 209, 223, 260
Flight simulation, 214

Flixborough disaster, 245
Flying formation, 178
Foamboard, 69
  models, 80
Fomcor, 322
Foreign languages, 34
Fortran, 217
FOV (*see* field of view)
Fractal geometry, 31
Frame-based knowledge systems, 264
Frames, 292, 294, 295
Frames of reference, 90
Front-end analysis, 99, 146, 151, 182
Fun, 248
Functional fixedness, 153
Functional requirements, 99, 124
Functions, 232
Fundamental design parameters, 242
Fuzziness, 294

Gas-dynamic lasers, 24
Gatekeepers, 239
Gear box design, 264, 268
General Dynamics Corporation, 195
General Motors Corporation, 38, 55, 194
Generational intuition, 307, 309
Generation of alternatives, 230
Generation of design alternatives, 287
Generation process, 316
Geometric modeling, 54, 192, 201, 203, 280, 282
Geometric modeling engines, 269
Geometric reasoning, 260
Geometric tools, 203
Gestalt, 291, 306
GLCM (*see* ground-launched cruise missile)
Global Positioning System, 25
Goal clarification, 177, 178, 179
Goddard Spaceflight Center, 19, 25
Goodness, 51, 53, 61, 83, 84
Government
  practices, 58, 277
  procurement, 57, 208
  regulations, 61

GRADE (*see* graphics for advanced design engineering)
Graham-Rudman-Hollings Act, 33
Graphical representations, 263
Graphics
 editors, 56
 for advanced design engineering, 204
 primitives, 263
 processors, 135
 workstations, 23
Ground-launched cruise missile, 195
Group design, 238
Grouping information, 79
Grumman Aircraft Corporation, 32
Guidelines, 164, 183, 222, 280
 for user acceptance, 250

Handbooks, 101
Hardware, 10
Head-aimed optical systems, 135
Head position trackers, 140
Heads-up displays, 145
Health hazards, 162
HEATHKIT amplifier, 69
Helmet displays, 270
Helmet-mounted displays, 145
Helmet-mounted tracking and display systems, 140
Heuristic knowledge, 264
Heuristic reasoning, 119
Heuristics, 9, 35, 192, 293, 298
High-energy laser weapons, 24
High resolution displays, 200
Holistic design, 123
Holistic thinking, 52, 307
Home appliances, 113
Horror stories, 160, 169
Hue, 133
Human circulatory systems, 108
Human cognition, 69, 71
Human-computer interaction, 212
Human engineering
 (*see* human factors)
Human engineering guidelines, 10

Human factors, 53, 56, 67, 89, 97, 133, 141, 149, 159, 160, 162, 164, 165, 167, 168, 208, 328
Human factors data, 223
Human factors engineers, 99, 100
Human information processing, 10, 11, 133, 300
Human-machine interfaces, 86, 91, 113, 114, 119, 145, 168, 192, 206, 270
Human-machine relationships, 114, 115, 120, 124
Human-machine symbiosis, 106
Human performance, 91, 116
 data, 92, 178, 181
 data bases, 243
 models, 282
Human problem solving, 77
Human-system design, 100, 102
Human-system interaction, 88
Human-system interfaces (*see* human-machine interfaces)
Hydrogen gas, 23
Hypothesis generation and test, 286

IAC (*see* information analysis centers)
IBM Corporation, 38, 309
IBM-709 computer, 23, 36
IBM computer systems, 218
IC (*see* integrated circuits)
ICAM (*see* integrated computer-aided manufacturing)
ICBM (*see* intercontinental ballistic missile)
Icons, 137
Idea generation, 182
Identification of information, 223
Ill-defined problems, 177
Ill-defined problem solving, 185
Ill-defined tasks, 69
Imagery, 183
Image understanding, 34
Implicit information, 122
Imprecision, 294, 298
Inactivists, 231
Incentives, 171

Subject Index   345

Inclinations, 9
Incrementalism, 37, 61, 62, 63
Individual differences, 248
Inductive skills, 287
Industrial design, 65, 69, 319
Industrial designers, 44, 46, 53
Industrial engineers, 208
Industrial psychology, 249
Industry, 165
Industry practices, 277
Infantry rifles, 9
Information
  access, 13, 16, 62, 281
  acquisition, 2, 89
  analysis centers, 282
  appropriateness of form, 15
  awareness, 281
  explosion, 200
  flow, 5, 58, 63, 85, 92, 101, 116, 132
  generation, 56
  management, 13, 32, 65, 78, 214, 215, 282
  manipulation, 62
  needs, 56
  processing, 56, 80
  processing characterisitics, 9
  processing models, 77
  reduction of uncertainty, 15
  requirements, 285, 286, 290, 299
  retrieval, 10, 13, 201, 221, 223, 282, 297
  retrieval systems, 11, 13, 14
  science, 106
  search, 10
  seeking, 222, 226
  seeking behavior, 87
  systems, 5, 106, 182
  task relevance, 15
  technology, 286
  transaction models, 71, 75
  transfer, 91
  transformation, 14, 107, 223, 282
  transformation processes, 46, 49, 62
  translation, 68
  utilization, 2, 5, 13, 16, 275, 280
  value, 15
Information Sciences Institute, 28
Innovation, 12, 48, 62, 63, 153
Innovative, 210
Input-output analysis, 300
Inquiry system models, 291
Inspectability, 188, 191
Inspection test requirements, 196
Inspector general, 38, 41
Instructional systems, 142
Instructions, 203
Instrumenting design activities, 54
Integrated circuits, 26, 141
Integrated computer-aided manufacturing, 188
Integration, 197, 200, 210
Intel, 27
Intellectual freedom, 209
Intelligence, 298
Intelligent human-machine systems, 122, 125
Intelligent machines, 114
Intelligent support systems, 16
Intelligent systems, 115
Intelligent workstations, 56
Intentions, 80, 325
Interactivists, 231
Intercontinental ballistic missiles, 23
Interdisciplinarity, 103, 104
Interdisciplinary cooperation, 193
Interdisciplinary education, 110
Interdisciplinary functioning, 259
Interface shells, 218
Internal models, 120, 123, 138 (*see also* mental models)
Interpretation of information, 89
Interpretive structural modeling, 295
Intuition, 88, 236, 291, 297, 305, 306
Invention, 70
Inventive opportunities, 80
Ion beams, 23
Irrationality, 307
ISI (*see* Information Sciences Institute)

Islands of automation, 200, 259

Japan, 26, 38
Jet engines, 37
Joint Chiefs of Staff, 38, 40
Joint Policy Coordinating Group on Logistics Research, Development, Test, and Evaluation, 194
Judgment, 52
Judgmental biases, 10
Judgmental intuition, 307, 309
Judgment style, 291

Keyboard design, 168
Keyboards, 11, 132
Knoesphere, 142
Knowledge, 9
  acquisition, 57
  engineering, 93, 155
  flow, 259
  representation, 212, 285, 292
  solicitation, 145
  with certainty, 307
Knowledge-based behavior, 297
Knowledge-based design, 204
Knowledge-based processing, 193
Knowledge-based support systems, 285
Kwajalein Missile Range, 24

Labeling information, 79
Language of design activities, 54
Large scale design integration, 21
Laser radars, 24
Laser weapons, 20, 22
Lawyers, 61
Learning, 298
Life cycle costs, 167, 171, 207, 210
Lincoln Laboratory, 23
Linear programming, 309, 311
LISP, 218, 263
Local area networks, 33
Lockheed Corporation, 204, 207, 209
Lockheed 1011 aircraft, 169
Lofting, 206

Logical reasoning tasks, 114
Logistics, 101, 167, 190, 194, 196, 197, 223
  support analysis requirements, 191
  system, 182
Loss of control, 15
Lotus 1-2-3, 14
LSAR (see logistics support analysis requirements)
Luminance, 133

M-1 tank, 161
MacDraw, 136
Machine intelligence, 19, 22, 29, 33, 113, 115, 139, 216 (see also artificial intelligence)
Machine operators, 258
Machine vision, 35
MacPaint, 136, 258
Macsyma, 263, 269
Magnetic circuits, 36
Magnetic confinement fusion, 32
Magnetic tracking systems, 135
Maintainability, 56, 61, 187, 193, 207, 208
Maintainers, 9
Maintenance systems, 182
Management, 55, 210
  acceptance, 100
  aids, 56
  functions, 199
  information systems, 106
  science, 55, 291
Man-machine interfaces, 168, 270 (see human-machine interfaces)
Man-machine systems, 2
Manning requirements, 169
Manpower, 168
  and personnel integration, 162, 181
  costs, 167
  requirements models, 191
MANPRINT (see manpower and personnel integration)
Manual control, 117
Manual control models, 120
Manufacturability, 195, 196, 246

Subject Index 347

Manufacturing, 133, 146, 165, 187, 189, 193, 200, 211, 258, 267, 271
  constraints, 322
  engineers, 208
  techniques, 85
Marine Corps, 232
Market demography, 85
Marketing, 12, 45, 277
Marketplace, 329
Markov processes, 309
Massachusetts Institute of Technology, 20, 32, 35, 194, 327
Master, 298
Master-slave relationship, 115, 121
Materials, 10
Materials science, 133
Mathematical equations, 204
Mathematical knowledge, 263
Mathematical models, 205
Mathematical programming techniques, 55
Mathematics, 55, 306
Means-ends analysis, 297
Means-ends representation, 292
Mechanical engineering, 54, 263, 267
Mechanical engineering design, 269
Mechanical ingenuity, 323
Media, 58
Media constraints, 322
Megamodels, 282
Mental effort, 248, 249
Mental images, 128
Mental models, 10, 120, 122, 127, 231, 234 (see also internal models)
Mental pictures (see mental models)
Mental representation (see mental models)
Mentor Graphics Corporation, 26, 194
Menus, 11
Message handling, 14
Metaknowledge, 298
Metal Oxide Semiconductor Implementation System Project, 28, 30
Meta-rules, 193
Microelectronics, 19, 26
Microprocessors, 27, 113

Microwave radars, 24
Middle management, 209
Military
  aircraft, 140
  force effectiveness, 160
  personnel, 164
  services, 57
  systems, 48, 98, 159
MIL-STD-1472, 178
MIL-STD 1472C, 240
Mindscape, 142
Mindsets, 171
Mind's eye, 130
Mindware, 138
MIPS, 30
MIS (see management information systems)
Missile site radars, 24
Mission
  analysis, 232
  analysis models, 282
  requirements, 160
MIT (see Massachusetts Institute of Technology)
Mockups, 179
Modeling, 13, 14, 23, 87
  methods, 282
  tools, 203
Model of design, 236, 290 (see also mathematical models)
Molecular structures, 141
Monarch, 30
MOSIS (see Metal Oxide Semiconductor Implementation System Project)
Motivation, 113, 319, 325
Motorola, Inc., 194
Movie industry, 31
Multi-attribute decision making, 55
Multi-attribute utility theory, 181, 185, 312
Multidisciplinarity, 104
Multidisciplinary access, 84
Multidisciplinary activity, 9
Multidisciplinary centers, 193
Multidisciplinary teams, 100, 223
Multidisciplinary utilization, 84

Multi-objective programming, 312
Multi-Plan, 13
Multiple launch rocket systems, 161
Multiprocessor architectures, 34
Murphy's Law, 239
Muscle fatigue, 11
Mutual cooperation, 122

NASA (*see* National Aeronautics and Space Administration)
NASA Ames Research Center, 36
National aerodynamic simulation facility, 36
National Aeronautics and Space Administration, 25, 57
National Science Foundation, 32
National Security Industrial Association, 194
National Semiconductor Corporation, 194
National Transportation Safety Board, 167
Natural language processing, 33
Nature of design, 7, 290
Nature of designers, 9
Navigation, 233
  satellites, 21
  satellite system, 25
Networking, 16, 201, 210
Network models, 55
Networks, 194
Non-destructive inspection, 191
Northwest Indians, 322
Novices, 115, 178, 298
NSF (*see* National Science Foundation)
Nuclear power plant, 86, 117
Nuclear power plant operators, 275
Numerically controlled machine, 29, 259
Numeric computation, 193

Observational methods, 211
Office design, 153
Office of Federal Procurement Policy, 98
Office of Management and Budget, 98
Office of the Army Deputy Chief of Staff, 43

Office of the Secretary of Defense, 20, 40
OFPP (*see* Office of Federal Procurement Policy)
Operability, 1, 61
Operational deployment, 290
Operations research, 55, 291, 306, 309
Operators, 9
Optimal feedback control, 55,
Optimization, 8, 52, 95, 191, 212
Option generation, 177, 196
Organizations
  attitudes, 245
  biases, 85
  choice model, 291
  considerations, 53
  criteria, 12
  design, 168
  functions, 199
  goals, 12
  issues, 4, 19, 37
  psychology, 249
  structures, 51, 62, 208
Orthographic sketches, 76
Overhead costs, 59, 60

Packard Commission, 38, 39, 40, 41, 277
PACs (*see* political action committees)
Parallax, 133
Parameters, 232
Partitioning information, 78
Part program libraries, 259
Part programmer, 259
Parts lists, 196
Pattern recognition abilities, 11
PC (*see* personal computers)
Peer group attitudes, 245
Penicillin, 128
Pentagon, 38, 40
Perceived level of discretion, 249
Perceived organizational attitudes, 249
Perceived peer group attitudes, 249
Perceived usability, 248
Perceived usefulness, 247, 281
Perceptual knowledge, 123

Subject Index

Perceptual-recognitional processes, 176
Performance
  capability, 235
  prediction, 169
  requirements, 50
Personal computers, 27
Personal data bases, 203
Personnel, 168
  requirements models, 191
  selection, 240
Perspective drawings, 71
Perspective sketches, 76, 80
Philosophy of science, 296
Physiology, 141
Planning, 35, 65, 73, 74, 271
Plasma stability, 23
Plot formatting, 203
Polaroid camera, 71
Policies, 207
Political action committees, 58
Political influences, 98
Politics, 98, 166, 326
POP-11, 271
POPLOG, 271
Power consumption, 207
Power plant control rooms, 45
Power plants, 245, 276
Practices, 206
Pragmatics, 54
Preactivists, 231
Precision, 299
Prejudices, 239
President's Blue Ribbon Commission on
  Defense Management, 38
Pride, 121
Princeton University, 23
Printed circuit boards, 196
Problem
  formulation, 8, 46, 55,57, 145, 146,
    154, 230, 290
  solving, 10, 80, 88, 106, 222, 292,
    300, 320
  space, 292
Procedural skills, 10

Procedures, 206
Process architectures, 289
Process control design, 151
Process control plant, 149
Process evaluation, 289
Process flexibility, 147
Procurement
  regulations, 12, 277
  requirements, 280
Producibility, 195, 200, 230
Product
  architecture, 289
  costs, 213
  evaluation, 289
  life cycle, 200, 203
  performance, 187
  quality, 200, 210, 213
  support, 200
Production, 50
Production engineering, 258
Production rules, 293
Production systems, 292
Productivity, 21, 210, 211
Programming
  environments, 56, 218
  languages, 132
Prolog, 263, 267, 271
Protocol analysis, 66
Prototypes, 289
  models, 179
  simulations, 201
Prototyping, 87, 214
Psychological issues, 51
Psychology, 122
  of design, 43, 53
  of problem solving, 69
  of system design, 97, 320
Psychomotor functions, 129
Psychophysics, 140
Public engineering, 166

Quality assurance engineers, 208
Questioning-answering system, 297
Queuing theory, 309

Radar, 108
  backscatter, 21
  scattering, 23
RAMCAD (*see* reliability and maintainability in computer-aided design)
Rapid decision making, 180
Rapid prototyping, 19, 28, 55, 56, 145, 150, 169, 184, 246, 289 (*see also* prototyping)
Rapid prototyping tools, 13
RAPT inference engine, 269
RAPT robot programming system, 269
R&D
  costs, 30
  directions, 53
  laboratories, 90
Reactivists, 231
Real-time simulation, 37, 214
Reasoning, 35
Recall, 11
Receiver-processor systems, 21
Recognition, 11
Recognitional processes, 180
Recombinant DNA, 20
Reconnaissance photography, 31
Redesign, 146, 196
REDUCE, 263, 269
Reduced instruction set computers, 30
Regulations, 93, 164, 165, 208, 222
Regulatory considerations, 58
Regulatory constraints, 85
Regulatory factors, 46
Regulatory pressures, 19, 38
Relational data bases, 156, 262
Relativity theory, 34
Relevancy, 299
Reliability, 133, 187, 193, 207, 208, 230, 235, 238, 299
Reliability and maintainability in computer-aided design, 194
Repairability, 188
Report formatting, 203
Representational modes, 75

Request for proposals, 50, 61, 99, 207
Requirements, 1, 12, 207
Requirements-driven design, 151
Requirements specification, 288
Research data, 83
Research subcultures, 89
Resistance to change, 23, 57, 60
Retinal disparity, 133
Retirement for cause program, 191, 195
Reverse engineering, 27, 161
RFP (*see* request for proposals)
RISC (*see* reduced instruction set computers)
Risk, 48, 61, 208
R&M (*see* reliability and maintainability)
ROBMOD polyhedral modeler, 270
Role playing, 100
Roles, 115
  of designers, 276
  of support systems, 14
Rule-based behavior, 297

Safety, 168, 208
  standards, 277
Sales, 45
Satellite Business Systems, 20, 25
Satisficing, 10, 37, 52, 84, 86, 88, 95, 180
Saturation, 133
Scenarios, 232
Scheduling, 271
Schemata, 292, 295
Scientific communication, 89
Scripts, 292, 294, 295
Sculpting, 69
Secretary of the Air Force, 188
Security control center, 152
Semantic content, 139
Semantic knowledge, 121, 123
Semantic networks, 292, 293
Semantics, 54
Semiconductor foundry, 28
Sensory experience, 214
Service life, 260
Ship, 30

Subject Index

Signal processing, 34
Simulation models, 168
Simulations, 14, 36, 87
Simulator design, 223
Simulator designers, 178
Simulators, 140
Sketching, 74, 80
Sketchy representations, 80
Skill-based behavior, 297
Skills, 9
Skunk works, 209
Small-scale integration, 26
SMD-FRAME construct, 50
Sneak circuit analysis, 195
Social aspects, 113, 120, 125
Social engineering, 106
Social goals, 59
Social interaction, 108
Social justice, 58
Social psychology, 249
Sociocultural characteristics, 307
Sociocultural contexts, 237
Sociology, 122
Software
   design, 150
   packages, 54
   system design, 33
   technology, 10
Solid geometry modeler, 269
Solid modeling, 81, 203
Sources of power, 119
SOW (*see* statement of work)
Space-based defense, 21
Space
   constraints, 324
   shuttle, 33
   station, 33
Spacesuits, 140
Spatial reasoning, 260
Spatial relationship reasoning engine, 269
Specialist's spectacles, 133
Specifications, 10, 101, 148, 207, 208
Speculation, 37
Speech understanding, 34

SPICE, 55,
Spread-sheet programs, 13, 282
Spread-spectrum, burst, time division
     multiple access format
     communication satellite, 25
Stability analysis, 55,
Standardization, 252
Standards, 85, 101, 208, 210, 222, 280
Stanford Center for Design Research, 218
Stanford University, 327
Staphylococci, 128
State estimation, 55,
Statement of work, 50, 277
Statistical mechanics, 34
STINGER, 161
Stochastic processes, 308
Stochastic process models, 55,
Strategic defense program, 39
Strategic defense system, 33
Stress, 11
   analyses, 246
Stress-based models, 291
Structural engineering, 133
Structural mechanics, 10
Subject matter experts, 87
Success stories, 169, 170
Supervisory control, 115, 117
   models, 117, 120
Supportability, 188, 189, 196, 200
Surface modeling, 203
Surgeons, 324
Surrogate imagery, 134
Symbiosis model, 117, 120, 125
Symbolic communication, 139
Symbolic representation, 31
Syntax, 54
Synthesis, 8, 104
Synthetic worlds, 270
System
   acquisition process, 164, 165, 170, 172
   analysis, 309
   architecture, 124, 289
   configuration, 224

design, 105
effectiveness, 1, 12, 52, 60, 83
engineering, 166, 194, 286
evaluation, 289
failure, 86
integration, 123
management, 115, 296
objectives, 147
performance, 55, 116, 159, 168, 171
requirements, 147, 150, 167, 225, 230, 232, 277
safety, 162
simulation languages, 10
specifications, 224
Systolic array, 30

TAC (*see* Tactical Air Command)
Tactical Air Command, 182
Tactile feedback, 270
Target acquisition performance, 233
Task
  allocation, 162
  analysis, 151
Taxonomic classification, 265
Taxonomic organization, 265
Technical information, 8
Technical orders, 196
Technical vocabulary, 224
Technological gatekeepers, 90
Technology
  transition, 39
  trends, 26
Tektronix, Inc., 194
Teleology, 308
Teleoperated surveillance platforms, 232
Telescope, 24
Ten Commandments, 326
Terrain data bases, 141
Testability, 190, 195
  analysis, 195
Test
  equipment, 9
  points, 196
Texas Instruments, Inc., 194

Theorems, 307
Theory of design, 211, 229
Third party buyers, 253
Thought-controlled matter manipulators, 132, 134, 138
Three-Mile Island, 86
Time constraints, 323
Top-down
  decomposition, 47
  design, 9, 147, 276
Touch feedback, 135
TOV (*see* teleoperated surveillance platforms)
Tower of Babel, 84
Toxic material, 324
Toy design, 320
Tracking and data relay satellite system, 25
Trainability, 187
Training, 14, 27, 162, 168, 218, 222, 226, 240
  device designers, 178
  devices, 178
  requirements, 201
  requirements models, 191
Trinity of ignorance, 311
Trust, 120
Turbojet engines, 37
Tutorials, 203
Tutoring, 14

ULCE (*see* unified life cycle engineering)
Uncertainty, 294, 298, 299
Unconscious analyses, 236
Unified life cycle engineering, 187
Universities, 105
University of California at Berkeley, 30
University of Illinois at Urbana-Champaign, 306, 319
University of Massachusetts, 312
University of Michigan, 29
University of Southern California, 28
UNIX, 271
Usability, 67
Users, 98, 310

Subject Index

acceptance, 15, 22, 245, 282
expectations, 85
friendliness, 138, 145
friendly, 15
friendly systems, 99
interfaces, 222, 270
involvement, 146
modeling, 270
needs, 275, 286, 287, 290
perceptions, 247
satisfaction, 290
User-system dialogue, 56
User-system interfaces, 222, 270 (*see also* human-machine interfaces)
Utilization of technical data, 83, 84

Value engineering, 237
Value of information, 87, 95, 222, 299
Value systems, 292, 310
VCASS (*see* visually coupled airborne systems simulator)
VCRs, 114
Venture capital, 38
Verbal protocol, 66
Very large-scale integrated circuits, 31, 189, 246
Very large-scale integration, 26, 30
Vibration, 234
Vibro-tactile simulators, 135
Videoplace, 140
Videotaped protocols, 74
Virtual cockpits, 140
Virtual design terminals, 140
Virtual rooms, 136
Virtual space, 127, 133, 139
Virtual space simulators, 140
Virtual terminals, 127
Virtual world generators, 134, 138
Virtual worlds, 127
Visicalc, 13

Visual accommodation, 133
Visual convergence, 133
Visual cortex, 129, 134
Visual fatigue, 11
Visual imaging, 131, 134
Visually coupled airborne systems simulator, 140
Visual representation, 65
Visual thinking, 127, 131
VLSI (*see* very large-scale integration)
VLSI circuits (*see* very large-scale integrated circuits)
Vocabulary, 54
Voice recognizers, 145
Volumetric modeling, 54
Volumetric relations, 70

Weapon system acquisition, 160
Weapon systems, 113, 120, 160, 188, 197
Wickedness, 308
Wideband radars, 24
Wind tunnels, 37
Wire frame models, 203, 269
Wire routing, 196
Word processing, 148
Working memory, 11
Workload, 185
Workplace management, 73, 74
Workshop on the Psychology of System Design, 3, 5, 19
Workshop themes, 7
World War II, 168
Wright-Patterson Air Force Base, 325

X29 aircraft (*see* advanced, forward swept wing, integrated technology aircraft)
Xerox Corporation Research Laboratory, 33

Yield, 28
Yield factors, 26